Design and Manufacture
An Integrated Approach

Other titles of interest to engineers from Palgrave Macmillan

Acoustics for Engineers, John Turner and A. J. Pretlove
CAD/CAM: Features, Applications and Management, Peter F. Jones
Engineering Mathematics, Third Edition, Ken Stroud
Form, Structure and Mechanism, M. J. French
Further Engineering Mathematics, Second Edition, Ken Stroud
Introduction to Engineering Materials, Third Edition, Vernon John
Introduction to Internal Combustion Engines, Second Edition, Richard Stone
Motor Vehicle Fuel Economy, Richard Stone
Tribology: Principles and Design Applications, D. Arnell, P. Davies, J. Halling and T. Whomes

Foundations of engineering

Dynamics, G. E. Drabble
Electric Circuits, P. Silvester
Electromagnetism, R. G. Powell
Fluid Mechanics, Martin Widden
Structural Mechanics, J. A. Cain and R. Hulse
Thermodynamics, J. Simonson

Design and Manufacture

An Integrated Approach

Rod Black

University of Portsmouth

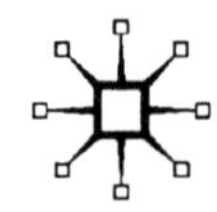

Published by
PALGRAVE MACMILLAN
Houndmills, Basingstoke, Hampshire RG21 6XS and
175 Fifth Avenue, New York, N. Y. 10010
Companies and representatives throughout the world

PALGRAVE MACMILLAN is the global academic imprint of the Palgrave Macmillan division of St. Martin's Press, LLC and of Palgrave Macmillan Ltd. Macmillan® is a registered trademark in the United States, United Kingdom and other countries. Palgrave is a registered trademark in the European Union and other countries.

ISBN 0–333–60915–8

This book is printed on paper suitable for recycling and made from fully managed and sustained forest sources.

A catalogue record for this book is available from the British Library.

Copy-edited and typeset by Povey–Edmondson
Tavistock and Rochdale, England

Transferred to digital printing 2002

Printed and bound in Great Britain by
Antony Rowe Ltd, Chippenham and Eastbourne

To the next generation, particularly Ben, Doug and Alex

Contents

Preface xv

Acknowledgements xvii

1 Concept to Solution – An Overview **1**

1.1 Introduction 1
1.2 New product introduction process 1
1.1.2 Background 1
1.2.2 Need 3
1.2.3 Engineering model 3
1.2.4 Design detail 3
1.2.5 Prototype 4
1.2.6 Field trials 4
1.2.7 Pre-production 4
1.2.8 Integrated approach 6
1.2.9 Control system 6
1.3 Summary 7
1.4 Questions 7

Part I The Basics of Design

2 Records and Communication: The Engineering Drawing **11**

2.1 Introduction 11
2.2 Engineering drawings 11
2.2.1 Equipment 12
2.2.2 Drawing layout 12
2.2.3 Sketching 12
2.2.4 Orthographic projection 15
2.2.5 Sections 18
2.2.6 Auxiliary projection 19
2.2.7 Types of drawings 21
2.2.8 Dimensions 23
2.2.9 Abbreviations and standard components 25
2.3 Maintaining records 26
2.3.1 Drawing identification 26
2.3.2 The working file 26
2.4 Summary 27
2.5 Questions 28

3 The Importance of Detail: Tolerances **30**
3.1 Tolerances 30
3.1.1 Introduction 30
3.1.2 Definition 30
3.1.3 Cost implications of tolerances 31
3.1.4 Presentation 32
3.1.5 Functional dimensions 32
3.1.6 Choice of datum 33
3.1.7 Tolerance allocation 34
3.1.8 Allocation examples 35
3.2 Limits and fits 38
3.2.1 Introduction 38
3.2.2 BS4500 38
3.2.3 Tolerance grade 38
3.2.4 Deviation 39
3.2.5 Syntax 39
3.2.6 Types of fit 40
3.2.7 Hole or shaft base? 40
3.3 Surface finish 41
3.3.1 Introduction 41
3.3.2 Assessment 42
3.3.3 Syntax 42
3.3.4 Cost 43
3.4 Geometric tolerances 44
3.4.1 Introduction 44
3.4.2 Syntax 44
3.4.3 Geometric tolerancing examples 46
3.4.4 When to use geometric tolerancing 49
3.5 Maximum material condition 50
3.6 Examples 51
3.7 Summary 53
3.8 Questions 53

4 Standard Components **54**
4.1 Introduction 54
4.2 Types of standard 54
4.3 Types of transmission 55
4.3.1 Introduction 55
4.3.2 Shafts, loads, keys and keyways 56
4.3.3 Belt transmissions 58
4.3.4 Chains and timing belts 62
4.3.5 Gears 62
4.3.6 Hydraulic transmissions 68
4.3.7 Electric transmissions 69
4.4 General standard components 71
4.4.1 Couplings 71
4.4.2 Clutches 73
4.4.3 Bearings 74

4.4.4 Seals 82
4.5 Joining devices 84
4.5.1 Threaded fasteners 84
4.5.2 Non-threaded fasteners 87
4.6 Summary 89
4.7 Questions 89

5 The Design Process **91**
5.1 Introduction 91
5.2 The process 91
5.2.1 Definition 92
5.3 Product design specification 92
5.4 Elements of the PDS 94
5.4.1 Origin of need 94
5.4.2 Operation 94
5.4.3 Life 96
5.4.4 Producer 98
5.4.5 Society 99
5.4.6 Customer 99
5.5 Information sources 101
5.5.1 Standards 101
5.5.2 Component manufacturers 101
5.5.3 Competitors' products 101
5.5.4 Library searches 101
5.5.5 Patents 102
5.6 Idea generation 103
5.6.1 Common obstacles 103
5.6.2 Improving creativity 104
5.7 Idea evaluation 108
5.7.1 Initial screening 108
5.7.2 Concept comparison 108
5.7.3 Criteria weighting 110
5.8 Summary 111
5.9 Questions 112

Part II The Basics of Manufacture

6 The Manufacturing Process **115**
6.1 Introduction 115
6.2 Material selection 115
6.2.1 Importance of materials 115
6.2.2 Factors influencing selection 116
6.3 Manufacturing processes 117
6.3.1 Shape modifying processes 117
6.3.2 Property modifying processes 120
6.3.3 Joining processes 121
6.4 Manufacturing system 121
6.5 Summary 121

7 Shape Modifying by Material Removal **122**
7.1 Introduction 122
7.2 Metal cutting 122
7.2.1 Cutting mechanism 123
7.2.2 Chip formation 123
7.2.3 Tool geometry 124
7.2.4 Tool materials 125
7.2.5 Cutting fluids 126
7.2.6 Tool wear 127
7.2.7 Tool life 128
7.2.8 Cutting economics 128
7.2.9 Cutting forces 129
7.2.10 Machine tools 132
7.3 Abrasive machining 136
7.3.1 Grinding 136
7.3.2 Honing 137
7.3.3 Superfinishing 137
7.3.4 Lapping 137
7.4 Other methods 137
7.4.1 Electrochemical machining 137
7.4.2 Electric discharge machining 137
7.4.3 Laser profiling 138
7.5 Summary 138
7.6 Questions 138

8 Shape Modifying: Retaining Material **139**
8.1 Introduction 139
8.2 Casting 139
8.2.1 Terminology 140
8.2.2 Sand casting 141
8.2.3 Shell moulding 142
8.2.4 Investment casting 143
8.2.5 Die casting 143
8.3 Powder forming 145
8.3.1 Powder manufacture 145
8.4.2 Compaction 145
8.4.3 Sintering 146
8.3.4 Hot isostatic pressing 146
8.4 Plastic deformation 146
8.4.1 Hot and cold working 146
8.4.2 Forging 146
8.4.3 Rolling 147
8.4.4 Extrusion 148
8.4.5 Sheet metal processes 148
8.5 Processes for non-metals 151
8.5.1 Plastics 151
8.5.2 Ceramics 153
8.5.3 Composites 154

8.6	Summary	154
8.7	Questions	155
9	**Joining Techniques**	**156**
9.1	Introduction	156
9.2	Non-permanent joints	157
9.3	Semi-permanent joints	157
	9.3.1 Mechanical	157
	9.3.2 Liquid–solid bonding (brazing and soldering)	157
9.4	Permanent joints	159
	9.4.1 Welding	159
	9.4.2 Adhesive bonding	164
9.5	Summary	165
9.6	Questions	167
10	**Property Modification**	**168**
10.1	Introduction	168
10.2	Heat treatment	169
	10.2.1 Structure of steel	169
	10.2.2 Stress relieving	170
	10.2.3 Annealing	170
	10.2.4 Normalising	170
	10.2.5 Hardening	171
	10.2.6 Tempering	171
	10.2.7 Surface hardening	171
	10.2.8 Case hardening	171
10.3	Surface finishing	171
	10.3.1 Mechanical	171
	10.3.2 Coating	172
	10.3.3 Friction surfacing	172
10.4	Summary	173
10.5	Questions	173
11	**Quality Control**	**174**
11.1	Introduction	174
11.2	Inspecting geometric features	174
	11.2.1 Inspection equipment	174
11.3	Inspecting non-geometric features	177
	11.3.1 Proof testing	177
	11.3.2 Visual inspection	177
	11.3.3 Dye penetrant	177
	11.3.4 Magnetic particle	178
	11.3.5 Radiographic	178
	11.3.6 Ultrasonic, acoustic	178
	11.3.7 Leak detection	178
11.4	Traceability	178
11.5	Process control and feedback	179
	11.5.1 Process capability	179

11.5.2 Statistical process control 181
11.5.3 Quality control charts 182
11.6 Summary 183
11.7 Questions 183

Part III Effective Integration of Design and Manufacture

12 Designing for Manufacturing Processes and Materials **187**
12.1 Introduction 187
12.2 Casting 187
12.2.1 When to use castings 187
12.2.2 Which casting process? 187
12.2.3 General design points 189
12.3 Forging 190
12.3.1 When to use forgings 190
12.3.2 Which forging process? 191
12.3.3 General design points 191
12.4 Powder metallurgy 192
12.4.1 When to use powder metallurgy 192
12.4.2 General design points 192
12.5 Material removal 193
12.5.1 When to machine 193
12.5.2 Which machining process? 194
12.5.3 General design points 195
12.6 Plastics 197
12.6.1 When to use plastics 197
12.6.2 Which plastic? 197
12.6.3 General design points 197
12.7 Ceramics 199
12.8 Composites 199
12.8.1 When to use composites 199
12.8.2 General design points 200
12.9 Summary 201
12.10 Questions 201

13 Designing for Joining and Assembly **202**
13.1 Introduction 202
13.2 Choice of process 203
13.3 Heat application 203
13.3.1 Welding 203
13.3.2 Brazing and soldering 207
13.4 Chemical application 207
13.4.1 Adhesives 207
13.5 Mechanical joints 208
13.5.1 Integral features 208
13.5.2 Threaded fasteners 212
13.5.3 Non-threaded fasteners 216

13.6 Automated assembly 216
13.7 Summary 218
13.8 Questions 219

14 Influences on Design and Manufacturing Choices 220
14.1 Introduction 220
14.2 Aesthetics 220
14.3 Ergonomics 222
14.4 Quantity 223
14.5 Safety 225
14.6 Strength, fatigue 227
14.7 Corrosion 229
14.8 Environment 230
14.9 Conflict, compromise 236
14.10 Summary 237

15 Analysis of Existing Designs and Manufacturing Processes 238
15.1 Introduction 238
15.2 Value analysis 238
15.3 Summary 247
15.4 Questions 248

16 Systems for Controlling Design and Manufacture 249
16.1 Introduction 249
16.2 Primary tasks 249
16.2.1 Company aims, objectives and strategy 250
16.2.2 Future products and processes 250
16.2.3 Develop and implement new products and processes 251
16.2.4 Obtain customer orders 251
16.2.5 Manufacture products and despatch to customers 252
16.2.6 Recover payment from the customer 254
16.2.7 Service the system 254
16.3 Information flows 256
16.3.1 Types of order 256
16.3.2 Special order 256
16.3.3 Standard product order 259
16.4 Control systems 261
16.4.1 Engineering changes 261
16.4.2 Efficiency measurement 262
16.4.3 Stock control 263
16.4.4 Material stock movements 264
16.5 Computer aided 265
16.5.1 Background 265
16.5.2 Product development 266
16.5.3 Manufacturing control 270
16.5.4 Computer integrated manufacture 273
16.6 Summary 273
16.7 Questions 274

17 The Business Context (Organisation and Costing) **275**
17.1 Introduction 275
17.2 Basics of costing 275
17.2.1 Profit 275
17.2.2 Income 275
17.2.3 Outgoings 276
17.2.4 Fixed and variable costs 276
17.3 Applications 277
17.3.1 Changes to an existing product 277
17.3.2 Introducing a new process or piece of plant 281
17.3.3 Introducing a new product 284
17.4 Quantity 287
17.5 Summary 288
17.6 Questions 288

Part IV Appendices

A Units and Definitions **293**
A.1 Introduction 293
A.2 Units 293
A.2.1 SI system 293
A.2.2 Imperial system 295
A.2.3 Unity brackets 298
A.2.4 Dimensional analysis 298
A.3 Equations of motion 299
A.4 Force, motion and work 300
A.4.1 Newton's laws of motion 300
A.4.2 Work, energy and power 300

B Data Tables **301**
B.1 Surface roughness 302
B.2 Relative costs of processes 303
B.3 Selected ISO fits: hole basis 304
B.4 Selected ISO fits: shaft basis 305
B.5 Properties of sections 306
B.6 Deflections and moments for simple beams 308
B.7 Friction coefficients 309
B.8 Corrosion: electrochemical series 310
B.9 Materials data (steels, non-ferrous metals, non-metals) 311

C Costing Calculations **314**

D Solutions **316**

E Equivalent Standards **320**

Bibliography and Further Reading **321**

Index **322**

Preface

In the early 1980s I worked for an engineering company, designing and manufacturing products. The design office contained the 'experts' in design, who created drawings for new products. When complete, these were passed to production engineering, located in a separate building remote from the design offices, where the 'experts' in production resided. They then modified the designs, so that they could be more easily manufactured. Much time was lost through the many conflicts over details that were 'vital' to the functioning of the final product and at the same time 'impossible' to manufacture. The solution was to combine the design and production engineering departments and locate them together, adjacent to the manufacturing plant. The design and production 'experts' were both involved in the product development process, working together as a team, rather than blaming each other for shortcomings in the products. The end result for the company was that new products could be introduced more quickly. In addition, they generally had fewer teething problems and tended to be more profitable. The moral is obvious.

The teaching of engineering tends to be very compartmentalised, with different aspects being taught by experts in various fields. Although there are many good reasons for this, I believe that we should never lose sight of the fact that in the real world, engineering problems rarely fit precisely into the compartments of knowledge as taught. It is therefore vital that the importance of integration across the subject boundaries is covered from an early stage in the teaching of engineering.

This book has been written for engineering students, recognising the need both to introduce basics skills and techniques *and* to show how they would be used in an integrated fashion.

Chapter 1 introduces the route by which an idea is converted into a product, using a case study to highlight the need for an understanding of the process. This is followed by the three main parts of the text.

The first part concentrates on the basics of design. The content includes the engineering drawing, tolerances, a range of standard components, the product design specification, and methods of generating and evaluating ideas. The second part concentrates on basic manufacturing techniques. These include material removal processes, casting, deforming, powder metallurgy, processes for metals and non-metals, joining techniques, processes for altering material properties, and an introduction to the concepts of quality.

The third part builds upon the basics, integrating aspects from the first two parts. Chapter 12, for example, looks at why a particular manufacturing process might be selected, and how the designer might take full advantage of the chosen process. Chapter 13 covers the application of joining techniques, discussing topics such as when and how should an interference joint be selected and designed, and how the use of automated assembly processes can influence the design. The next chapter discusses, largely through case studies, how a variety of influences result in engineering compromises between design and manufacture. Chapter 15 shows, again through the use of case studies, how value analysis can be used to improve both the component design and the manufacturing process. Chapter 16 takes an overall view of an organisation designing and manufacturing products, examining the systems used, how control is maintained, and how computers can be used to help. The final chapter examines the business context, costing systems, and how decisions might be made when selecting a new piece of manufacturing plant, or introducing a new product.

Recognising that units are a common source of error for students, Appendix A includes definitions, conversion factors, and sections on unity brackets and dimensional analysis.

Revision questions are included, where appropriate, at the end of each chapter. Solutions can be found either in the text, as directed, or in appendix D. Throughout there are various references to British Standards. Equivalent, or similar, International Standards are listed in appendix E.

The book builds on much that I have learned working as an engineer over the past two decades, but has only been possible because of the help, co-operation, suggestions and tolerance from a variety of colleagues, too many to mention individually. I would, however, like to thank, in particular, John Bishop for his general help and advice, and specifically for his contribution, with the assistance of 3M, to the section on adhesives. I would also like to single out Bill Puttick for his helpful comments and advice on the chapter on metal cutting. Many thanks, Bill. Last, but not least, I must also thank my family, who have shown remarkable tolerance whilst I concentrated on 'the book', ignoring them for far too long.

ROD BLACK

Acknowledgements

The author and publishers would like to thank the following for the use of copyright material:

Figures 4.17, 4.18, 4.19, 4.20, 4.21 reprinted with the permission of SEED (Sharing Experience in Engineering Design) and McGraw-Hill Book Company Europe, from their publication, K. Hurst, *Rotary Power Transmission Design*, 1994.

Figure 5.5, Table 5.2 reproduced with the permission of the Institution of Electrical Engineers from *A Guide to Design for Production*, published by The Institution of Production Engineers, 1984.

Figure 4.38 reprinted with the permission of SKF (UK) from their *General Catalogue 1989*.

Figures 3.21, 3.22, B.2 and Tables B.3, B.4 are extracts from British Standards and are reproduced with the permission of BSI. Complete copies can be obtained by post from BSI Sales, Linford Wood, Milton Keynes MK14 6LE.

1 Concept to Solution: An Overview

1.1 Introduction

Why do companies employ engineers? Obviously it is to solve any engineering problems they may have. If this is true, it assumes that companies have a deep-seated interest in solving such problems. As this seems unlikely, we need to ask why a company might want the solution to an engineering problem. The lifeblood of a company is profit. Remove it and the company will die; increase it and the company will thrive. Engineers are employed for the same reasons as other professionals – to help improve and maintain company profits. It therefore follows that if all the employees are working towards the same goal, the more closely they cooperate with each other, and the more likely they are to be successful.

Consider a company making and selling products. If the product is well received by the market and is able to generate good profits, both the product and the company will be successful. At some stage, the product must have been designed and manufactured. Design and manufacture are two elements of engineering. However, it would be wrong to separate them totally, for they are highly dependent on each other. Close integration of the design and manufacturing functions is essential.

Chapters 2 to 11 will discuss the basics of Design and Manufacture, but before launching into the detail, we will start by examining the process of introducing a new product and demonstrate the need for an integrated approach between the various engineering areas of expertise.

1.2 New product introduction process

1.2.1 Background

Many companies start with an idea from an enthusiastic entrepreneur who manages to put his idea into practice, thereby generating a successful product. In the early stages the company is usually very small, possibly even only a one-man business. The lines of communication are very short and do not cause problems. As the company grows, so does the number of staff involved, and with them grows the potential for communication difficulties. The need for formalising procedures increases, and the skills for running the company gradually change. In many cases the founder of the company finds these changes restricting to the entrepreneurial spirit that started the company, and leaves. With or without the originator, the company must continue to introduce new products if it is to survive in an ever more competitive world. Some form of system will be required if the process is to be completed successfully.

The best way to demonstrate the need for a formal approach to the new product introduction process is by considering an example.

This case study involves a company producing a range of laboratory centrifuges. The company started in the 1940s in the UK and quickly established a worldwide reputation for a range of products whose name was synonymous with quality and practicality. The company grew over the years until by the mid-1970s it had several hundred employees.

One of its staple products was a simple bench-top centrifuge first introduced around 1950. The body was an alloy casting containing a vertically mounted electric motor with its output shaft pointing upward. On this shaft was fitted a rotor containing 4 test-tube holders, each mounted on a trunnion. At rest the tube holders hung vertically, and when the motor was run they would swing to a horizontal attitude, so that centrifugal force would force the denser constituents of the test-tube contents to the bottom of the tube. A single on/off switch was fitted to the front of the casting, which was topped off with a hinged lid. The speed of the centrifuge could be checked by inserting a contact tachometer onto the centre of the rotor via a hole in the lid.

The simplicity of the design, its low cost and reliability resulted in large sales on a worldwide basis. The product sold so well that the company felt only minimal further development was needed. A minor update to the body was introduced in 1963. Eventually competitors began to realise that they could produce a more desirable alternative, and proceeded to do so. Sales of the original were maintained at first, but then market share started to fall.

The competition updated their products by adding some aesthetic appeal and safety features such as a locking lid. The result was falling sales of the original, starting with the export markets. As the UK market continued to thrive, excuses for the sluggish export sales were given: poor exchange rate, unsatisfactory overseas agents, different market conditions abroad, etc.

Eventually the competitors' products appeared on the UK market, with the result that the original's sales were badly hit. Alarm bells rang, and the company at last realised that a replacement product was needed urgently. The brief to the design department was short and to the point: create a new bench-top centrifuge with some style, a locking lid, with a low unit cost, to be available with the minimum of delay. The urgency was so great that no time was spent preparing a detailed specification (PDS – Chapter 5); the only defined requirement was to replicate the performance of the original machine whilst adding the features mentioned.

The Research and Development department used a team of enthusiastic engineers who produced a prototype that even exceeded the basic brief, by adding a timer and a larger rotor offering increased carrying capacity. The proposal, which was rapidly accepted, incorporated a number of interesting features.

The body introduced a novel form of construction, being made from structural foam. This offered a number of advantages. It was not labour intensive in its manufacture, it allowed a good deal of freedom in terms of styling, and the low number of pieces kept the assembly operation to a minimum. As the company did not have any in-house expertise in the use of structural foam, manufacture of the bodies was subcontracted to a specialist.

The increased carrying capacity was simply achieved by equipping the machine with a larger rotor. The increased inertia of this rotor put greater demands on the electric motor driving it. (Remember, the rotor was fitted directly to the output shaft of the motor.) To cope with this increased load the motor drew more current, and thus tended to overheat. The designers anticipated the problem and used the rotor as a fan to draw air across the windings and hence provide cooling. (See Figure 1.1.)

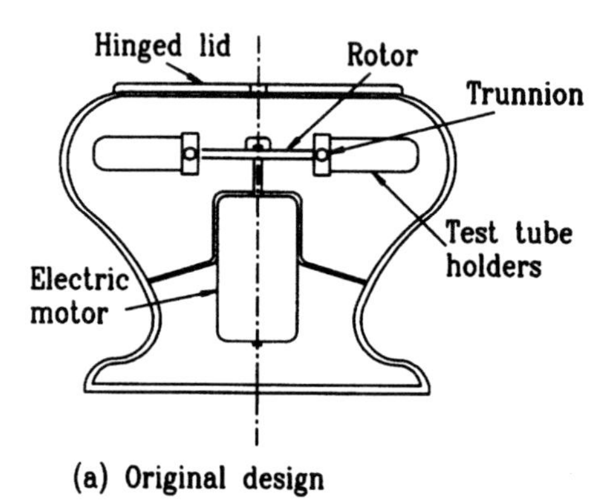

(a) Original design

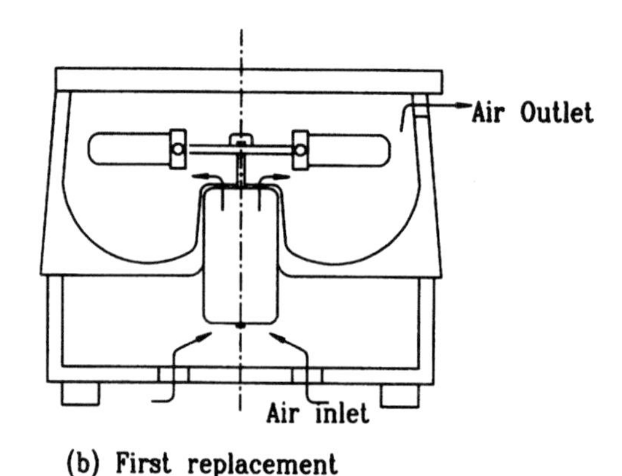

(b) First replacement

Figure 1.1 *Centrifuges with rotors spinning*

The new design appeared to more than satisfy the design brief, so was put into production with the minimum of delay. The initial reaction from customers was very favourable, and sales soared above expectations. However, after a few months problems started to appear.

The high volume of sales put unexpected pressures on the specialist suppliers of the body. Warranty claims also started to increase, particularly over the body. Over a period of time, areas of the structural foam were showing signs of deterioration caused by the vibrations of the machine in operation. Modifications in the form of metal inserts were introduced, but these simply added to the production problems of the supplier. As a result the cost of the body increased, adding to the already high cost of warranty claims.

A number of complaints from customers raised fundamental problems with the design. Most customers used the centrifuges to separate samples that would deteriorate if heated. The innovative idea of using the rotor to draw the cooling air over the motor resulted in warm air being passed over the samples, thus raising their temperatures. The motor employed was a DC electric motor that used carbon brushes. As the brushes gradually wore down, the air flow carried the carbon dust into the bowl area of the centrifuge and deposited it on the structural foam surface. This white surface rapidly attained a dirty appearance – not an endearing feature in a hygienic laboratory! To compound these problems the white surface was not easily cleanable, particularly if a blood sample was spilt.

The exhaust cooling air exited at the rear of the machine via a vent just below the lid, about 300mm above its base. The designers had tested the centrifuges on benches placed against the walls of their test area. Under these conditions the exhaust caused no problems. However, many customers used peninsular benches with technicians working on both sides. Not unreasonably, the person situated close to the exhaust was not impressed.

As if these complaints were not enough, the noise levels were too high, with the final straw being the failure of some of the locking mechanisms for the lid. Overall, an excellent design solution had, within a matter of months, turned into a disaster.

What went wrong? Detail mistakes had been made in a number of areas, but most (if not all) of these could have been avoided if the company had had a formal procedure for introducing new products.

At this point we will introduce the fundamentals of a new product introduction process, and then return to the case study and see what difference its presence could have made.

1.2.2 Need

The process must start with the identification of a need. This need must consist of two elements, the first being the existence of potential customers, and the second being the potential to satisfy the needs of the producing organisation. These aspects leading to the creation of a product design specification will be discussed more fully in Chapter 5, so for the moment we will concentrate on what happens once some conceptual ideas as to an appropriate solution have been devised.

1.2.3 Engineering model

Although most concepts are developed initially on paper, there comes a point where the concept must be proved to work in the physical world. This is the role of the engineering model which, although it may only be a very crude representation of the final product (indeed, it may only replicate a few essential aspects of the final product), must prove (or disprove) the principles involved. For example, a car manufacturer may produce a mock-up to test the effects of a new front-seat tilting mechanism on ease of access to the rear seat of a car. This may be a plywood representation of the car side, floor and door opening, with the only true vehicle parts being the new front and rear seats. This could be called an engineering model and used to test the practicality of the proposed concept. In many cases more than one engineering model may be required to evaluate alternative concepts and then possibly to refine the chosen solution.

1.2.4 Design detail

Many designs will incorporate elements similar to existing products. For these aspects the preliminary stages of the detail design work can take place in parallel with the concept proving stage. Once the engineers are satisfied that the concept is technically feasible the detail design can proceed in earnest. This detail work will involve in-depth decisions as to the size and type of bearings, motors, surface finish, tolerance requirements, etc. It will also include decisions as to the method of manufacture, the materials to be used, the need for special purpose machinery or tooling, whether existing components can be adapted, the advisability of making in-house or buying in, and so on. It is vital at the earliest possible stage to involve all within the organisation who can have a positive input. In other words, there is a need for an integrated approach.

1.2.5 Prototype

Today there are very few products where both performance and reliability are not crucial to the product's success. A poor reputation in this area, even if it only applies to the first few production batches, can be extremely hard to live down. A number of prototypes are normally built to generate some data on these aspects at the earliest possible stage. These will more closely represent the final product than the engineering model, but will almost certainly be hand-built rather than built with the production tooling, which will not at this stage be ready. It is very important that as the prototypes are being built, any change that the fitter implements during the build is recorded and then reflected in the drawings. The same, of course, applies to any modifications put in place during the prototype testing programme. The number of prototypes made depends largely on the costs involved. These costs are a mixture of the physical cost of producing the prototype and testing it, the time taken and the risks associated with minimising the testing. Note that there is also a risk of testing too much, in that this may delay the product's introduction date to the extent that a competitor steals the market.

1.2.6 Field trials

As the programme proceeds, final prototype models may be demonstrated or even lent to specified customers for final evaluation. These are called field trials and can prove invaluable in picking up those small omissions that could have otherwise marred the product's reputation. No matter how sophisticated an in-house test environment may be, it can do no more than simulate final use. It is fairly common for car companies, for example, to lend a particular model with, say, a new type of engine, to a fleet user such as a police force to get feedback as to the customer's opinion of the product.

In parallel with these tests and trials, the detail of the design and manufacturing processes is being refined and amended in light of feedback from the tests and trials. Production tooling will be being procured and the production facility made ready.

1.2.7 Pre-production

Once the production tooling and associated assembly lines have been installed, a pre-production batch is run through the system. This is usually a small batch, the primary purpose being to iron out any final production problems. Often early models are put through an accelerated life test so that the producer can have the earliest possible warning of any potential problems. Any reliability data generated from these will usually have greater validity than that generated from the tests on the prototypes, but of course will not be available until much later in the new product introduction process.

Finally the production stage is reached and the product is launched. From this point, for a successful product, the engineering input can be reduced but is rarely stopped altogether, as there will always be room for improvements to both the product and its production process.

The detail and emphasis of a new product introduction programme will, of course, depend on the organisation and the product, but the procedure as described and summarised in Table 1.1 will apply in most cases.

Now we will return to the case study and see how many of the faults could have been avoided if even the fundamental stages had been adhered to.

The first mistake the company made was in not recognising that the original product had a finite life; at some point someone was almost certain to

Table 1.1 *Product introduction process*

Stage	*Activity*
NEED	Market surveys
SPECIFICATION	Concept design
	Idea generation and evaluation
ENGINEERING	
MODEL	Prove concept
	Preliminary design
PROTOTYPE	Detail design
FIELD TRIALS	
	Refine the design
PRE-PRODUCTION	
PRODUCTION	
	On-going enhancements
SOLUTION TO A NEED	

introduce a better product. This mistake was then compounded by a misunderstanding of how the product was used by the customers. Problems such as the tendency to heat blood samples, generate excessive levels of noise, and blow the exhaust gases in the face of a technician when used on a peninsular bench, could all have been avoided by better knowledge of the market requirements. The product that was introduced met the needs that the company assumed the customer had, rather than the actual needs.

No formal product design specification (PDS) was produced, simply an instruction given that the new machine had to be aesthetically pleasing, with a locking lid and be available with the minimum of delay. No reference was made to reliability, ease of cleaning, the need for quiet operation, etc. The machine that emerged exceeded the basic brief by also including a tachometer, and having an increased carrying capacity, and without reference to any customers was put straight into production. The engineers had produced a machine that met their interpretation of the customers' needs and the company's requirement for a short development time. Had a full PDS been generated, the search for data required (as we shall see later in Chapter 5) would have forced the designers to define in more detail the specific needs of the customers and hence avoid some of the fundamental problems.

Although problems such as the excessive sample temperature rise on this particular machine would have been difficult to predict, the requirement to avoid heat build-up should have been specified in the PDS. This in itself would not have removed the problem, but it should have identified a characteristic to be tested at the engineering model stage.

Even if some of the problems had slipped through the net, the use of field trials should have identified virtually all the problems mentioned so far. They would, however, have had to be exceptionally extensive to have uncovered the problems that arose from the failure of the lid locking mechanism, or the deterioration of the structural foam body. These should have come to light with a programme of extended life tests on a number of prototypes. Of course, testing until a 100% level of confidence in the product had been achieved would not be practical. However, the duration of the tests and number of models tested should have been such that statistically the level of risk was understood by those responsible for sanctioning the start of production. This is an important concept that we will return to later.

For the moment we will return to the company and examine how they resolved the problems of the structural foam machine. The most important element is that the disaster of this product made them recognise the error of their ways. They introduced a product planning department responsible for ensuring that future products were directed towards the customers' needs. They also introduced a new product introduction procedure containing all the stages discussed above. A replacement for the machine was urgently needed but, to the company's credit, it surveyed a large sample of customers, and generated a formal PDS, the relevant sections of which were shown to selected customers. A formal programme of testing, including engineering models, prototypes and pre-production models, was undertaken. Some of the testing was done in-house and some with field trials. The final result was a product that was well received by customers, met the company cost requirements, and over a period of some years proved a very worthy product, thereby recovering the reputation of the company.

1.2.8 Integrated approach

We have concentrated on the benefits of a formal new product introduction process in terms of minimising problems that may arise after the unit is in production. The procedure does this partly by highlighting the stages and tests that need to be undertaken. In fact it does more than this; it encourages the very necessary liaison between those parties involved in the introduction process. For example, one of the parties, the customer, had partly been ignored. The need for integration goes much further than simply between the customer and the company; there must be very close liaison between all the relevant areas within the organisation. The prototypes will be built by skilled fitters whose level of engineering expertise will be much greater than that of the assembly workers who will eventually produce the final product. The designer must work closely with the fitter to identify any components which need to be 'fitted' together rather than assembled. The manufacturing engineer may also be involved at this stage, to get a closer view of potential problems that may affect the final assembly. If he (or she) does identify potential difficulties, the design may need to be modified or possibly an alternative assembly technique devised. Such changes may have knock-on effects on other aspects of the product. These may be physical, may delay the introduction date or add to the development costs. Generally, the earlier such modifications are identified, the better. It therefore follows that the earlier that close liaison between all those involved in the process can take place, the more likely a successful product will emerge.

Building the prototype is simply one example of the need for liaison between different areas; this need percolates through most aspects of the product introduction process. The marketing department should be involved with any new features that may be added (Are they really needed? Will the customer pay more if they are included?). Buyers, materials specialists, production staff, designers and test engineers should perhaps all be involved if a new material or technique is proposed.

1.2.9 Control system

In general, although the process may start as a one-person exercise in the marketing department, it will soon mushroom into requiring a much wider level of participation. At this point it is worth reminding ourselves of the reason for introducing a new product. It is to ensure the continuing health and growth of the company, i.e. to make a profit. This can obviously only be achieved by producing a product that the customer is willing to buy but, equally obviously, it must be possible to produce the product at a cost that will enable sufficient profits to be generated. It follows that any new product introduction process must maintain control of costs and timescales just as rigorously as the physical aspects. The costs include both the development or introduction costs and the actual production cost of the final product.

At the start of the programme there are estimates or targets covering the unit cost of the final product, the costs of tooling, the time to introduce the product, the reliability and functionality of the product, etc. As the programme progresses these estimates become firmer. For a given product, the greater the time and money spent on its development, the greater the level of confidence will be that the product will meet its unit cost, functionality and reliability targets. However, it also follows that the extra development costs will have to come from somewhere, i.e. the profits from the product. It is no good having an excellent product that is unable to contribute to company profits because its development costs were too high. A compromise has to be reached and judgements made in the knowledge of the level of risk.

Normally a new product introduction programme would have a number of critical review points where decisions would be made to proceed (or not) to the next review point. The first of these points could be called a marketing appraisal. This could include marketing information such as sales estimates (volume and income), identification of markets, competition, a PDS, and estimates for the costs of proceeding with the introduction programme. This information would equate to data on potential profitability. If the figures looked satisfactory a decision would be made to finance the next stage.

The next review point could be technical feasibility, where some preliminary design work and an engineering model could generate a potentially suitable concept. The more detailed knowledge gained would enable the estimates for both the final unit cost and introduction costs to be refined, so that a decision to move to the detail design stage, build a prototype, and possibly commit some tooling to be produced would be taken.

At each of the subsequent review points the costs, reliability data and market information would all be updated and refined, so that although the authorisation to the next review point would normally involve an ever increasing commitment, the decision would be made with an ever more accurate set of information.

The precise number of review points will vary according to the type of organisation and complexity of the product. These will be discussed further in Chapter 17.

1.3 Summary

Although in this chapter we have concentrated on the introduction of new products, the same principles apply equally well to modifications or changes to existing products. On completing this chapter you should:

- Be able to recognise the potential problems that can arise when introducing a new product and appreciate the need for an integrated approach that involves close liaison with a wide number of skills within the organisation.
- Understand that, although decisions made have to be taken with less than 100 per cent knowledge of the facts, a systematic approach should reduce the risks to an acceptable level.

1.4 Questions

Now use the following questions to check your understanding of this chapter. The relevant section of the chapter is shown in brackets.

1. Name the principle stages in the development of a new product. (1.2)
2. What do you understand by the term 'engineering model'? (1.2.3)
3. What is the primary purpose of a pre-production run? (1.2.7)
4. Should customers be involved in the product development process? If so, at which stages? (1.2.2, 1.2.6)

Part 1
The Basics of Design

2 Records and Communications: The Engineering Drawing

2.1 Introduction

Introducing a new product, or an improvement to an existing one, is normally a very straightforward process for a one-man business, where the designer, builder and salesman are one and the same person. Today, very few engineering products are produced this way. The product complexity and sophisticated controls needed in the producing organisation (to remain competitive), have resulted in organisations splitting into departments specialising in particular areas of the business. These may include design, production engineering, purchasing, testing, sales, etc. The new product introduction process as introduced in Chapter 1 indicates that for almost all organisations a number of departments and people are going to be involved in the process. It is essential that an efficient means of communicating the details of the engineering artefact is used.

The graphical form is ideal for quickly conveying complex ideas in an unambiguous form. Anyone attempting to write a complex description without reference to diagrams will soon appreciate the saying 'a picture is worth a thousand words'. If a set of rules and procedures can be agreed and adopted for the methods by which the 'pictures' are to be set out, and if sufficient views are drawn, any assembly can be described completely and exactly without ambiguity, no matter how complex the assembly may be. This chapter does not pretend to be a full drawing manual, but will explain the basics of these rules. Access to a copy of BS308, or a copy of PP7308 to hand, would normally be needed to cover the rules in greater detail when completing a drawing.

Although the graphical format is extensively used, the written form also has to play its part. The last section of this chapter will cover the need to maintain records and the use of a working file:

2.2 Engineering drawings

2.2.1 Equipment

As with the written word, engineering drawings can either be produced manually or with the help of a computer. The use of CAD (Computer Aided Design) will be covered later, in Chapter 16, so for the moment we will concentrate on manual drawings.

Original drawings are normally produced on a transparent medium, either tracing paper or a film made of a material such as polyester. Tracing paper is cheap, but it can stretch or shrink with temperature and humidity changes. The film is more expensive but compensates by being much more stable. When complete, paper copies of the drawing can be made by passing the transparent medium through a dyeline machine. The resulting copies (prints) can then be distributed throughout the organisation as required, with the original remaining in the drawing office.

Storage can become a problem in a prolific drawing office where the larger sizes of drawing sheets are used. One technique to save on storage space is to photograph each drawing and mount the resulting 35mm slide on a card. This is called microfilming. The card can be inserted into a special viewing machine, where an image of the drawing is back-projected onto a screen. These machines often also have the facility to produce a print of the drawing.

Drawing paper or film comes in a series of standard sizes.

Note that each size is a multiple of the next smaller size. This means that any size of drawing can be folded to an A4 size, and will therefore be able to fit into a standard folder or report.

A0	841mm × 1189mm
A1	594mm × 841mm
A2	420mm × 594mm
A3	297mm × 420mm
A4	210mm × 297mm

The basic equipment needed to produce a drawing is:

- Clutch pencil *(preferably two, one with 0.3mm and the other 0.7mm leads)*
- Compass
- Adjustable set-square *(not required if the drawing board has a draughting machine)*
- Eraser
- Eraser shield
- Drawing scale
- Plastic templates for small radii
- PP7308, or access to BS308

2.2.2 Drawing layout

As the engineering drawing is simply a means of transferring information between individuals, it follows that standardising the format should minimise any unnecessary confusion.

Title block: All drawings must have a title block, which should be placed in the lower right-hand corner and contain a minimum of

- Draughtsman's name
- Date completed
- Projection used (see section 2.2.4)
- Scale
- Title
- Drawing number

Placing this basic information in the same place on all drawings means that prints can be folded to a common A4 size so that this reference information is at the same place on the folded drawing.

Scale: Formal drawings should be to a uniform scale. The drawing can be full size 1:1 or should be to one of the recommended scales:

1:2, 1:5, 1:10, for smaller than full size
2:1, 5:1, 10:1, for larger than full size.

The scale should always be stated in the title block. However, it is bad practice to measure a dimension on a drawing physically, as the paper print is almost certain to have shrunk or stretched. It is good practice to add a note to the drawing: 'DO NOT SCALE. IF IN DOUBT, ASK'.

2.2.3 Sketching

Although most engineering drawings are formal in layout there are many occasions when it is very useful to generate a simple freehand sketch to emphasise the point being made. Such occasions could include discussions with suppliers, a shop-floor operative, or one's boss.

Although a great deal of skill is required to produce a high-quality sketch, by using some simple techniques most of us should be able to generate basic sketches that assist in the explanation of a particular point. For example, the creation of a freehand circle can often be simplified by first sketching a square whose sides define the size of the circle. Knowing that the circle must pass through the mid points of each of the sides of the square helps to maintain the correct proportions.

The most useful sketch is a 3D view of the object. Probably the most commonly used format is an *isometric sketch*. An object, such as the cube in Figure 2.1, is positioned with a vertical edge facing the viewer and then tilted through just over 35° so the top surface is also visible. The

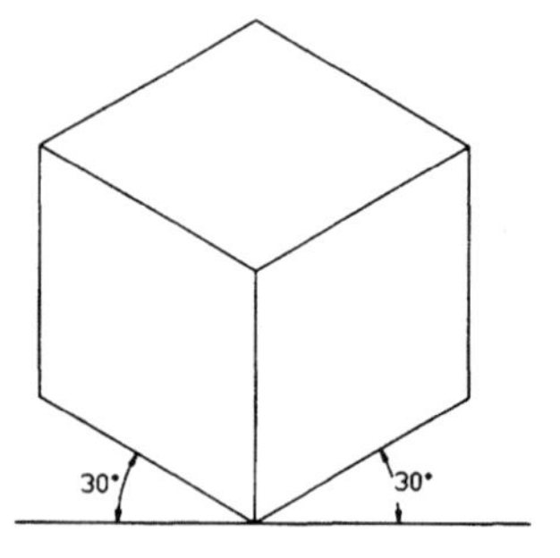

Figure 2.1

lower edges of the cube lie along what are called the isometric axes. These each form an angle of 30° to the horizontal. The success of a sketch depends to a large degree on getting these axes correct. Isometric graph paper (where the axes are defined) can be a great help for the beginner.

Creating isometric views of objects with straight edges is fairly easy. Circles provide a greater challenge. The steps in Figure 2.2/3 describe a simple technique that can be used to sketch an approximate isometric circle. In reality, it is a 3D variant of using a square to help draw a 2D circle.

1. *Draw an isometric view of the top face of a cube whose side length equals the diameter of the required circle, i.e. an isometric square.*

2. *Draw construction lines joining the mid points of opposite sides, and add a diagonal joining the two corners with the* smaller *angles.*

3. *Using one of the corners (A) with the* larger *angles as the centre, draw an arc from B to C. (Remember, you are creating a sketch, so this should be completed freehand, imagining that the arc centre is at A.)*

4. *Now add two more construction lines, joining the ends of the arc BC with its centre A. The intersection of the lines with the diagonal creates points D and E.*

5. *Using D as the centre, draw an arc radius DB as shown on Figure 2.2.*

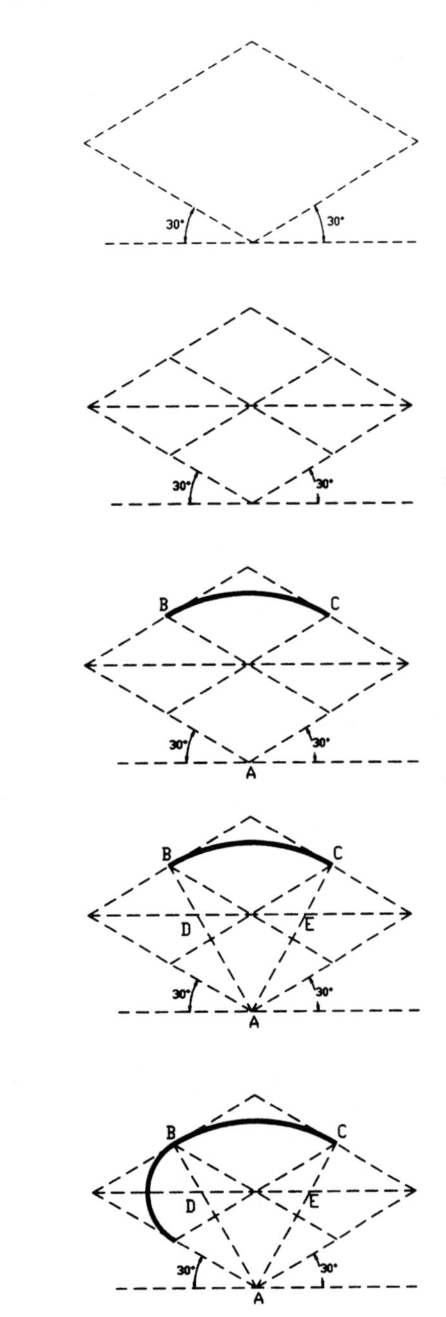

Figure 2.2

6. *Using E as the centre, draw an arc radius EC as shown.*

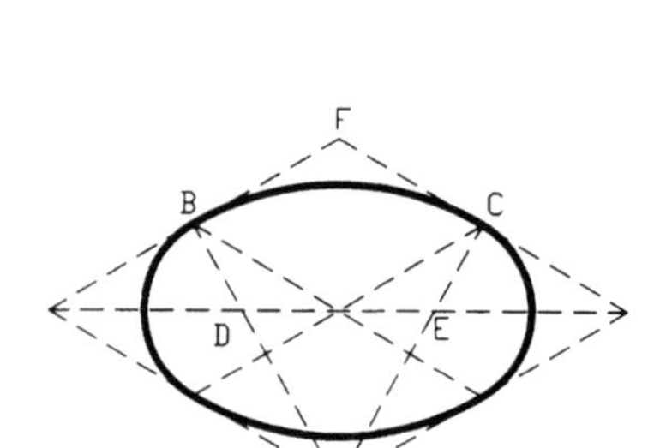

7. *Finally, to complete the isometric circle, repeat step 3 using the remaining corner with the* large *angle F as the arc centre (see Figure 2.3).*

Figure 2.3

This technique can be applied equally successfully to any side of an isometric cube, provided the diagonal is always drawn joining the corners with the *small* angles. It can also be used to create arcs representing features such as filleted corners and edges.

Generating a reasonable sketch of a simple shape such as a cube is fairly easy. Most items that we need to sketch can be divided into cube-type shapes. The principle is to sketch lightly the 'rectangular boxes' enclosing the various features roughly to scale, before adding the detail. Figure 2.4 shows how this can be done.

Sketches can be improved by the addition of suitable shading, and by recognising those features that need highlighting, and those that could perhaps be ignored.

There are, however, severe limitations to the usefulness of sketches in describing a component precisely. Try defining the size of all the features, the positions and sizes of all holes, etc. on a sketch and its limitations will become apparent. In fact, unless the item is very simple, a number of sketches are almost certain to be needed (to show those hidden faces). A rigorous technique that can be learnt, rather than depending to a degree on artistic skill, is needed. Such a technique is *orthographic projection.*

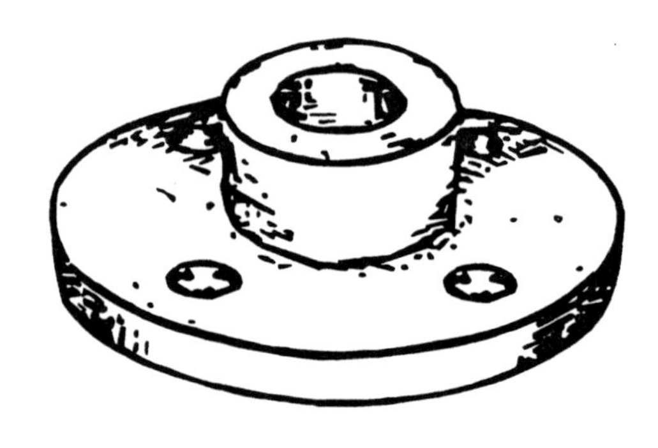

Figure 2.4 *Sketch*

2.2.4 Orthographic projection

In order that drawings can be understood easily and with a minimum of confusion, a set of rules, BS308, covering their presentation, has been generated.

Take a simple component such as a die. Say we wish to define the location of the spots on one of its faces. The simplest way to do this is to draw a square representing the face, and draw in the spots in the correct positions. It therefore follows that if we wish to fully define all faces of the die we need to draw 6 diagrams, each representing a different face. These 6 diagrams alone would not fully define a die. We need additional information to define the relative positions of the faces, e.g. the 'six' face is always opposite the 'one' face. Orthographic projection provides us with a simple set of rules for defining the relative positions of the faces of the die.

Orthographic projection involves viewing an object on orthographic planes (i.e. planes that are at right angles to each other). Imagine such planes as in Figure 2.5 (a). We will call the top right-hand segment the first angle quadrant.

Now if we place the item to be drawn in this quadrant, view it from above and draw what we see on the plane directly below the object, we will have created what is called a plan view. If we view the object from the front and project what we see onto the plane directly behind the object, we will have created a front view.

For the end view to be shown we need to imagine a third orthographic plane, and view the object as Figure 2.5 (c).

Now assume the views on each plane are permanent, remove the die, and unfold the planes so that they lie flat. The result is a *first angle orthographic projection* of the die. Although only three views are shown, the remaining sides of the die could be shown by projecting them onto the other three orthographic planes that would have formed a box enclosing the die. Note that all the views are to scale and that the edges line up as appropriate.

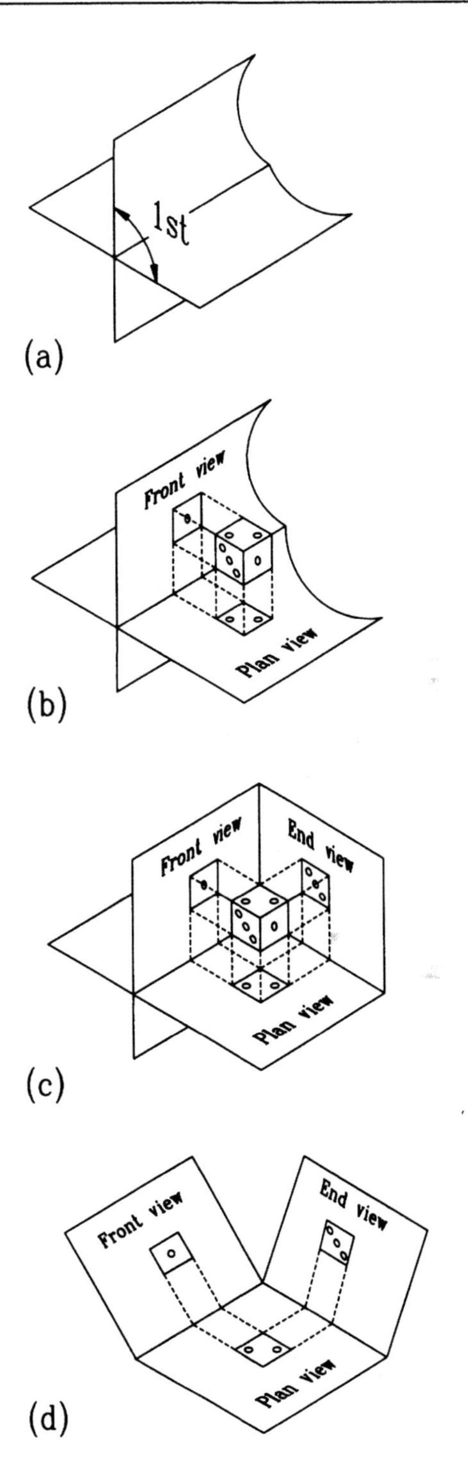

Figure 2.5 *Development of first angle projection*

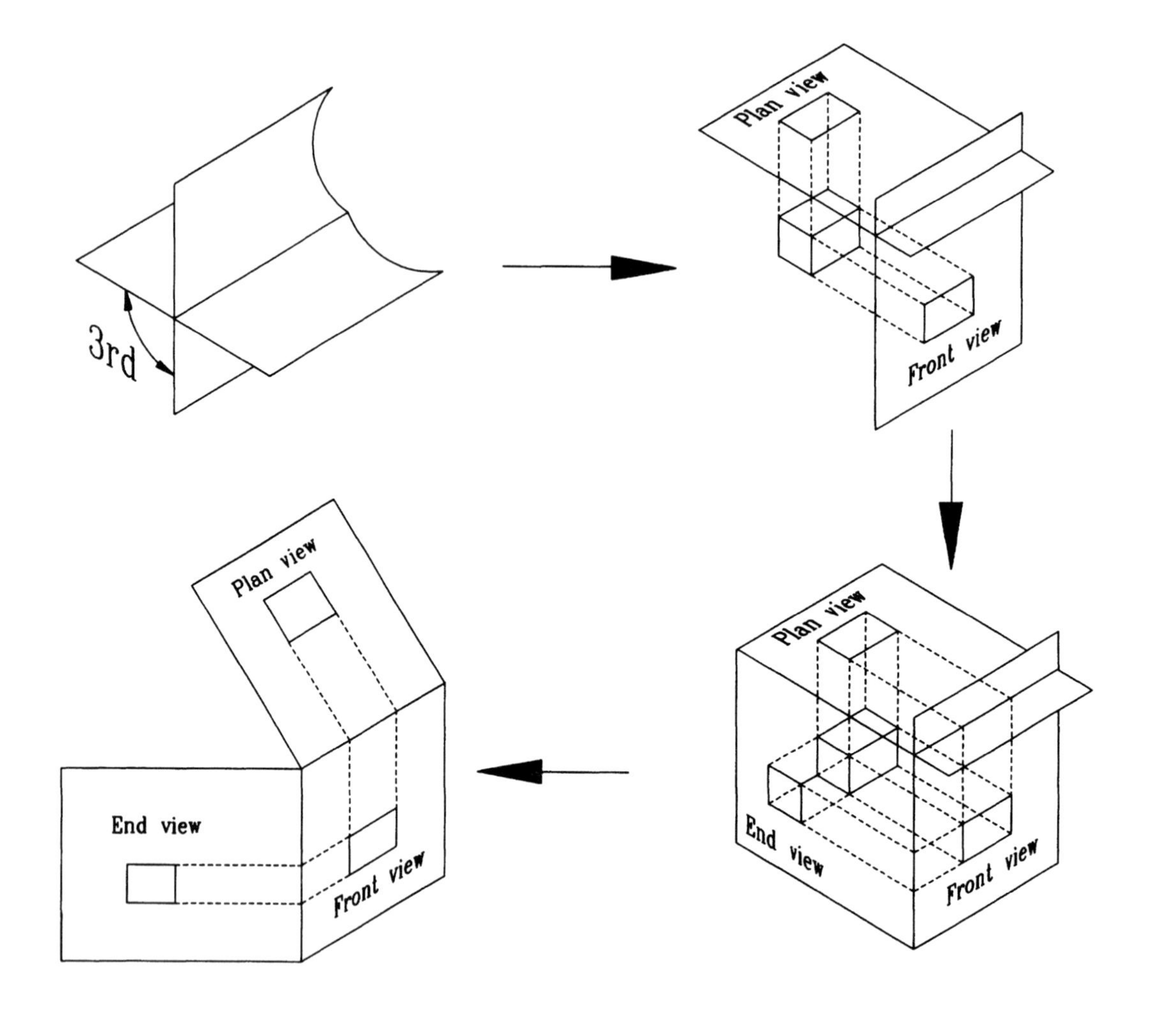

Figure 2.6 *Development of third angle projection*

There is another equally useful version of orthographic projection called *third angle projection.*

Let us return to our two orthographic planes and place a box in the segment called the third angle (Figure 2.6). This time imagine the planes are the sides of a glass box and that we are viewing the box through the planes. We then draw the view of what we see on the appropriate side of the glass box. As before, the drawings are projected directly from the object and should be both to scale and in correct alignment. Again, the end view is generated by adding a third side to the glass box. Finally, as before, the box is removed and the 'glass box' unfolded to reveal the drawing, this time in third angle projection.

With use these systems soon become second nature to the engineer.

In both forms of projection the individual views are the same, but their relative positions on the paper are not. Figure 2.7 shows a simple component in both first and third angle projection. You should take particular note of the differences.

Most engineering organisations will standardise on one form of projection, with US firms

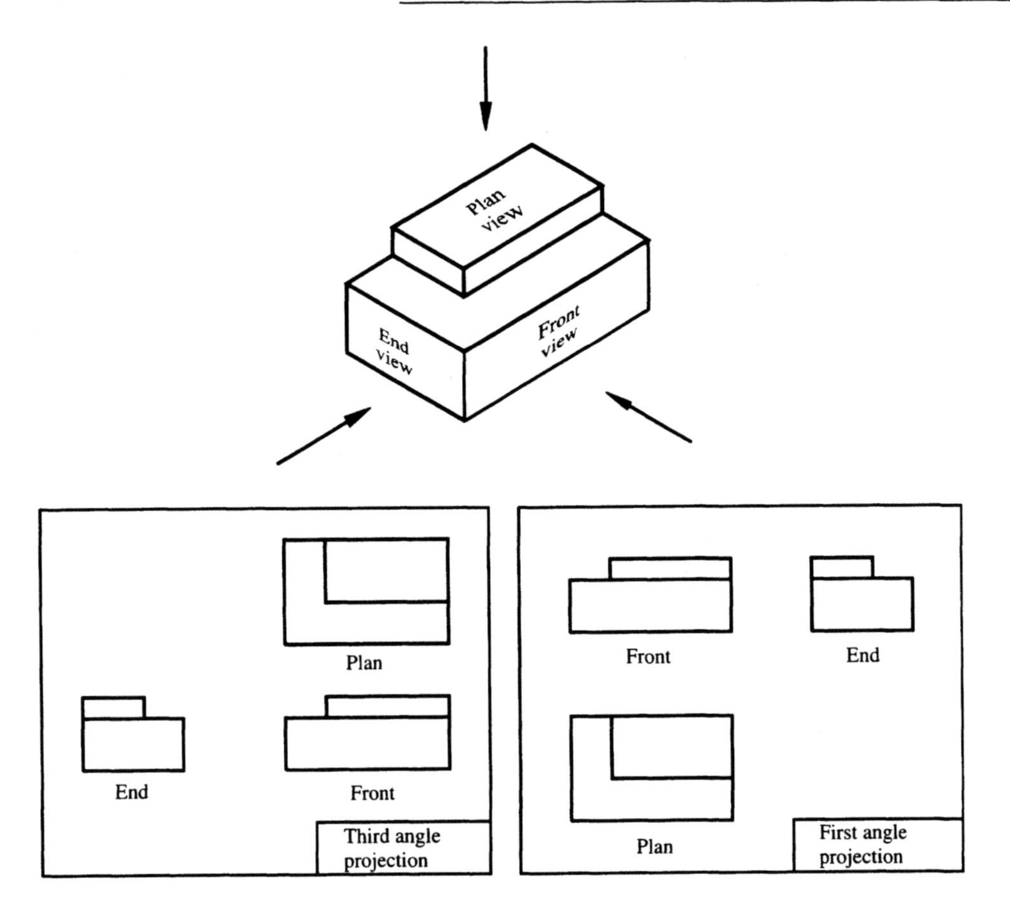

Figure 2.7 *Comparison of first and third angle projections*

having a preference for third angle and the European organisations first angle. To avoid any unnecessary confusion, all drawings must state which form of projection they are using. This can be done either in words or by using the appropriate BS symbol, as in Figure 2.8.

Each of the examples given so far has shown three views. Indeed, for many components, three views will prove sufficient. However, you should appreciate that either projection system will allow the generation of up to six orthographic views. (The 'glass box' fully enclosing the component would have six sides.) In practice this is rarely necessary. A good drawing will not include any more views than are actually needed.

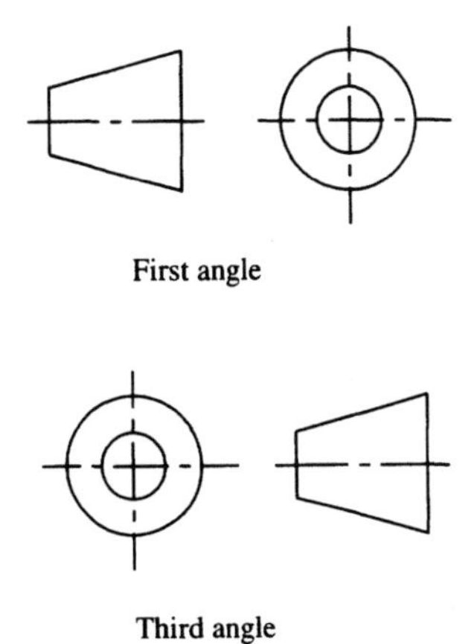

Figure 2.8 *Projection symbols*

2.2.5 Sections

Hidden detail: The projection systems as described only show the outside of a component, whereas in many cases vital details are hidden from view. Such hidden detail may be shown by using thin dashed lines. The bracket (Figure 2.9) would be represented by the drawing Figure 2.10).

Sections: Although the use of dashed lines works well with the bracket shown, if there is a lot of hidden detail the drawing can be complex and difficult to interpret. In these cases, sections should be taken. An imaginary plane is used to cut away those parts that are hiding the detail that needs to be shown. If the cutting plane cuts through solid elements of the component this is normally indicated on the drawing by hatching these areas (see later for exceptions to this). The remaining part (or section) is shown using the normal projection rules. For example, on our simple bracket example the cutting plane AA has been used to reveal more clearly the detail of the lug in Figure 2.11. Note that the sectioning plane has been identified by a thin chain line, thickened at the ends where the identifying letters are shown. The section must also be identified, e.g. Section AA. These drawings have been produced using first angle projection. Note the use of the appropriate symbol. Also note that the axis of the hole has been identified in each view with a thin chained line. This is normal practice on engineering drawings.

Hatching: Hatching is used to indicate those parts of a component cut by a sectioning plane. The hatching lines are normally thin continuous lines drawn at 45° with equal spacing to suit the drawing scale, preferably more than 4mm apart. A single component should have common hatching, but adjacent components must have different hatching patterns. The patterns can differ by varying the angle and spacing of the hatching lines. Indeed, on a complex section this is one way the engineer can tell one part from another.

Although there is a general rule that all material cut should be hatched, there are some notable exceptions. When the sectioning plane passes longitudinally through fasteners such as nuts and bolts, and features such as shafts, webs and ribs-hatching is not required.

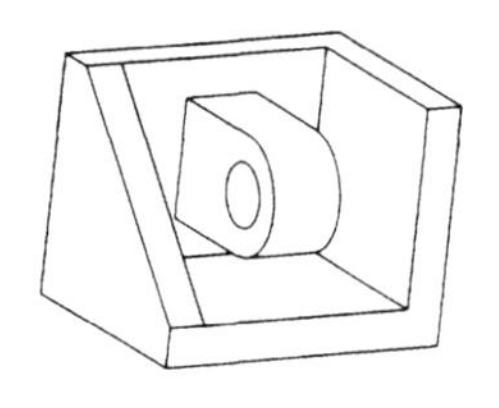

Figure 2.9

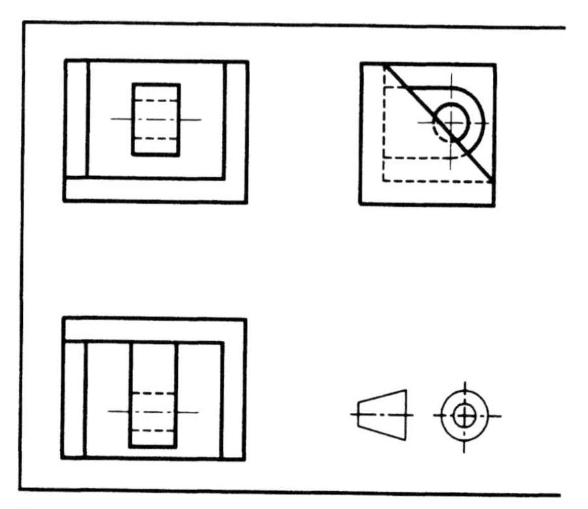

Figure 2.10 *Hidden detail*

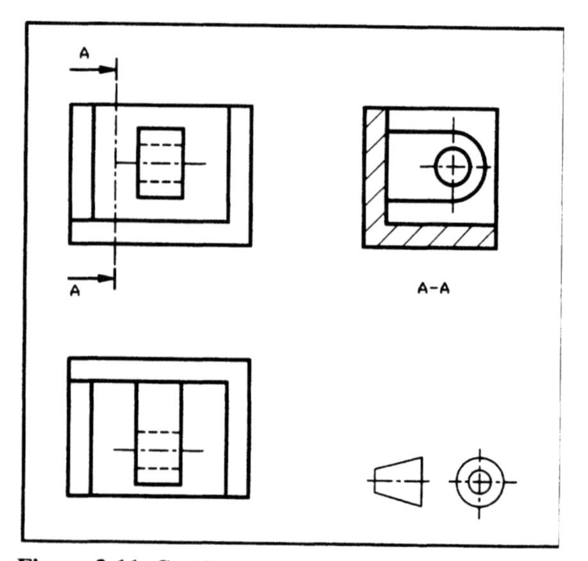

Figure 2.11 *Sections*

2.2.6 Auxiliary projection

Orthographic projection works well with many components, but consider a component such as the cast bracket in Figure 2.12. This item has an angled face containing some holes that need to be detailed. If simple orthographic projection is used the holes will appear as ellipses, thereby being both difficult to draw and to understand. The most useful view would be one at right angles to the angled face. What is needed is an auxiliary view, i.e. another view projected onto an auxiliary plane that is parallel to that of the angled surface.

The process of generating the auxiliary view is best described by taking a simple rectangular block (Figure 2.13) and creating the extra view (Figure 2.14) by using the following rules:

1. We need two orthogonal views, plan (view A) and front (view B) to create the auxiliary view (C).
2. Using the corner numbers as in Figure 2.13, label view A and B (Figure 2.14), the numbers for the hidden corners should be in brackets.
3. Draw projection lines at the required angle from view B to enable view C to be drawn.
4. Draw a reference line AA between views A and B at right angles to the projection lines, preferably touching a point on view A.
5. Draw reference line CC at right angles to the projection lines between view B and view C (the view to be drawn).
6. Measure the distance of corner 1 on view A from the reference line AA.
7. Plot this point along the relevant projection line (view B to view C), the same distance from reference line CC.
8. Repeat for the remaining corners.
9. Lightly join the plotted points, taking care to join only corners that are joined on the original. Decide which lines are hidden and complete.

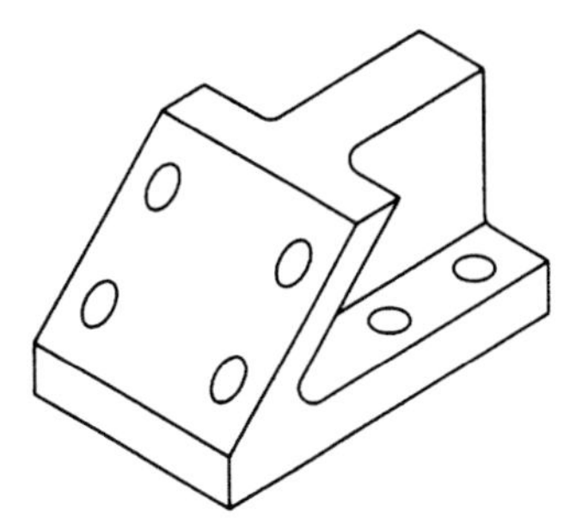

Figure 2.12 *Cast bracket*

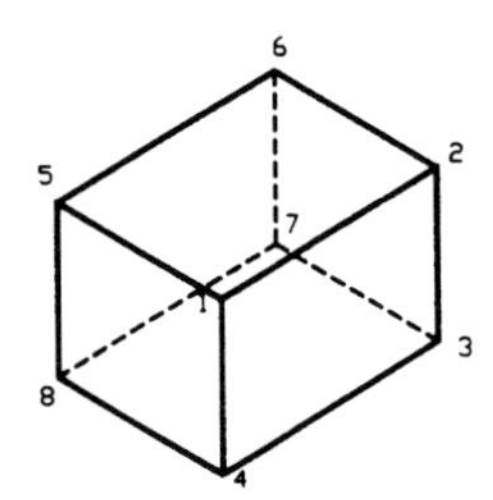

Figure 2.13

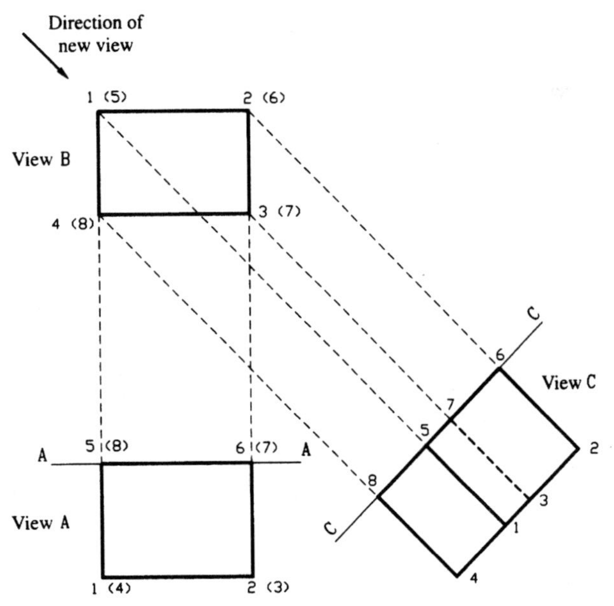

Figure 2.14

The reader should note that in Figure 2.14, although view C can be taken at any angle to view B, these two views are still orthogonal. It is therefore possible to use these two views to create another auxiliary view, Figure 2.15. Note the corner numbering stays the same, but the labelling of the reference lines and the views has been changed so that the above rules can still apply. We are, in effect, rotating the rectangular block and recording the views seen. Try picking up a rectangular object such as an eraser, and viewing it from different angles to check the views shown.

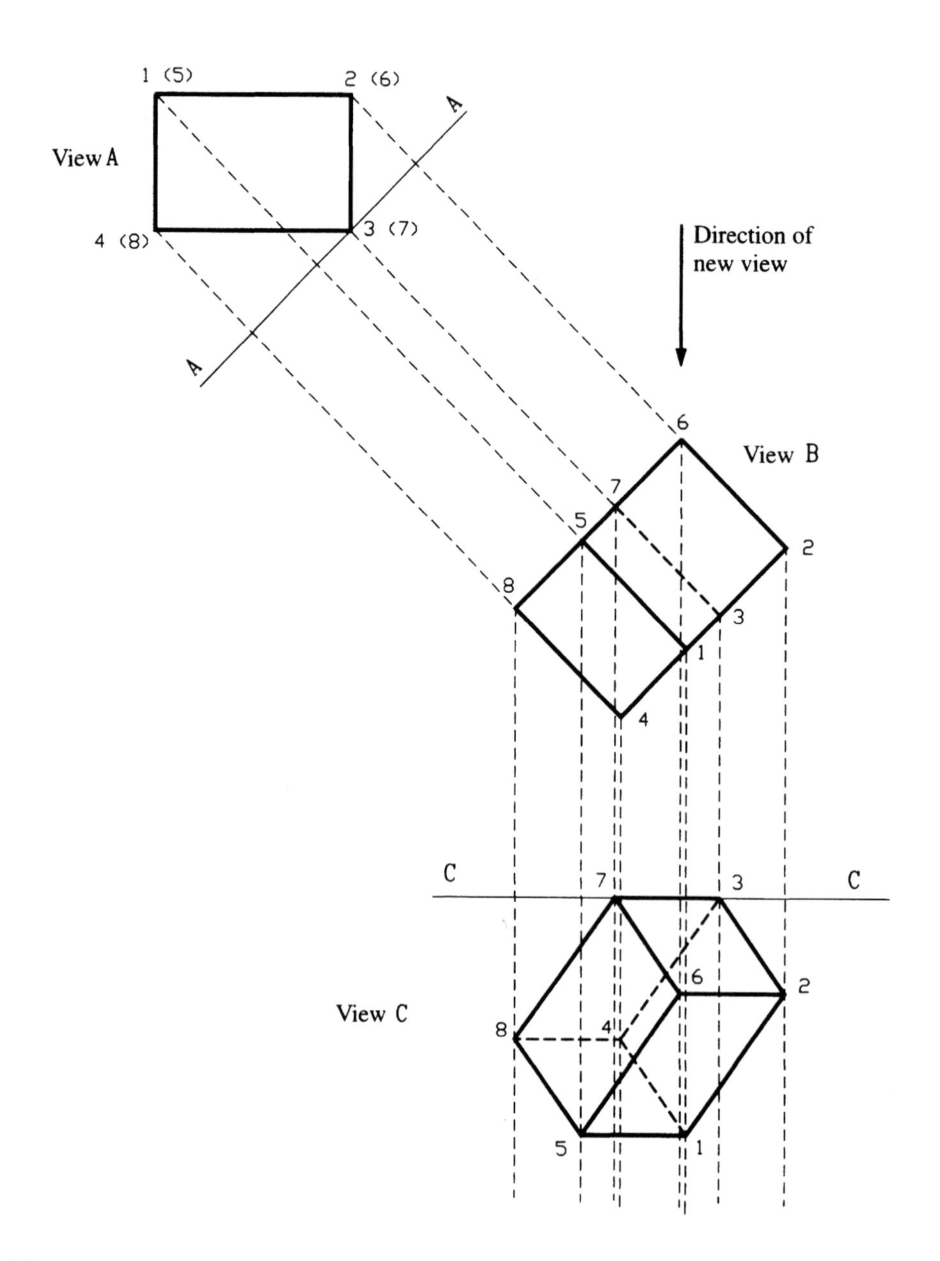

Figure 2.15

2.2.7 Types of drawings

Detail drawings

In general, each engineering item that is to be manufactured will need a separate drawing. These are called detail drawings and should normally include all the necessary information required to manufacture the component (see Figure 2.16). This information must include:

- All necessary dimensions and tolerances
- Raw material details (size, form, type)
- Surface texture requirements
- Any finishing processes (e.g. painting).

Details of these will be described in later chapters. In some cases, more than one drawing of a component is needed. For example, an item produced from a casting will have a casting drawing showing the essential details of the 'as cast item'. A separate machining drawing will also be needed, dimensioning only those details requiring machining. This drawing will identify the casting as its raw material. Remember that the purpose of the drawing is simply to pass information on. The machining drawing is the instruction sheet for the machinist, and therefore should not include dimensions of as-cast elements that are of no interest to him or her.

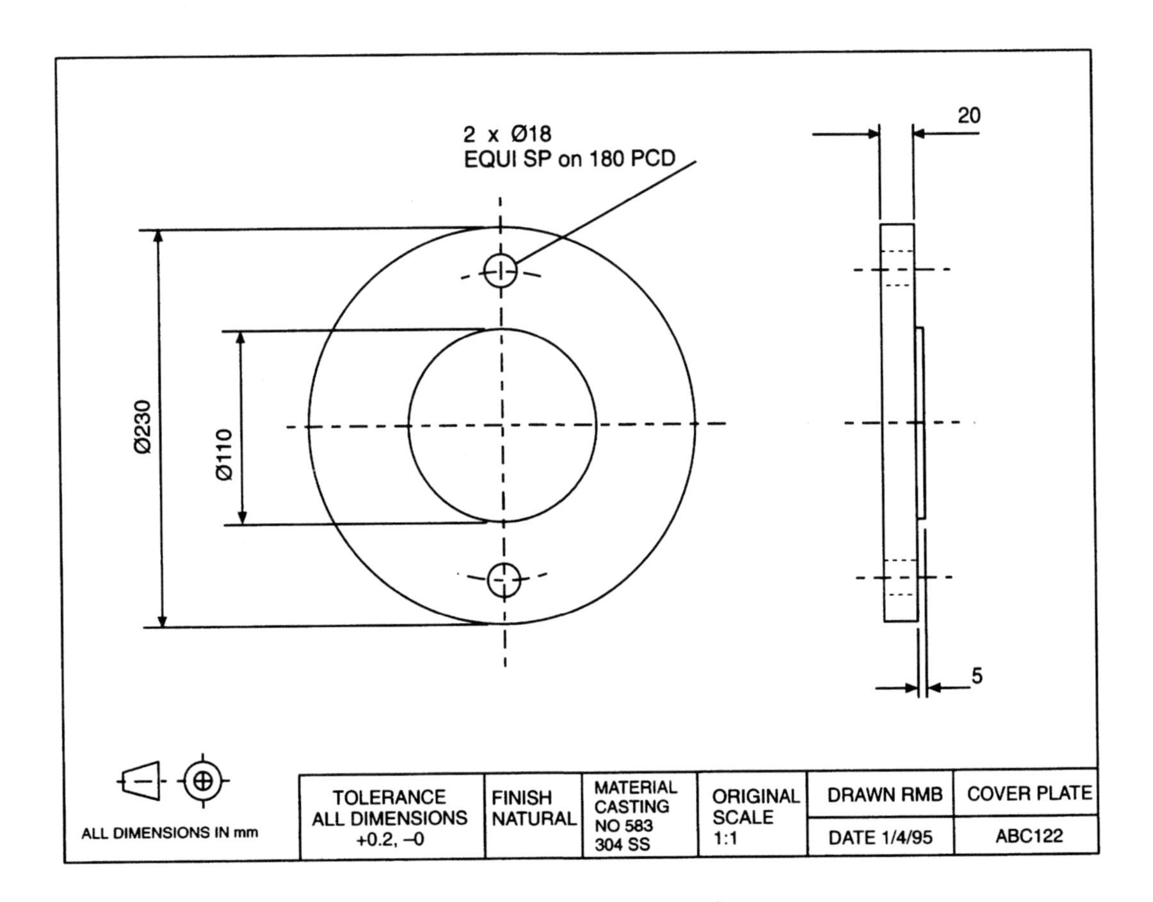

Figure 2.16 *Drawing detail*

Assembly drawings

In many cases the component will form part of a larger assembly. If this is so, an assembly drawing (sometimes called a general assembly or GA) will be required (Figure 2.17). The purpose of this drawing is to show how the parts fit together. The designer would normally generate a layout drawing as part of the design process to help in the evolution of the components. The assembly drawing will be a formal development of the final layout.

It does not normally need to include dimensions, other than those that might be needed during the assembly process, such as a clearance that may need to be set. However, it must identify all the individual components needed. It does this via a parts list which should contain a minimum of 4 columns:

- Item number
- Number off
- Item description
- Item reference number.

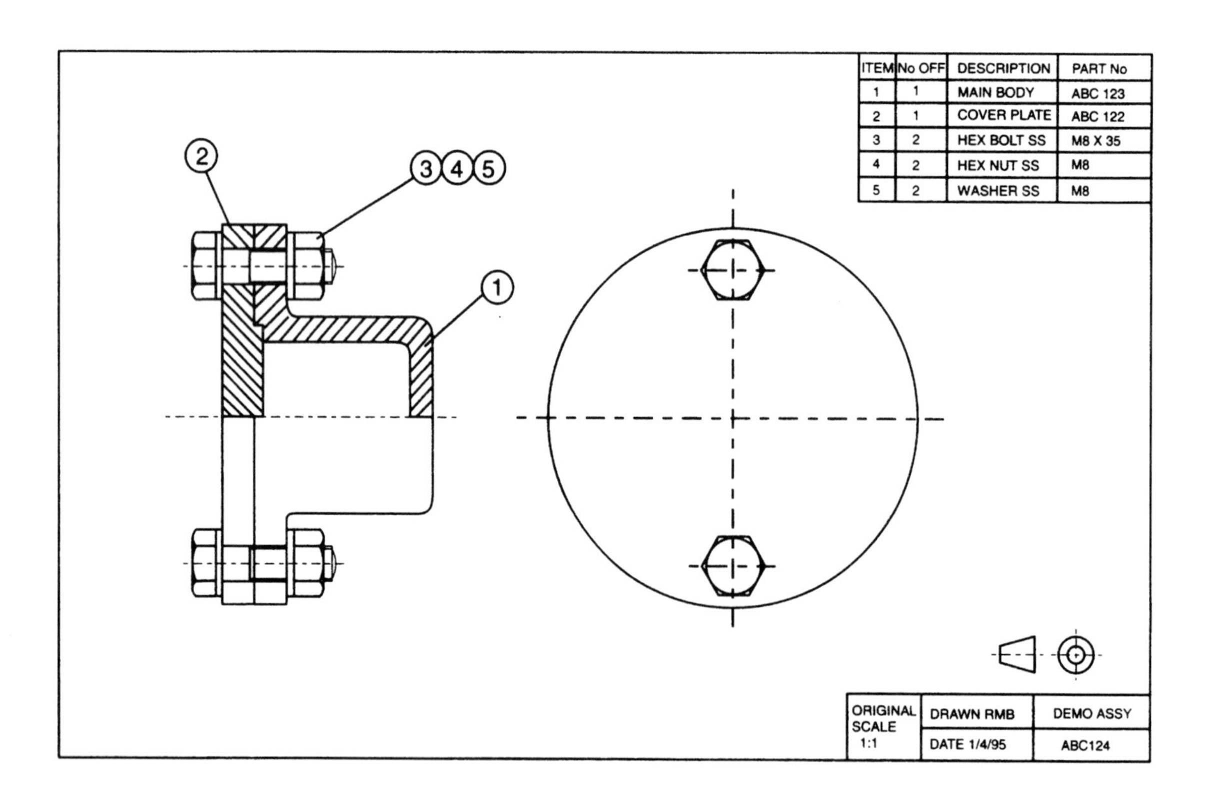

Figure 2.17 *Assembly drawing*

2.2.8 Dimensions

So far we have only covered the representation of the shape of the component on the drawing. Even though this should be to scale, the drawing's usefulness is very limited unless relevant dimensions are added. There are two main aspects to dimensioning: how to show the dimensions, and which features need to be dimensioned. The second of these is covered in Chapter 3. We will now introduce the basics of adding dimensions.

Sizes are added by using dimension and projection lines, as in Figure 2.18. Dimensioning should obey the following basic rules:

- Keep dimensions outside the component boundary.
- Avoid crossing of dimension and projection lines.
- Do not include units on each dimension: add a general note to the drawing (All Dimensions in mm).
- Centre lines can be used as projection lines, but should never be used as dimension lines.
- Use centre lines (chain-dashed) to indicate centres and axes of holes, as well as features of symmetry.
- Hole positions are indicated by the location of the hole centre.
- Hole size is always given by diameter, never radius. Use Ø to indicate diameter.
- Do not over-dimension, as this can lead to confusion and errors. Each dimension should only be indicated once on the drawing. Auxiliary dimensions are the exception to this rule, and are only added for clarification. They are not toleranced and should be shown in parentheses.

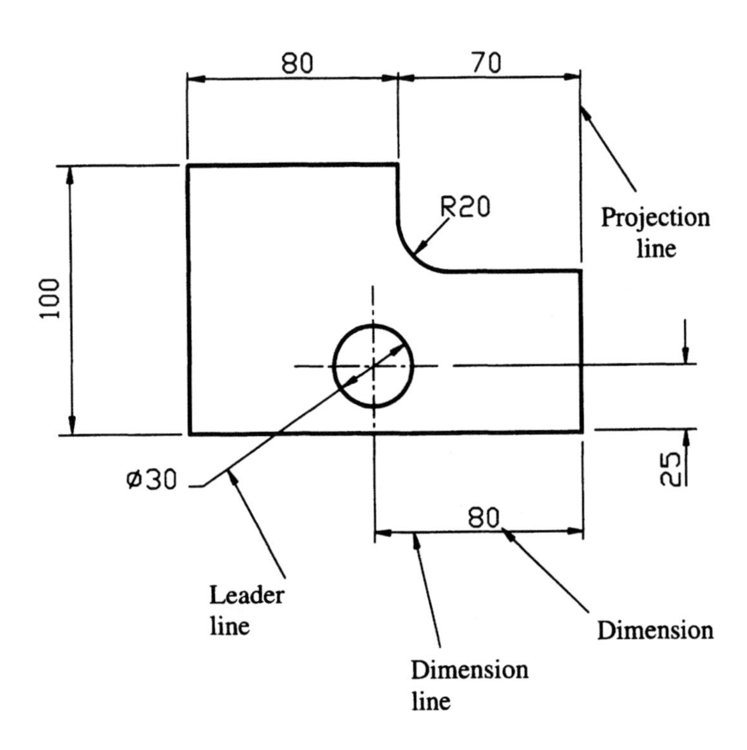

Figure 2.18

Figure 2.19 indicates a number of acceptable alternative ways of dimensioning circles and diameters. Never lose sight of the purpose of the drawing: to communicate specific information. Choose the dimensioning method that appears simple and easy to understand.

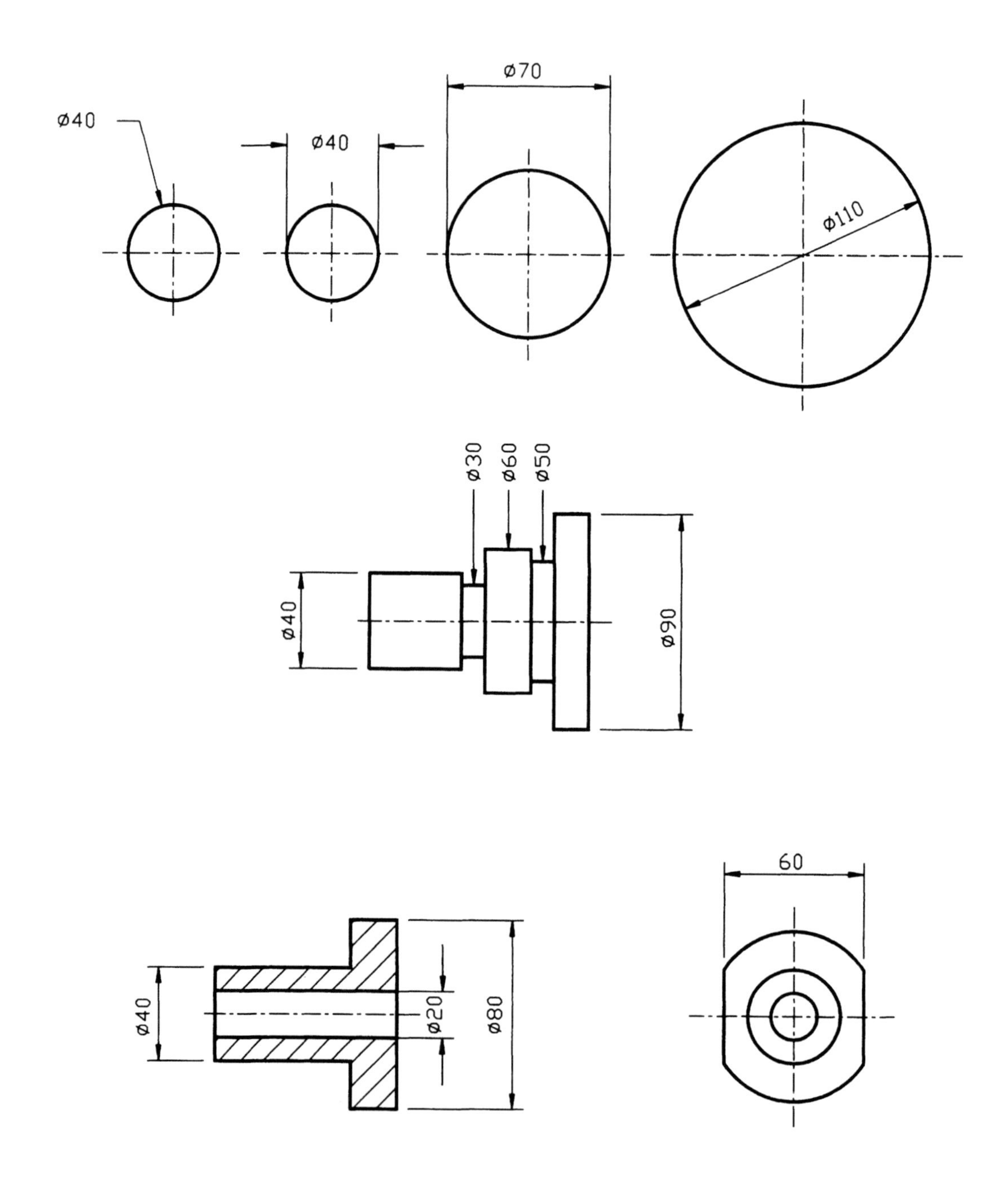

Figure 2.19

2.2.9 Abbreviations and standard components

BS308 recognises that certain features and words appear frequently on drawings, and allows abbreviations and drawing shortcuts to be used in a number of cases. Some of the acceptable abbreviations are listed below. See PP7308, or BS308 for a full list.

Assembly	ASSY
Centres	CRS
Centre Line (on drawing)	℄
Centre Line (in note)	CL
Countersunk	CSK
Counterbore	CBORE
Cylindrical	CYL
Diameter (in note)	DIA
Diameter (on dimension)	Ø
Drawing	DRG
Equally Spaced	EQUI SP
Hexagonal	HEX
Head	HD
Material	MATL
Maximum	MAX
Minimum	MIN
Pitch Circle Diameter	PCD
Radius (in note)	RAD
Radius (on dimension)	R
Required	REQD
Sheet	SH
Spotface	SFACE
Standard	STD
Thread	THD
Typically	TYP
Undercut	UCUT
Weight	WT
Volume	VOL

Nuts, bolts and screw threads are amongst the most common features to appear on engineering drawings. Figure 2.20 indicates acceptable proportions of hexagonal bolt heads. Note that nuts are usually a little thinner than bolt heads, so 0.7D would be more appropriate for a nut thickness.

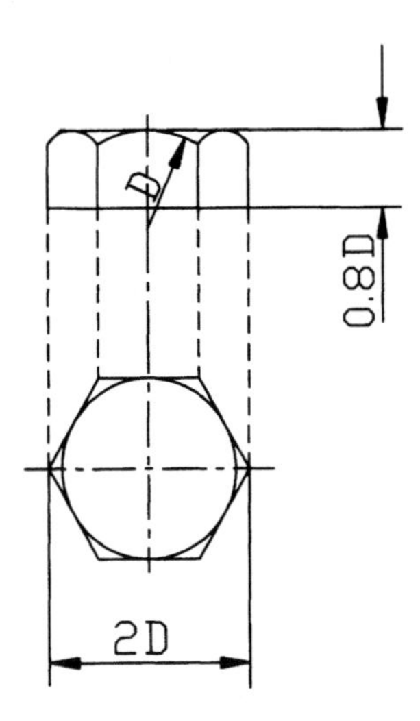

Figure 2.20

Standard threads are shown in Figure 2.21. Note the treatment of the hatching on the internal thread. Note also that the tapped thread does not exist for the whole depth of the hole. This is to allow clearance for the tap (see Chapter 13). Detail drawings for standard items such as nuts and bolts are not normally required. They will be shown on the assembly drawing and called up in the parts list, where the type of head, thread and length will be specified. Students should study a copy of BS308 (or PP7308), where acceptable representations of other standard components such as rolling element bearings, springs and gears are shown.

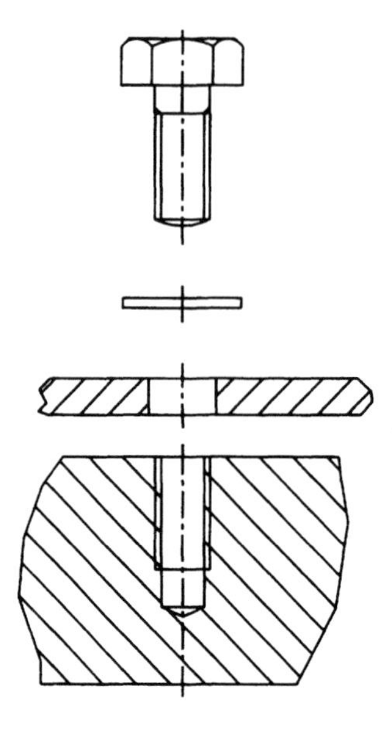

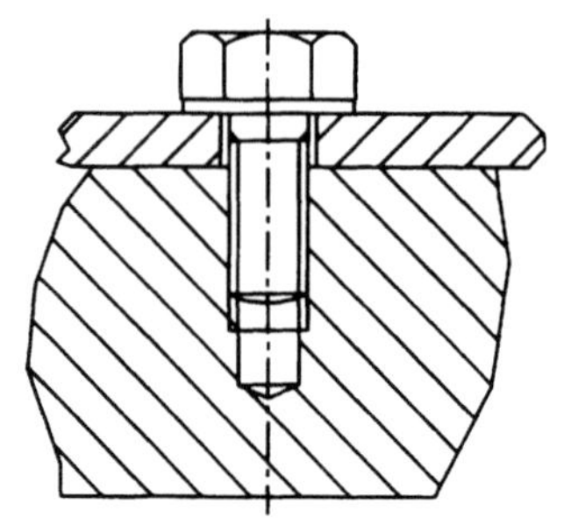

Figure 2.21

2.3 Maintaining records

2.3.1 Drawing identification

The simplest form of drawing identification is by the drawing number. This can range from just taking the next number in a book listing drawing numbers and titles, to sophisticated numbering systems incorporating codes indicating the type of component, other assemblies used on, carry-over component from, etc. One of the simplest but very useful additions to the drawing number is the paper size: A1, A4, etc. Although seemingly trivial this makes locating a past drawing much easier.

Design is an iterative process that in some ways is never complete, as the relentless march of technology is forever generating opportunities for improvement and enhancement. The result is that engineering drawings are not static, but are subject to alterations. A system of recording and identifying the changes is needed to avoid confusion.

A drawing is normally identified by a drawing number and title. We now need a third identifier, the issue number. Each alteration to the drawing will result in the next numerically higher issue number being added. The shop-floor foreman will then be able to check that he has the latest version of the drawing. However, this needs to be taken a stage further, by recording the actual modification. For simple changes, these can be recorded via a short note on the drawing. More complex modifications may refer to a change number on the drawing, details of the actual change being recorded separately and referenced via the change number.

Issue and change numbers may be adequate for the designer to keep control of modifications, but once the product is in production, any changes may have far-reaching consequences. Means of controlling and recording engineering changes such as these will be covered in Chapter 16.

2.3.2 The working file

The engineering drawing is a vital element in the design process. It represents the conclusion of an element of the designer's task. It records and communicates that conclusion. However, it gives very little information as to the reasons behind the 'conclusion'. These are contained within another important document, the *working file*.

Even though this document may never be used directly in the manufacturing process, there are a number of reasons for its importance:

1. Within most environments, the design engineer is likely to be working on a number of projects. To operate efficiently he/she needs to maintain an up-to-date record of the status of each task.
2. Changing circumstances within an organisation can often result in a project being put 'on hold' for several months. When the pieces are picked up later the designer will be expected to resume progress from the point that work stopped. This will only be possible if a comprehensive working file has been maintained.
3. It is not unusual for a design engineer to inherit a project from a colleague who has left the organisation, or who has been promoted to a different post. Again, the working file will prove vital.
4. Despite the best efforts during development and testing, many new products suffer teething problems. These can include manufacturing difficulties, failure to meet cost targets, actual product failure, or supply problems. The designer may well have to justify his or her selection of a particular material, choice of bought out component, surface finish, tolerance value, etc.
5. Finally, if legal proceedings are taken against the designer, he or she may well have to prove the competence of the design. Without a comprehensive working file this could prove difficult.

Now that we have established the need for working file we must define its contents:

(a) The working file should be a regularly maintained record of all progress on the project concerned. It would normally be largely hand-written, with appropriate sketches, diagrams, etc., but must be sufficiently legible, and contain adequate detail to be understood by a third party.

(b) All options and alternatives considered must be included. The rejected options, together with the reasons for their rejection, are important in that they will avoid unnecessary repetition. This applies equally well to design options, material selections and even, in some cases, to the choice of supplier.

(c) Calculations should be clearly laid out, with all assumptions and simplifications stated. If standards, or codes of practice, have been used, this should also be recorded.

(d) Records of any trials or tests on prototypes or with supplier's products must be included.

(e) Engineering inevitably involves compromise between a wide variety of interests. For example, a company with a machine shop running with spare capacity may introduce a policy of using in-house manufacture wherever possible. The available plant may affect some of the engineering choices. The reasons for these choices should also be in the file.

To summarise, the working file is a regularly maintained record of the design engineer's contribution towards the project or product.

2.4 Summary

On completing this chapter, you should:

- Recognise the need to communicate via illustrations, which can include both simple sketches and formal engineering drawings.
- Appreciate the need for a common standard (BS308), and at least acquaint yourself with PP7308.
- Understand the fundamentals of orthographic projection, including the difference between first and third angle, sections, hatching, auxiliary projection, the basics of annotating dimensions, the content of detail and assembly drawings, and the appropriate abbreviations for standard items.
- Be aware of the need to record engineering progress, both via relevant updates of drawings and also via a working file.

2.5 Questions

The best way to reach a level of competence in creating and reading engineering drawings is by practice, so you are recommended to create some drawings of simple everyday items to build up your level of confidence. Always keep a copy of PP7308 to hand to ensure that you are following the correct procedures. Try the following questions to check your understanding at this stage, referring to the sections in brackets if you find difficulty with the answer. (Solutions for the drawing exercise questions, 8 to 11, can be found in Appendix D.)

1. Name the recommended scales for engineering drawings. (2.2.2)
2. Each drawing should have a title block. What is the minimum information that should be in the title block? (2.2.2)
3. What is the difference between an Assembly drawing (General Assembly) and a Detail Component drawing? (2.2.7)
4. An Assembly drawing should have a parts list. What information is kept in the parts list? (2.2.7)
5. Sometimes (during a meeting with non-engineers, for example) it is more appropriate to generate a sketch than a formal engineering drawing. Sketch an isometric cube with a hole in each of the three visible sides. This should be done freehand, but use isometric paper if it helps. (2.2.3)
6. There are two forms of orthographic projection, first and third angle. Show that you understand the difference by generating two drawings of a die, one in each projection system, and note the differences. (2.2.4)
7. How many orthographic views would be needed to fully define the die in first angle projection? Would the same number of views be required in third angle projection? (2.2.4)
8. Figure 2.22 contains isometric views of 5 separate objects. Draw, for each object, a minimum of 3 orthogonal views, in first angle projection. Select your views so that each of your drawings fully defines the item. An orthogonal drawing needs to be to scale, so before you start select suitable dimensions for the length of each side, etc., keeping the proportions as in the isometric views. (2.2.4)
9. Understanding an engineering drawing means that you must be able to view an orthogonal drawing and visualise the shape in 3D. Figure 2.23 shows engineering drawings of two items. Test your ability to visualise the items in 3D by creating an isometric sketch of each. (2.2.2)
10. Demonstrate your understanding of the difference between first and third angle

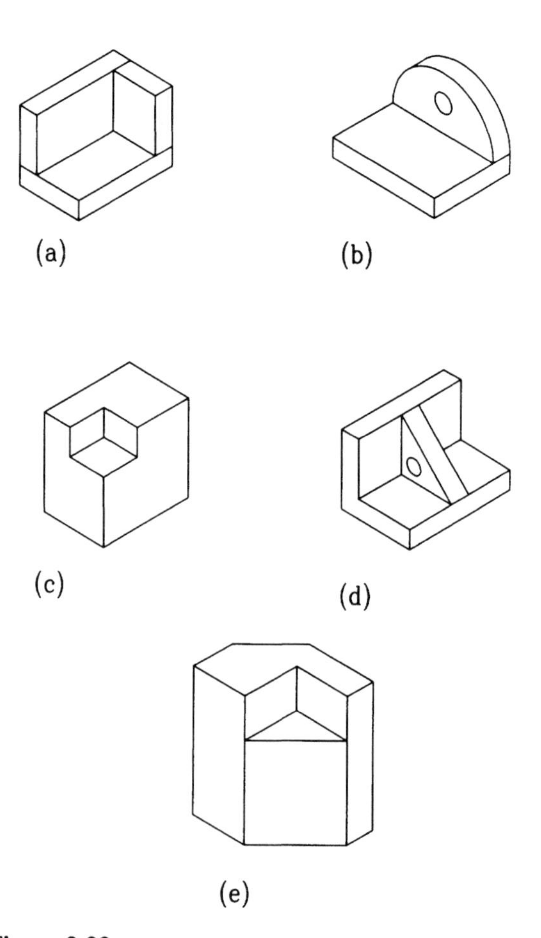

Figure 2.22

projection by converting each of the drawings in Figure 2.23 to third angle. (2.2.4)

11. Repeat question 8, but this time use third angle projection. (2.2.4)
12. When would an auxiliary projection be needed? Try generating an auxiliary view of a simple object such as a matchbox, and check the correctness of your paper view by viewing the object from the same angle. (2.2.6)
13. The size of a hole can be given either by dimensioning its radius or its diameter. True or false? When can the radius be shown? (2.2.8)
14. What should be contained in the working file, and why is it needed? (2.3.2)
15. Figure 2.24 shows a simple assembly drawing of bushed support housing for a shaft. Create your version of this drawing, replacing the upper left view with that of section A–A as indicated. Before you start, select suitable dimensions, so that you retain the proportions in Figure 2.24. (2.2.5)

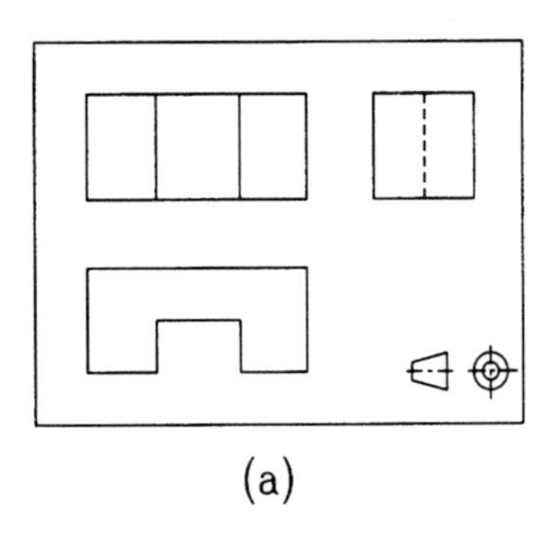

(a)

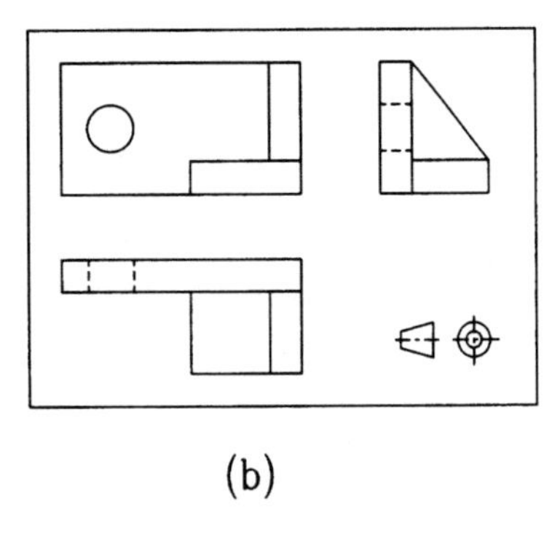

(b)

Figure 2.23

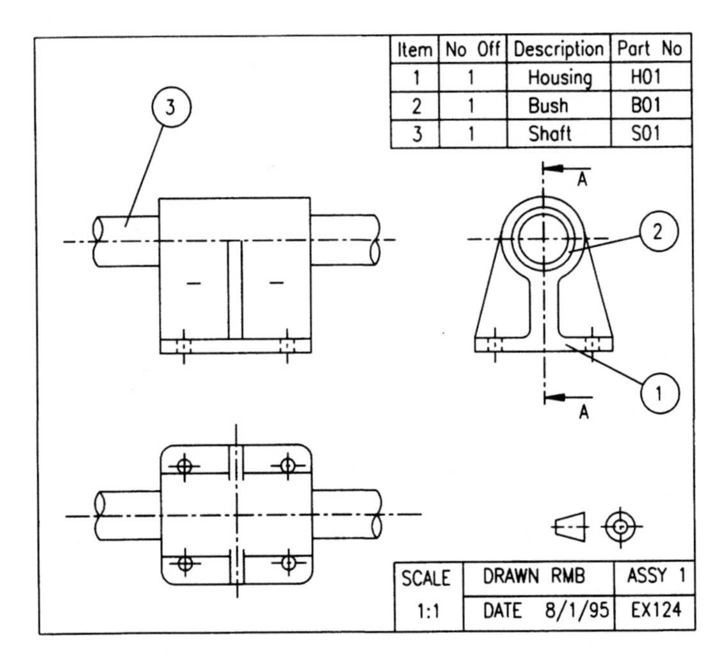

Figure 2.24

3 The Importance of Detail: Tolerances

3.1 Tolerances

3.1.1 Introduction

If it were possible to manufacture components to precise dimensions, an engineer's life would be greatly simplified. It would only be necessary to specify one dimension for each feature of the component, and provided it was correctly specified and manufactured, a perfect fit would be guaranteed. Unfortunately life is not like that. Precise absolute measurements do not exist in the real world. When asked the time at midday you are likely to reply '12 o'clock' even though the actual time will probably be either just before or just after 12 o'clock. What we really mean by our reply is within a minute or two of midday. Our reply is not a precise absolute measurement. We accept that it has an error or tolerance. The same is true of component dimensions. For example, although a diameter may be specified as 55mm, without any further information this would indicate that any value larger then 54.5mm or smaller than 55.5mm would be satisfactory. The dimension is said to have a tolerance of 1mm.

3.1.2 Definition

The tolerance is the stated permissible variation in the size of a dimension, and is the difference between the upper and lower acceptable limits.

Why do we need tolerances?

An assembly of components will only function correctly if its constituent parts fit together in a predictable manner. As it is not possible to manufacture to an exact size, the designer has to decide how close to the ideal size is satisfactory. In the early days of engineering, components would be made with wide tolerances and would require modification or 'fitting' as they were assembled. This fitting often needed a high level of skill. Today, advances in engineering mean that tighter tolerances are possible. As a result, most manufacturing now involves the manufacture of interchangeable parts, giving the following benefits:

(a) The assembly of components requires less skill, as the need for adjustment to size is eliminated. Assembly is therefore cheaper, and quality easier to maintain.

(b) Servicing a defective part is made simple. The faulty component can be replaced by another manufactured to within the same range of dimensions.

(c) Parts can be made in large quantities, in many cases with less demand on the skill of the operator. This will almost certainly involve a high level of investment in the provision and use of special purpose machines, tools, jigs, fixtures, gauges, etc. However, high production volumes allow this investment to be spread over many components, so that the individual component cost is much less than if manufactured on an individual basis by skilled craftsmen.

As with every rule there are exceptions. Sometimes the costs of achieving full interchangeability are greater than the benefits available. For example, motor-car pistons are often made to the closest economic tolerances achievable by the manufacturing process and then sorted into suitable sets. The volumes in this case are such that full interchangeability would not be the cheapest solution.

3.1.3 Cost implications of tolerances

As we have seen, there are considerable benefits available from being able to manufacture to close tolerances, but there is also a penalty – cost (see Figure 3.1). There are three main characteristics that affect the cost of achieving a given tolerance:

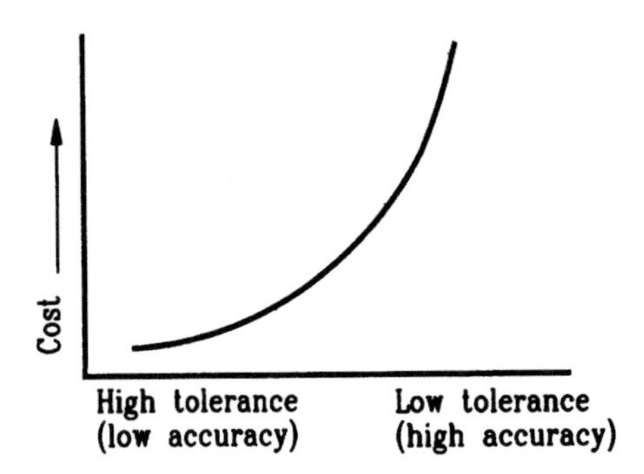

Figure 3.1 *Tolerance v. cost*

(a) *Tolerance size*: In very general terms, the smaller the tolerance the greater the cost to achieve that tolerance on a given component. Why?

The tolerance may affect the type of raw material form. For example, a sand casting cannot be held to the same dimensional accuracy as the more expensive investment casting. It may also affect the choice of machine required to perform the operation.

The required tolerance is likely to influence the choice of machine to manufacture the component. A new machine tool is likely to be capable of producing a component to within tighter tolerances than an older machine, possibly suffering from worn spindle bearings. However, the higher capital value of the newer machine will almost certainly result in a higher running cost. Again, the effect of tighter tolerances is higher costs.

Closer tolerances are also likely to require more manufacturing operations, more intensive inspection, and probably higher levels of scrap.

(b) *Component size*: The larger the component, the more it will cost to achieve a given tolerance. Large components are more difficult to handle and require bigger, more expensive machines to produce them. Any scrap produced is also likely to be of higher value.

(c) *Detail of the feature*: For a given tolerance on a given component, some features are easier to achieve than others. For example, it is generally easier to produce an outside diameter on a component that can be rotated than a hole, on one that cannot. (This may not be true for very large items.) This is because the revolved component can fit in a chuck on a lathe where the turned diameter may be set to any value by adjusting the lathe feed. A hole on the other hand, will require different drills and possibly different reamers for different diameters. This leads to high tool and setting costs.

Knowing the general areas to avoid is one thing, but the designer must have a means of checking his costs. It is very difficult to generalise on costs as different organisations have different ways of costing their processes. For example, one company that the author worked for had its own foundry, so the costing system encouraged the use of castings. It had a surplus of turning machines, but few milling machines. Turning therefore tended to be cheaper than machining-centre work.

Some companies may have a special expertise in a particular process. This process may be economic for them, but very expensive for their competitors. Most design offices have their own manuals, giving guidance re: manufacturing capability and cost. The prudent designer will, of course, maintain close liaison with the production engineers and process planners.

Unfortunately, none of this is available to most students. Instead, PD6470 gives a guide to the relative costs of achieving tolerances and surface finishes on different components with different machines.

3.1.4 Presentation

Almost all dimensions on a drawing are subject to tolerances (the exception being auxiliary dimensions, which are added for clarity rather than production quality control). To simplify, the drawing tolerances are normally only noted directly on critical dimensions. The remaining dimensions are usually covered by a general tolerance defined in the drawing notes: e.g. *Tolerance* ± 0.1 *except where otherwise stated.*

The tolerances of the critical dimensions can be shown in one of three ways (Figure 3.2):

1. *Direct*: The maximum and minimum acceptable values for the dimension are given. This format may help an inspector checking the component, and possibly the engineer as he allocates the tolerances.
2. *Uni-lateral*: The diameter given is the Maximum Material Condition (MMC), together with the maximum allowable departure from this condition. (Note: MMC for a shaft is the largest size, and for a hole is the smallest size.) Tolerancing in this way is helpful to the machinist, who must remove enough material to at least reach MMC.
3. *Bi-lateral:* The size given is the nominal (mid-way) value together with the tolerance as a $\pm$ value.

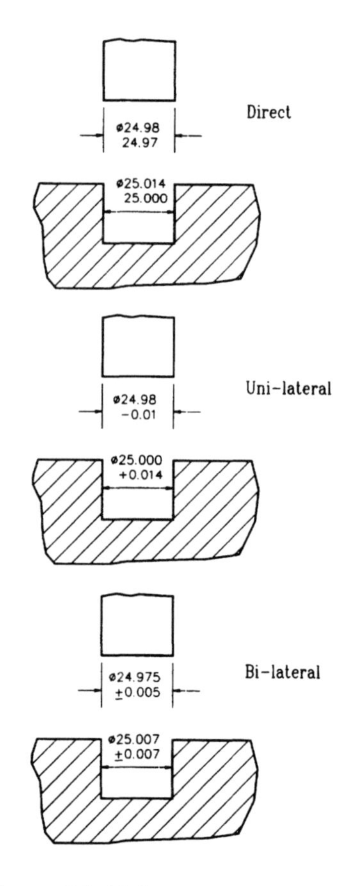

Figure 3.2 *Tolerance presentation*

3.1.5 Functional dimensions

The addition of dimensions and tolerances to a drawing will inevitably add to the production and inspection costs of the component. Care must therefore be taken to ensure that the dimensions and their associated tolerances are defined in such a way as to simplify the manufacturing process, hence keeping the costs to a minimum. This is done by concentrating on the critical or functional dimensions.

A functional dimension is one that directly affects the function of the product.

As an example, take the assembly of a roller such as may be used in a desk-drawer runner (Figure 3.3).

Now consider the pin, which is retained by means of a tight (inter-ference) fit in the main support. Its prime function is to support the roller whilst allowing it to rotate. If we examine the axial dimensions of the pin we only find one critical or functional dimension.

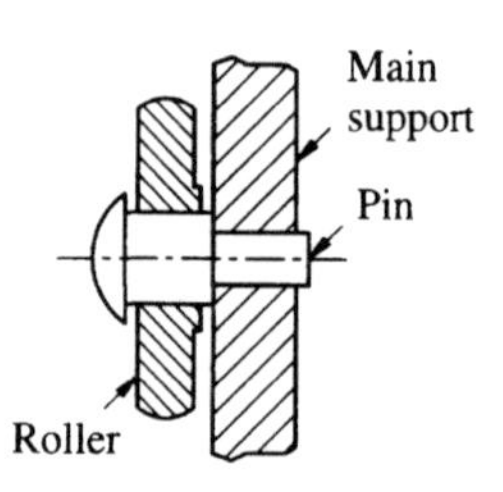

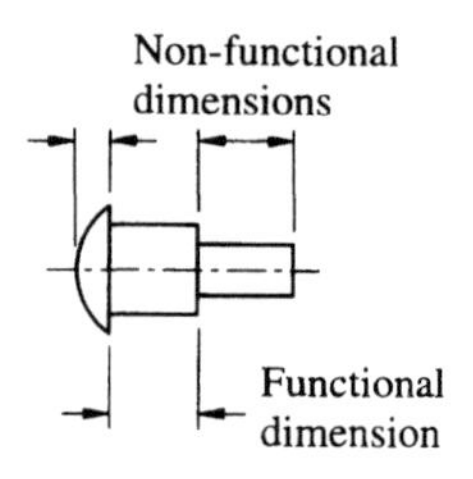

Figure 3.3 *Functional dimensions*

3.1.6 Choice of datum

In general, using functional dimensions as datums will prevent the use of unnecessarily tight tolerances. Normally other datums will only be chosen if they allow a simpler method of manufacture or inspection.

Figure 3.4 demonstrates this. A widget is to be fitted to the body of a larger component via an adaptor. To function correctly, the distance between the outer surface of the body and the outer surface of the widget (i.e. the offset) is critical. The widget is an existing component with a width tolerance as shown.

The offset will be affected by a combination of dimensions of the widget and the adaptor. The total tolerance allowed for the offset is 0.1, but as the widget has a tolerance of 0.04, this leaves only a possible maximum of 0.06 to be allocated to the adaptor.

If the end face of the adaptor is taken as the datum (Figure 3.5), both of the dimensions shown will affect the offset. Sharing the available tolerance equally between these dimensions results in two tolerances, each of 0.03.

However, by using a functional dimension as a datum (Figure 3.6), only one dimension on the adaptor will affect the offset. This dimension can therefore utilise all of the available 0.06 tolerance. The other dimension shown is not functional and can have its tolerance relaxed to, say, 0.5.

Note that both ways of dimensioning will achieve the functional purpose equally well, but the tighter tolerances on the first example are almost certain to incur a cost penalty over the example using the functional surfaces as a datum. This rule should be used as a general guideline. There may be cases where other datums may prove more helpful for manufacture or inspection. Close liaison between the designer and his or her manufacturing colleagues is essential.

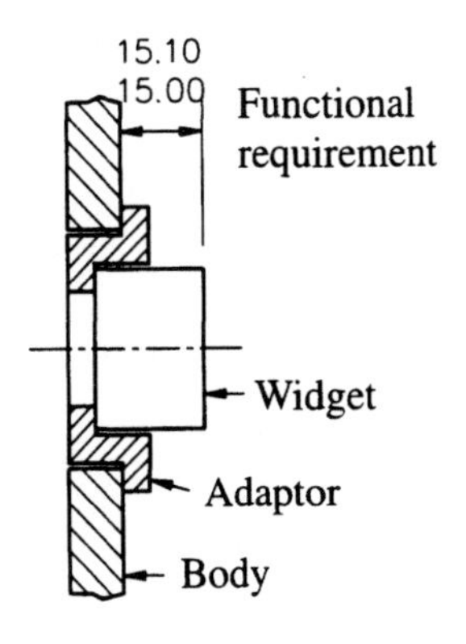

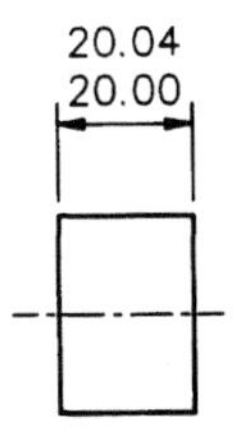

Figure 3.4 *Functional requirements*

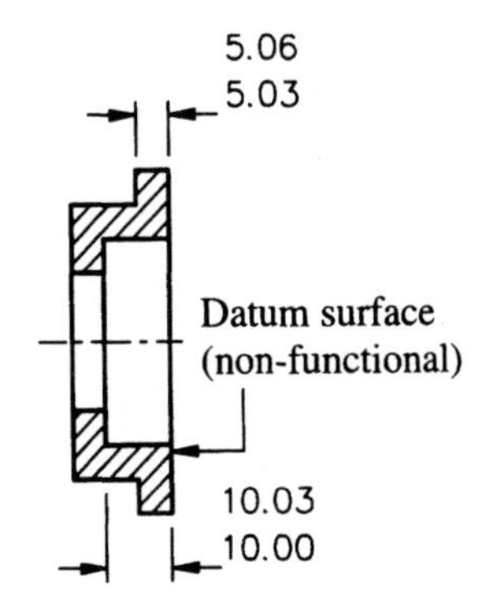

Figure 3.5

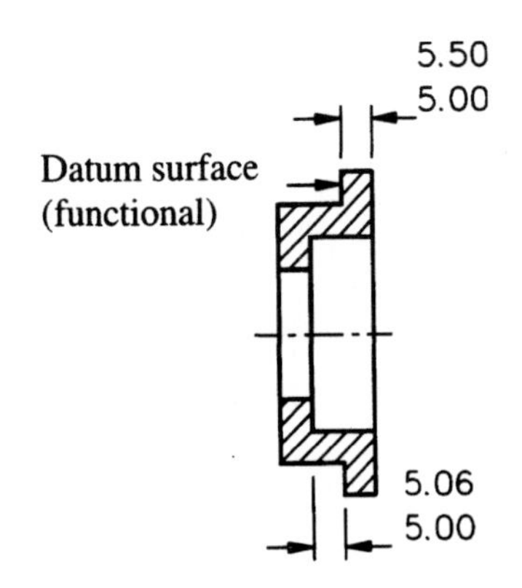

Figure 3.6

3.1.7 Tolerance allocation

General case

Although a drawing must contain enough dimensions to ensure that the component is fully defined, particular care must be taken to ensure that it is not over-dimensioned. If more than the minimum number of dimensions are shown, some will be redundant and are almost certain to cause confusion, particularly when tolerances are involved.

For example, consider a rectangular plate with cut-outs on two sides, dimensioned as in Figure 3.7.

At first sight, the drawing appears satisfactory: the overall height as measured on the left-hand side is 70, the same as can be calculated from the dimensions on the right-hand side. However, as it is possible to calculate the overall height in two completely independent ways, one or more of the dimensions must be redundant.

Now see what happens if a general tolerance of say $\pm$ 0.5 is added to the drawing. Assume all the dimensions on the left are at their minimum and those on the right are at their maximum permitted values. The overall height according to the left-hand dimensions is 68, but when calculated from those on the right is 72. As both answers cannot be correct there must be an error. There is – the drawing is over-dimensioned.

One of the dimensions must be omitted so that it can 'float', or adjust its value, to give an unambiguous answer for the overall height. This dimension is called the dependent dimension. Although it cannot be specified on the drawing it may well affect the component function. The designer needs to consider how its value is affected when selecting the other dimensions and their tolerances.

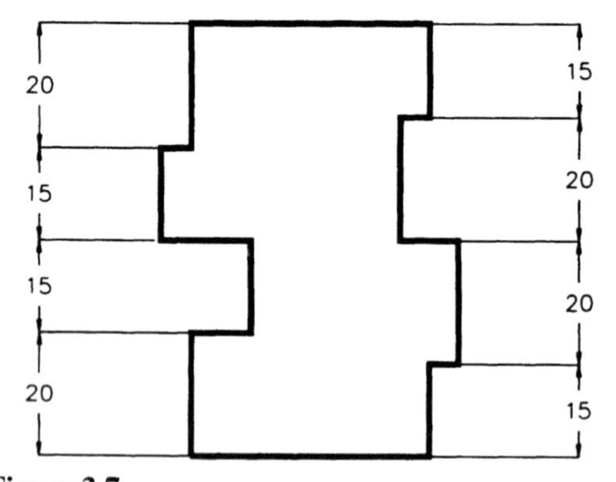

Figure 3.7

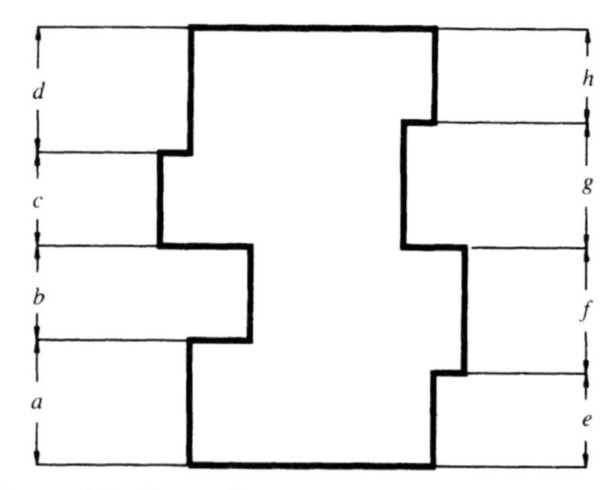

Figure 3.8 *General case*

Now consider a general case where the dimensions have been replaced by the letters a to e (Figure 3.8). The limits of each dimension are represented by subscripts, 1 being the maximum and 2 being the minimum values.

One dimension cannot be specified directly: the dependent dimension. In this case we will choose e as the dependent dimension. Its value can be calculated from the other dimensions as follows:

$$e = a + b + c + d - f - g - h$$

The maximum and minimum values are therefore given by:

$$e_1 = a_1 + b_1 + c_1 + d_1 - f_2 - g_2 - h_2$$
$$e_2 = a_2 + b_2 + c_2 + d_2 - f_1 - g_1 - h_1$$

Subtract one from the other

$$e_1 - e_2 = a_1 - a_2 + b_1 - b_2 + c_1 - c_2 + d_1 - d_2 + c_1 - c_2 + f_1 - f_2 + g_1 - g_2 + h_1 - h_2$$

Now examine the result:

$(e_1 - e_2)$ is the tolerance on e

$(a_1 - a_2)$ is the tolerance on a

$(b_1 - b_2)$ is the tolerance on b

etc.

The equation could therefore be written as:

$$\text{tol}(e) = \text{tol}(a) + \text{tol}(b) + \text{tol}(c) + \text{tol}(d) + \text{tol}(e) + \text{tol}(d) + \text{tol}(f) + \text{tol}(g) + \text{tol}(h)$$

or, in other words:

The tolerance on the dependent dimension is equal to the sum of the tolerances on all the dimensions that affect the dependent dimension.

This is an important theorem that will significantly help the designer in the task of allocating tolerances.

Although, as demonstrated above, the redundant dimensions can make a drawing ambiguous, if such dimensions could provide useful information, they may be included as auxiliary dimensions. These are shown in parentheses and are not toleranced. An acceptable method of dimensioning the vertical features of the plate is as shown in Figure 3.9.

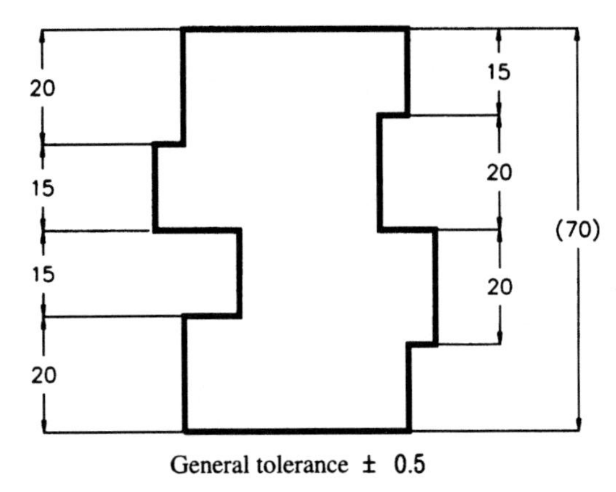

Figure 3.9

3.1.8 Allocation examples

Tolerances can affect both the cost and function of a component. An otherwise good design could be ruined by poorly allocated tolerances. To be successful, design needs attention to detail, especially in the allocation of tolerances. The following examples demonstrate some of the principles involved.

Example 3.1

Consider a piece of round bar with a flat machined on it. It has already been designed and has the dimensions as in Figure 3.10.

As can be seen, it is not over-dimensioned, and sufficient dimensions are present to define the component size. So far so good. Now, consider how the component is to be made. The diameter has a wide tolerance, so standard bar stock could probably be used. The first operation will be to cut the bar to length. However, the overall length is not given on the drawing and as each of the linear dimensions has close tolerances the operator will probably follow the sequence of three operations as shown in Figure 3.11.

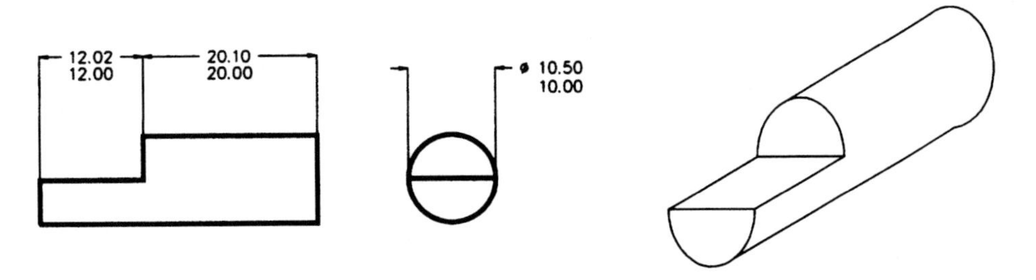

Figure 3.10

If the overall length had been specified on the drawing the last manufacturing operation (trim end) could be omitted, saving both time and material. However, if c (Figure 3.12) is specified, either a or b must be omitted (i.e. become the dependent dimension) to avoid any over-dimensioning problems. In this case we will choose b as the dependent dimension because it has the wider tolerance.

Remember: *The tolerance on the dependent dimension is equal to the sum of the tolerances of all the dimensions that affect the dependent dimension.*

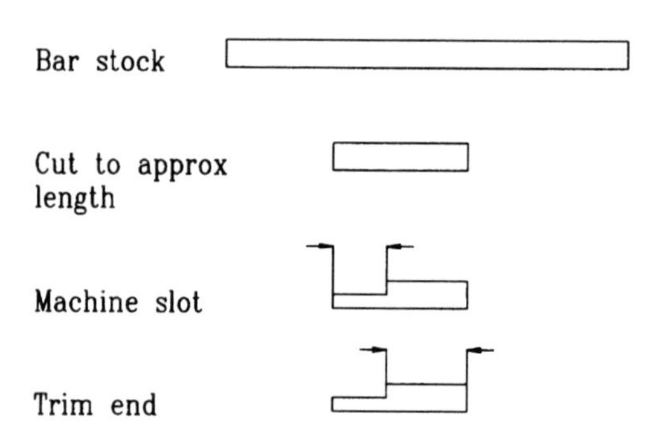

Figure 3.11

$$\text{tol}(b) = \text{tol}(a) + \text{tol}(c)$$
$$0.1 = 0.02 + \text{tol}(c)$$

Hence

$$\text{tol}(c) = 0.08 \qquad (1)$$

The relationship between a, b and c is given by

$$b = c - a$$

Using the subscripts 1 and 2 to represent the maximum and minimum values,

$$b_1 = c_1 - a_2$$
$$b_2 = c_2 - a_1$$

giving

$$c_1 = 20.10 + 12.00$$
$$c_1 = 32.10$$
$$c_2 = 20.00 + 12.02$$
$$c_2 = 32.02$$

Now check by calculating the tolerance on c and compare with (1) above.

$$\text{tol}(c) = c_1 - c_2$$
$$= 32.10 - 32.02$$
$$= 0.08$$

As the check has proved satisfactory the drawing can be dimensioned as in Figure 3.13.

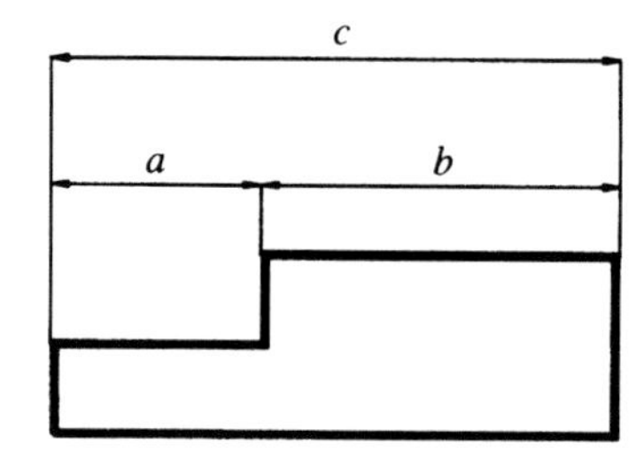

Figure 3.12

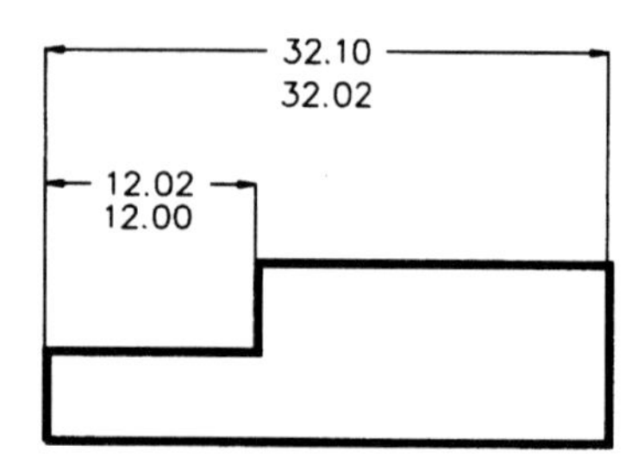

Figure 3.13

Example 3.2

In the previous example, one dimension was simply replaced with another by calculating a suitable tolerance for the new dimension. Sometimes the tolerance on the existing dimension needs to be adjusted.

Consider a short piece of bar cut from a larger strip and then drilled (Figure 3.14). The position of the hole is controlled relative to each end. The problem is similar to the previous example in that the first operation is to cut the bar to length, but the length is not specified on the drawing.

To do this we need to specify c (Figure 3.15) and eliminate either a or b. The tolerance is the same on each so it does not matter which we eliminate. Choose a to be the dependent dimension, i.e. the one we will eliminate (Figure 3.15):

$$\begin{aligned} \text{tol}(a) &= \text{tol}(b) + \text{tol}(c) \\ 0.1 &= 0.1 + \text{tol}(c) \end{aligned}$$

Hence

$$\text{tol}(c) = 0$$

As we cannot have a zero tolerance, the tolerance on b must be reduced to, say, 0.05, thus allowing a tolerance of 0.05 for c.

Note that reducing the tolerance on a dimension will not adversely affect its ability to function, as the new dimension will still be within the constraints of the old.

The relationship between a, b and c is given by

$$a = c - b$$

Using the subscripts 1 and 2 to represent the maximum and minimum values,

$$\begin{aligned} a_1 &= c_1 - b_2 \\ a_2 &= c_2 - b_1 \end{aligned}$$

giving

$$\begin{aligned} c_1 &= 10.10 + 15.00 \\ &= 25.10 \\ c_2 &= 10.00 + 15.05 \\ &= 25.05 \end{aligned}$$

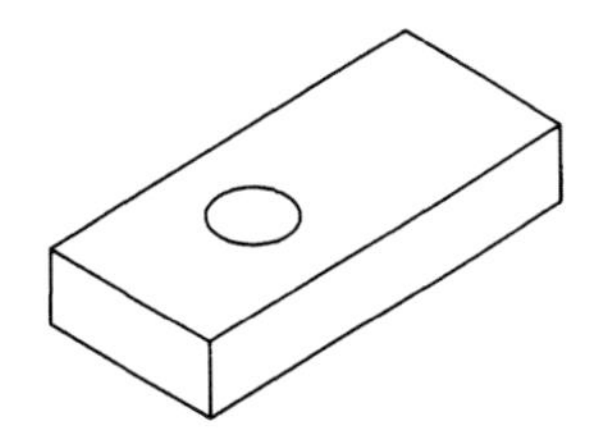

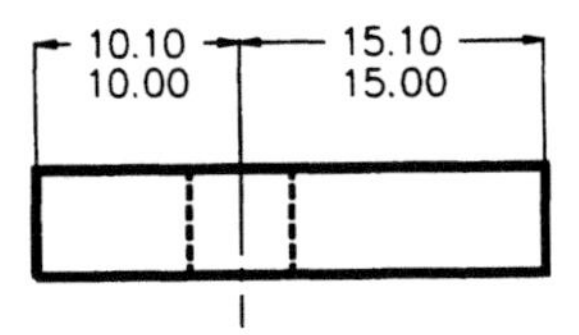

Figure 3.14

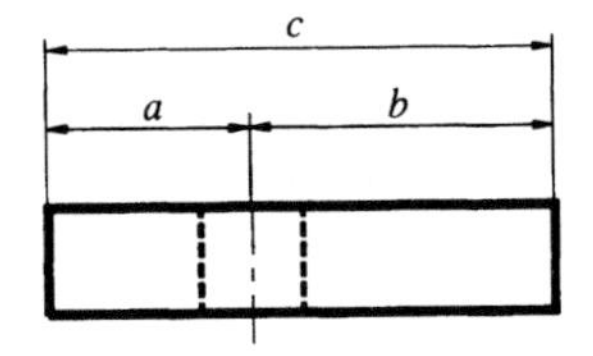

Figure 3.15

Now check by calculating the tolerance on c and compare with the allocated value of 0.05 above.

$$\begin{aligned} \text{tol}(c) &= c_1 - c_2 \\ &= 25.10 - 25.05 \\ &= 0.05 \end{aligned}$$

As this checks with the allocated value, the new values for b and c can replace the old a and b (see Figure 3.16).

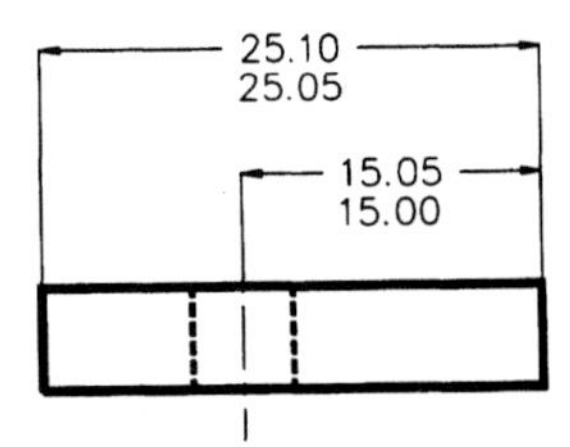

Figure 3.16

3.2 Limits and fits

3.2.1 Introduction

We have seen that controlling the tolerances on dimensions is necessary to ensure that parts are interchangeable. However, a good design requires not only that the components are interchangeable, but also that they fit together in such a way that they can function correctly. The way components fit together will be determined by a combination of the chosen dimensions of each component and the tolerances on those dimensions.

Imagine a shaft that is able to rotate freely supported in a nylon bush. The bush in turn is mounted in a housing. To achieve its function, the bush must be held in the housing so that it does not move. The dimensions of the outside diameter of the bush and the inside diameter of the housing are therefore selected so that there is a tight fit between these components. On the other hand, the diameter of the shaft must always be less than the inside diameter of the bush, so that it is able to rotate freely. We have described here two types of fit: an interference fit (the bush in its housing) and a clearance fit (the shaft in the bush).

There are three basic classes of fit:

(a) Clearance: The hole is always larger than the shaft.

(b) Interference: The shaft is always larger than the hole (prior to assembly).

(c) Transition: The tolerances on the hole and shaft diameters are such that either a clearance or interference fit could result.

3.2.2 BS4500

Within these classes there is a multitude of ways in which components can fit together. BS4500 provides a system for defining such 'fits'.

A chosen fit should exhibit the same functional characteristics over a range of sizes. For example, a running fit should enable a shaft to rotate freely. However, the clearance necessary to achieve this will not be the same for a shaft of 250mm diameter as it would be for a shaft of 25mm diameter. BS4500 provides a system in which a fit characteristic (e.g. running fit) can be achieved for different size conditions. The fits are the result of two characteristics – tolerance and deviation.

3.2.3 Tolerance grade

The tolerance of a dimension is its permissible size variation, i.e. the difference between the upper and lower acceptable values of the dimension. In practice, the tolerance increases with size. A tolerance grade relates the actual tolerance to component size. The ISO (BS4500) system uses 18 tolerance grades: IT01, IT0, IT1, IT2, . . . IT16. The list below links the grades with some of the manufacturing processes required to produce them. In the case of the finer grades, some applications have also been included.

IT16 Sand casting (approx), flame cutting
IT15 Stamping (approx)
IT14 Die casting, rubber moulding
IT13 Press work, tube rolling
IT12 Light press work, tube drawing
IT11 Drilling, milling, slotting, planing, metal rolling or extrusion
IT9 Worn capstan or automatic, horizontal or vertical boring machine
IT8 Centre lathe turning and boring, reaming, capstan or automatic in good condition
IT7 High quality turning, broaching, honing
IT6 Grinding, fine honing
IT5 Ball bearings, machine lapping, diamond or fine boring, fine grinding
IT4 Gauges, fits of extreme precision produced by lapping
IT3 Good quality gauges, gap gauges
IT2 High quality gauges, plug gauges
IT1 Slip blocks, reference gauges
IT0 Very fine reference gauges, horology
IT01 Extremely fine reference gauges, precision horology

3.2.4 Deviation

The term *deviation* is used to indicate the position of the size limits relative to the nominal (basic) size. To provide a wide range of engineering fits the ISO (BS4500) system includes 27 deviations. They are indicated by letters: upper case for holes and lower case for shafts.

For a hole, an 'H' designation sets the lower limit of hole diameter (Maximum Material Condition) at the basic size. Letters alphabetically prior to H (G, E, D etc.) progressively move the MMC to values greater than the basic size. Letters after H (K, N, P etc.) progressively move the MMC to values less than the basic size.

For a shaft, an 'h' designation sets the upper limit of shaft diameter (Maximum Material Condition) at the basic size. Letters prior to h (g, e, d etc.) progressively move the MMC to values less than the basic size. Letters after h (k, n, p etc.) progressively move the MMC to values greater than the basic size.

In the case of two components fitting together (e.g. a shaft and a hole) the basic size is the same for both items. (See Figure 3.17.)

The fit will be affected by the deviation of both components, so to simplify things one of the deviations is set to zero (i.e. MMC of either the hole or the shaft is set to the basic size).

If the hole deviation is set to zero (H designation used) the system is said to be *hole based.* If the shaft deviation is set to zero (h designation used) the system is said to be *shaft based.* (Also see section 3.2.7.)

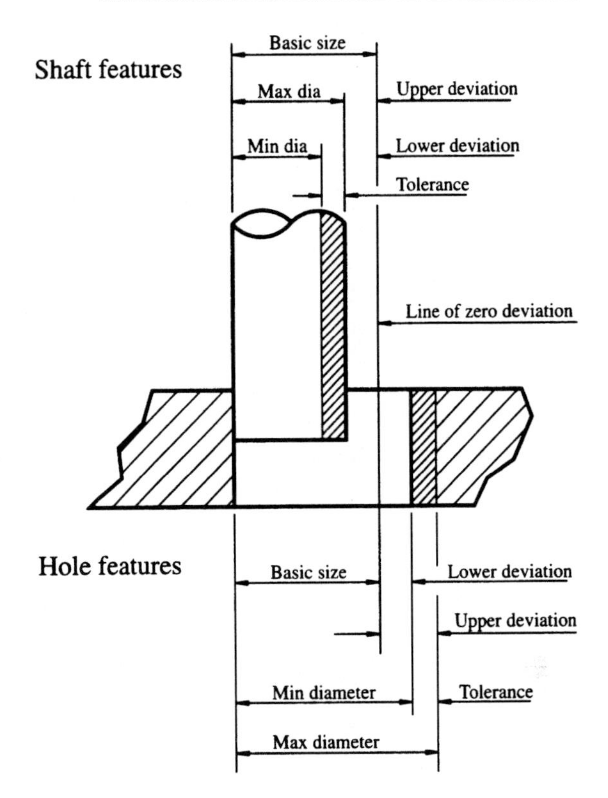

Figure 3.17

3.2.5 Syntax

The method of describing a fit in terms of the tolerances and deviations is:

(Hole dev)(Hole tol grade) (shaft dev)(shaft tol grade)

e.g. H7 k6

3.2.6 Types of fit

There is a vast choice of fits available from all possible combinations of the tolerance grades and deviations in the standard. In practice, however, most engineering applications can be satisfied from a more limited range of fits. BS4500 provides tables covering selected fits which should satisfy most applications. The following comments expand upon the use of some of these fits.

- *H7 f7 Normal running:* Common, good quality fit used with oil or grease lubrication where no substantial temperature difference occurs, e.g. gearbox shafts, engine tappet guides.

- *H7 g6 slide:* More expensive, only used for continuous running bearings if ideal conditions exist (i.e. light load and accurate alignment with good lubrication). Normally used for accurate sliding components or an easy location fit, e.g. gear pump spindle bearings, sewing machine needle bar guide.

- *H7 h6 Location:* At MMC there is zero clearance. This can be used as a precision sliding fit if good lubrication is present, e.g. hydraulic spool valve, exhaust valve guide.

- *H7 k6 Push:* Transition fit used for location where a slight interference is desirable to eliminate vibration, e.g. shaft coupling or end cap with bearing.

- *H7 p6 Light press:* Secure fixing for non-ferrous parts that can be assembled or dismantled without over stressing, e.g. bronze bush in a ferrous housing.

- *H7 s6 Heavy press:* Permanent assembly of ferrous parts, e.g. thin cylinder lining in a cylinder block.

If you are using standard components such as bearings, the supplier's catalogue will often give information as to the appropriate fit that should be used. In such cases you would be well advised to heed their recommendations. In the absence of such guidelines you should first decide whether to use the hole or shaft base system (section 3.2.7), then choose one of the selected fits from Figure 3.18, and refer to the tables in Appendix B for the actual tolerance values.

3.2.7 Hole or shaft base?

Except in the case of very large sizes, holes are produced using 'fixed size' tools such as drills and reamers. A large number of hole diameters would require an equally large number of such tools. It is relatively easy, however, to adjust the diameter of a shaft produced on a lathe, which is effectively an adjustable-size tool.

Using a standard hole for each nominal size and varying the mating shaft diameter to produce the required fit has the advantage of limiting the number of 'fixed size' tools needed. This is the *hole based* system and is the preferred system.

The *shaft based* system would normally only be used in the case of a common shaft to which a number of different components (with different fits) are assembled, thus enabling a single shaft diameter to be employed.

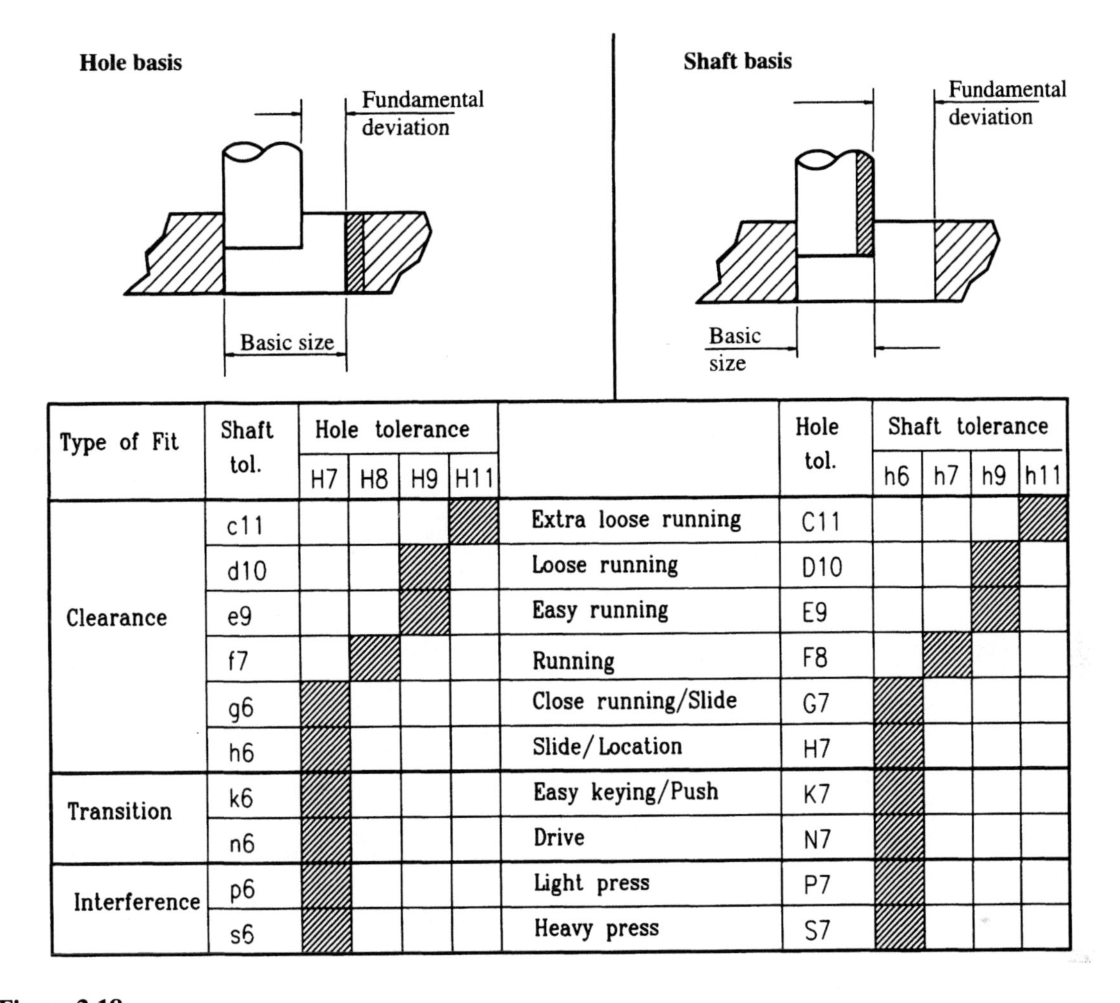

Type of Fit	Shaft tol.	Hole tolerance					Hole tol.	Shaft tolerance			
		H7	H8	H9	H11			h6	h7	h9	h11
Clearance	c11				■	Extra loose running	C11				■
	d10			■		Loose running	D10			■	
	e9			■		Easy running	E9			■	
	f7		■			Running	F8		■		
	g6	■				Close running/Slide	G7	■			
	h6	■				Slide/Location	H7	■			
Transition	k6	■				Easy keying/Push	K7	■			
	n6	■				Drive	N7	■			
Interference	p6	■				Light press	P7	■			
	s6	■				Heavy press	S7	■			

Figure 3.18

3.3 Surface finish

3.3.1 Introduction

All surfaces have characteristics relating to size, form and texture. The first two characteristics largely determine the ability of components to fit together and are controlled by size dimensions and tolerancing (including geometric tolerances, see section 3.4). Surface texture is also important as it can affect friction, wear, fatigue resistance, aesthetics, suitability for hygienic applications, etc.

The actual texture is largely the result of the inherent limitations of the manufacturing process. For example, a ground surface would be smoother than a milled surface, which in turn is likely to be smoother than a sand cast surface. For a given process the surface finish can also be affected by machine vibration, cutter deflection, choice of cutting speeds and feeds, fixture design, etc.

As with limits and fits, much knowledge exists as to the suitability of particular levels of surface roughness for particular applications, so a means of assessing and specifying surface texture is required.

3.3.2 Assessment

BS1134 defines a number of parameters to describe a surface finish, but the most commonly used is R_a or CLA (Centre Line Average).

The R_a value is evolved by taking the actual surface profile and generating an arithmetic mean reference line, positioned such that the total hatched area above the line is the same as that below the line (Figure 3.19).

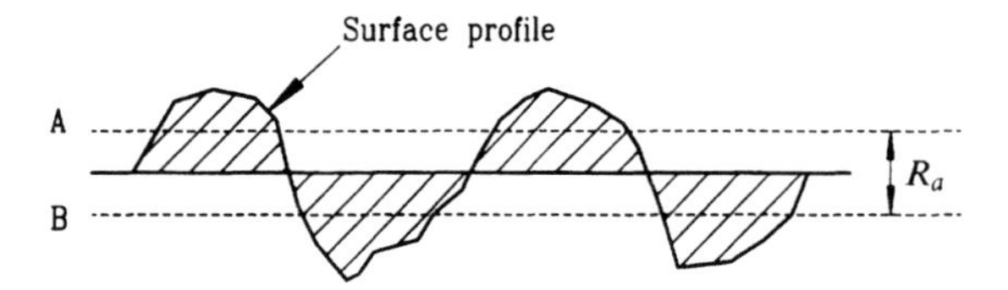

Figure 3.19

Line A is now drawn, representing the average departure of the actual surface from the reference line, considering only the surface that lies *above* the reference line. Line B represents the average departure of the surface from the reference line considering only the surface that lies *below* it. The R_a value is the distance between the lines A and B.

The finish can be physically measured in a number of ways, often by using a device called a 'Tallysurf'. This device pulls a stylus across the surface and gives a readout of the R_a value, usually in µm. Note that the readings are likely to be affected by the direction in which they are taken, and the length over which the study is made.

3.3.3 Syntax

The relevant R_a value is added to the symbol, as in Figure 3.20. This represents the maximum permitted roughness value. Two values may be added if there is a requirement to control the surface finish between maximum and minimum levels. Figure 3.20 (a) in-

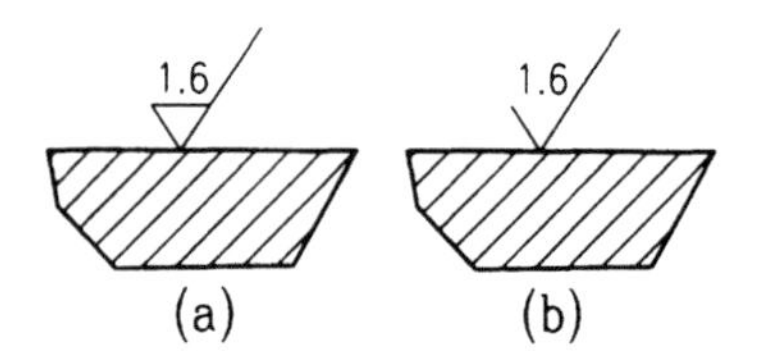

Figure 3.20

dicates the surface must be machined and (b) indicates the required finish but does not specify machining.

As with tolerances, it is common practice to indicate all the machined surfaces, but only include an R_a value on those surfaces considered to be critical. A general rate will give the value of the surface finish on the non-specified areas.

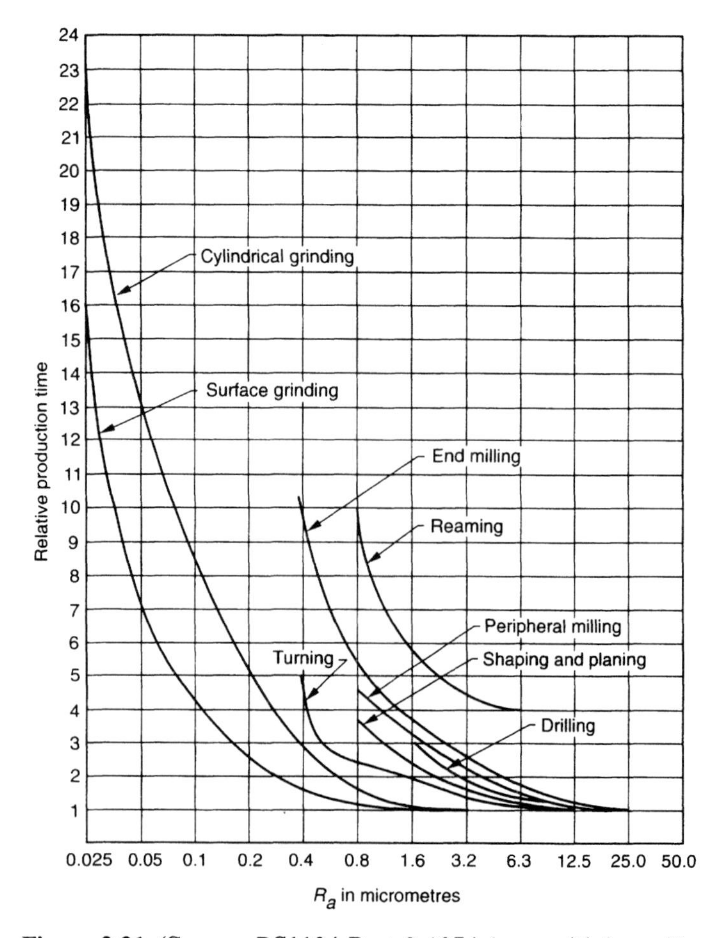

Figure 3.21 (Source: BS1134 Part 2 1974 (now withdrawn))

3.3.4 Cost

Again, as with tolerances, high-quality surface finishes are expensive and should only be specified where necessary. Where possible, the inherent surface finish generated by the manufacturing process specified (or that required to achieve the tolerance) should be used. Note that changing to a grade only one level finer can result in a considerable cost increase.

Figure 3.21 from BS1134 gives an indication of the increases in production time resulting from an increase in the smoothness of the surface finish required.

Figure 3.22 (taken from BS1134; Part 2: 1990) indicates typical ranges of surface finish available from some common manufacturing processes. Remember, these are typical values only. In special cases higher or lower values may be obtained.

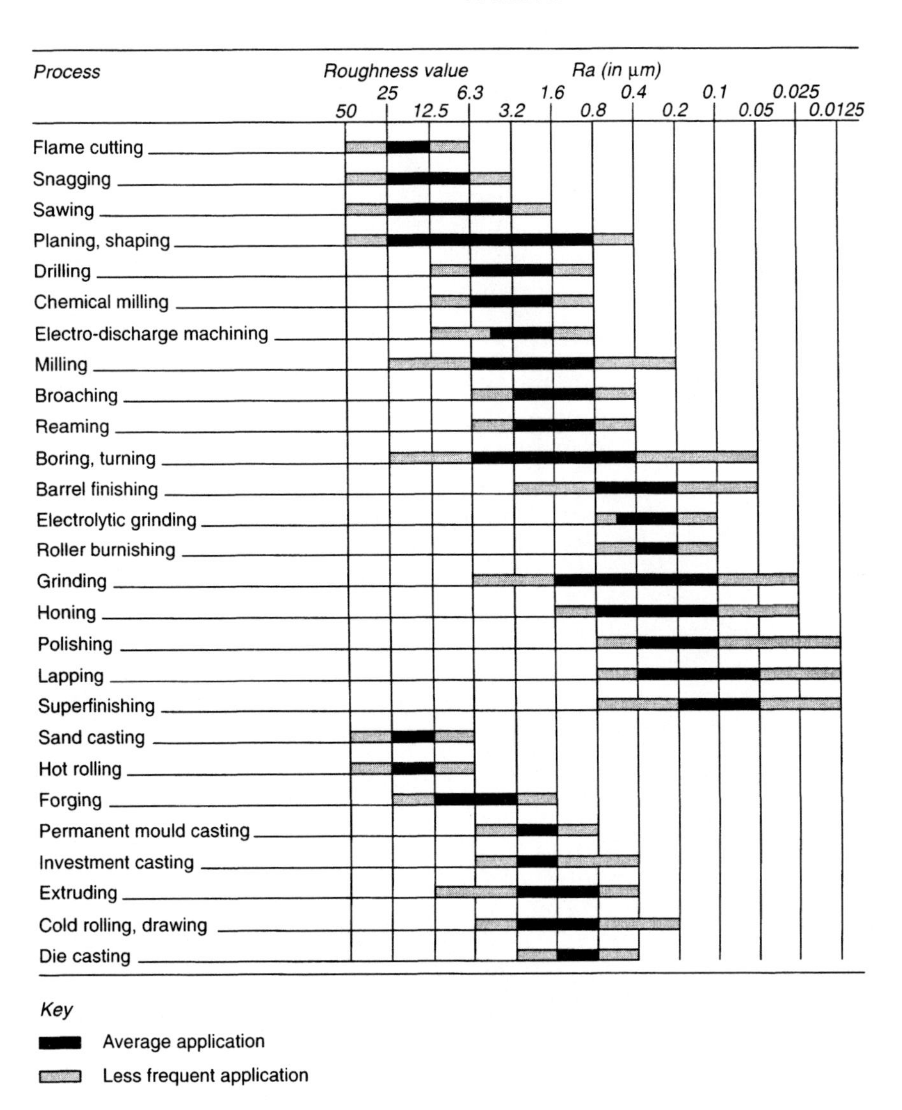

Figure 3.22

Source: BS1134: Part 2: 1990.

3.4 Geometric tolerances

3.4.1 Introduction

A dimensional tolerance only defines the difference between the upper and lower limits of a dimension; it ignores any departure from true form, i.e. roundness, straightness, etc. Geometric tolerancing specifies the maximum variation of form or position of a feature by defining a tolerance zone within which the feature is to be contained.

To avoid the need for descriptive notes, geometric tolerances are specified by a system of symbols and tolerance frames which define the feature in question and the size and type of tolerance. We will start by introducing the system for adding this information to the drawing, and then show a number of examples.

3.4.2 Syntax

Tolerance frame

The frame is as below and can have two or more boxes.

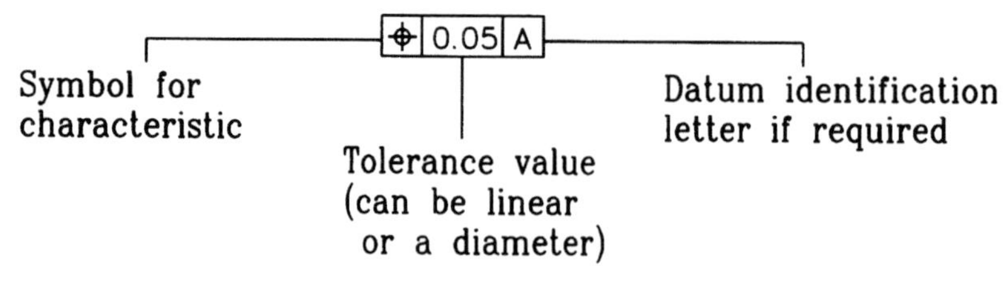

Figure 3.23

Feature identification

The tolerance frame is connected to the toleranced feature by a leader line terminating in an arrow. The position of the arrow defines the feature. The angle of the leader line is also important. The width of the tolerance zone is in the direction of the arrow.

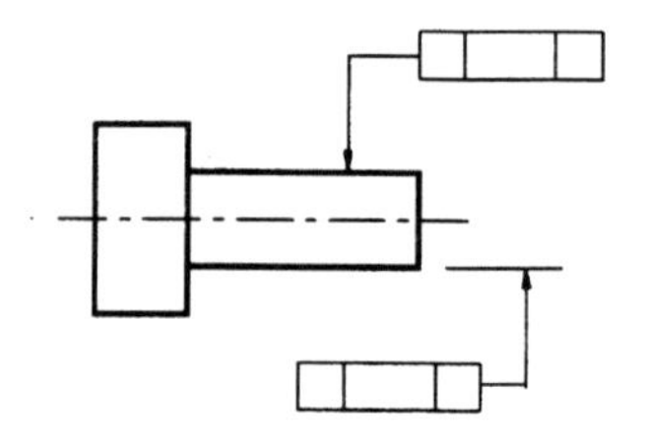

Figure 3.24

A surface, face or edge is identified by an arrow head which touches the feature or an extension of the feature (Figure 3.24).

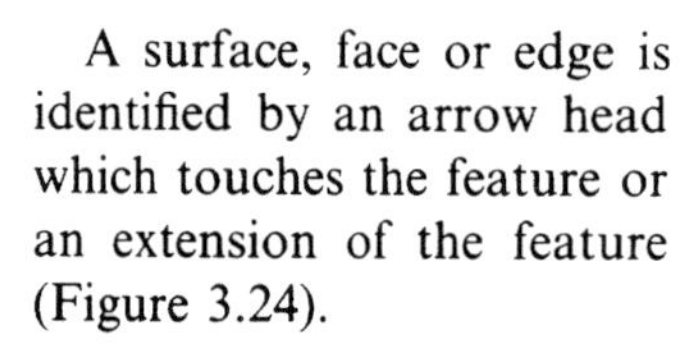

An arrow head touching a centre line indicates the common axis (or median plane) of all features with this axis (or median plane) (Figure 3.25).

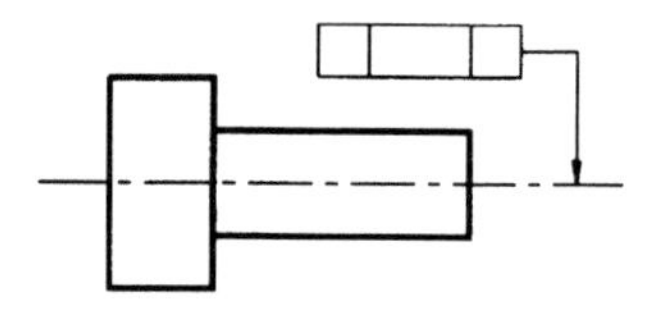

Figure 3.25

If the arrow head touches the feature at a dimension line, the axis (or median plane) of that feature alone is indicated (Figure 3.26).

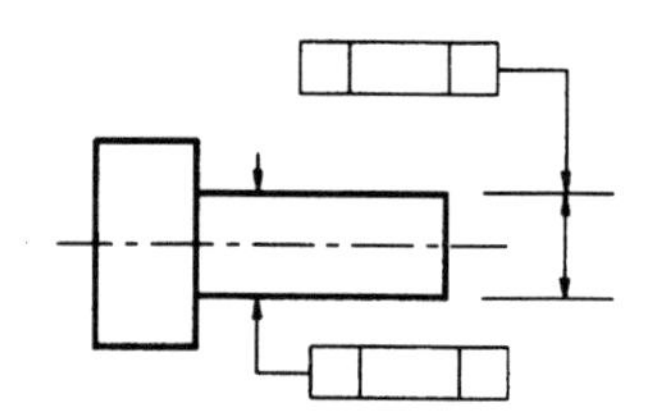

Figure 3.26

Datum identification

A datum feature may be a plane, surface, or an axis. It can be identified directly or by a capital letter in a box connected to the feature by a leader line ending as below, with a solid triangle, following the same rules as for feature identification (Figures 3.27 and 3.28).

If needed, more than one datum feature can be referenced by placing a hyphen between the two identifying letters, thereby indicating a single datum established by the two datum features (Figure 3.29).

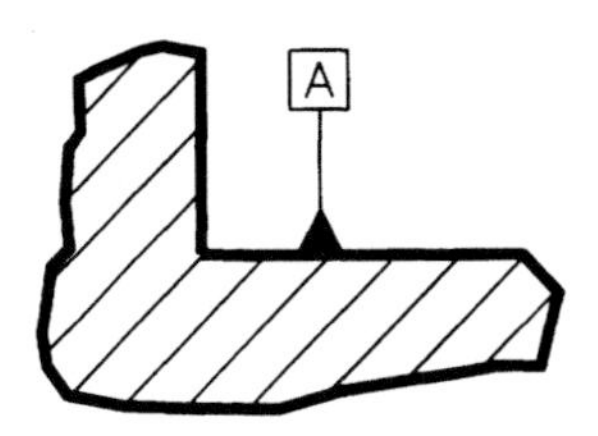

Figure 3.27

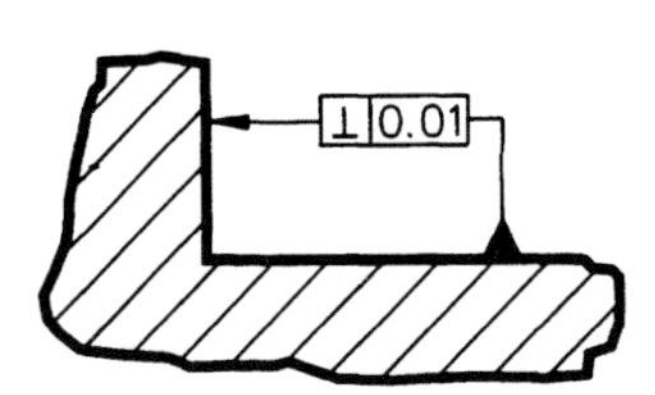

Figure 3.28

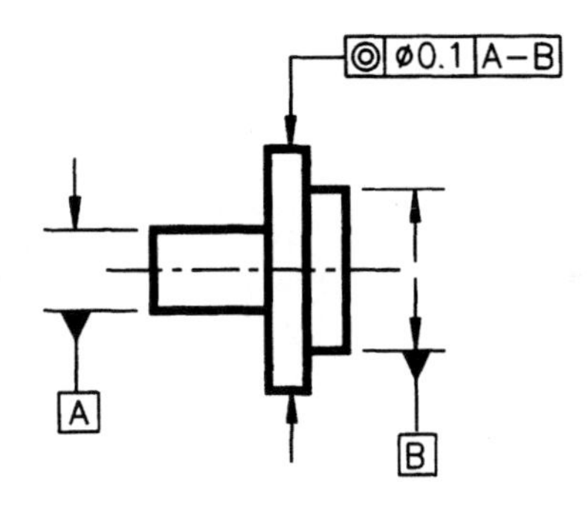

Figure 3.29

Symbols for toleranced characteristcs

TYPE OF TOLERANCE	CHARACTERISTIC	
FORM	Straightness	—
	Flatness	▱
	Roundness	○
	Cylindricity	⌭
	Line profile	⌒
	Surface profile	⌓
ORIENTATION	Parallelism	//
	Squareness	⊥
	Angularity	∠
LOCATION	Position	⌖
	Concentricity	◎
	Symmetry	⌯
COMPOSITE	Run-out	↗

Figure 3.30

Theoretically exact dimension	□
Datum feature identification	▲
Maximum material condition	Ⓜ
Circular/cylindrical tolerance zones	ø

Figure 3.31

3.4.3 Geometric tolerancing examples

(a) Tolerances of form. These do not need to be referenced to a datum.

Characteristic	*Example*	*Interpretation*
Straightness Controls the straightness of a line on a surface, an edge, or the axis of a feature. The tolerance zone may be the area between two parallel straight lines or a pair of cylinders.	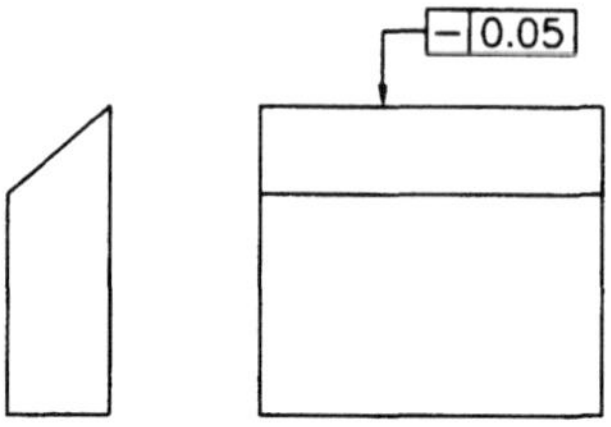	The edge must lie between two parallel straight lines 0.05 apart in the plane of the drawing. 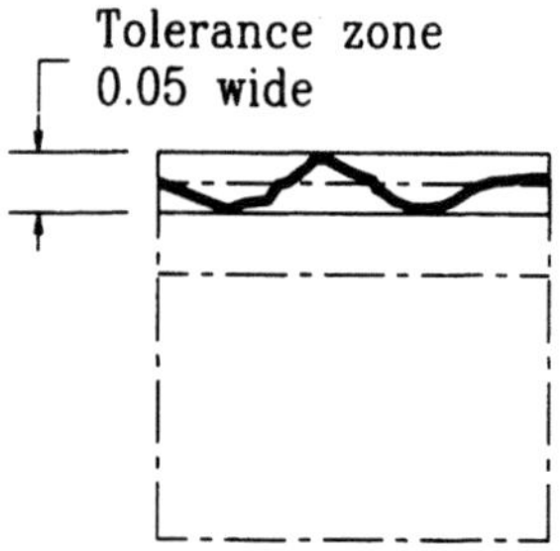
	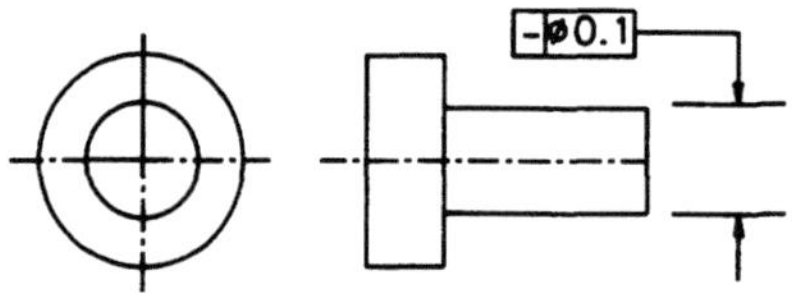	The axis of the dimensioned cylinder must lie within a tolerance zone formed by a cylinder of diameter 0.1. 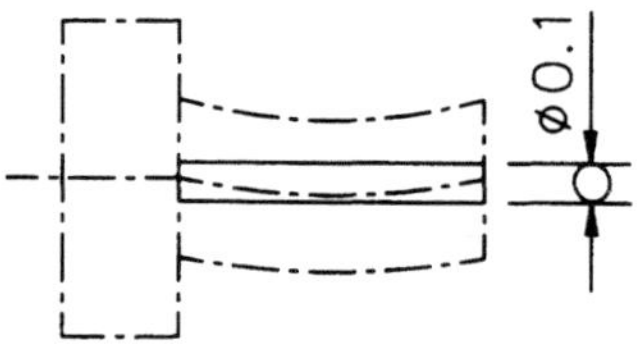
	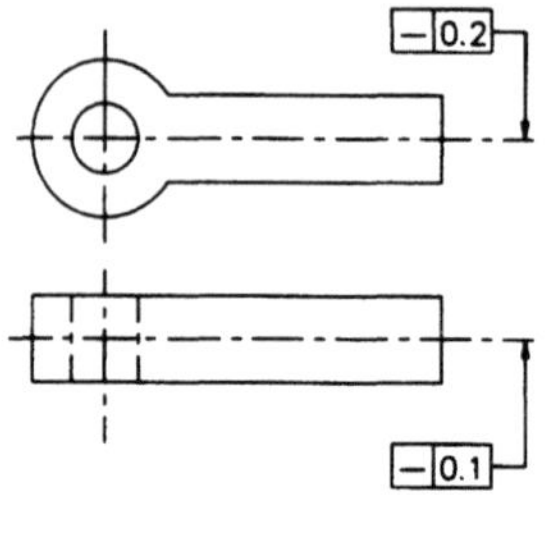	The axis of the whole component must lie within a parallelepiped 0.2 wide in the horizontal plane and 0.1 wide in the vertical plane.
Flatness Controls the deviation of a surface from the true plane, its tolerance zone being the space between two parallel planes. If required, notes such as NOT CONCAVE can be added.	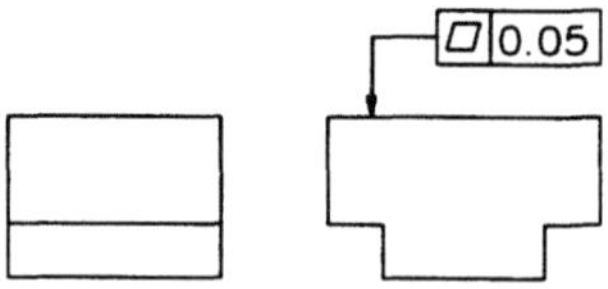	The surface must lie between two parallel planes 0.05 apart. 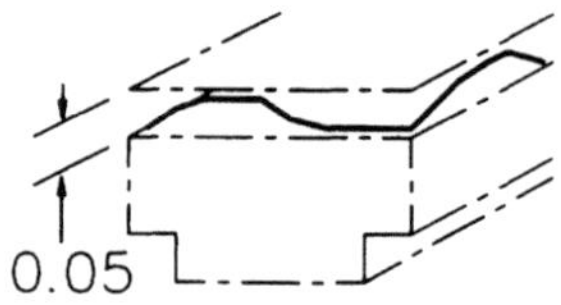

Characteristic	*Example*	*Interpretation*
Roundness Controls the errors in form of a circle in the plane in which it lies. It may refer to a section of a solid of revolution (sphere, cylinder etc.) the tolerance controlling the roundness of the cross-section. Roundness is not concerned with the position of the circle.	○ 0.04 ○ 0.04	In both of these examples the toleranced feature is any cross section at right angles to the axis of the designated surface. The tolerance zone is the space between two concentric circles 0.04 apart. Tolerance zone 0.04 wide Possible form
Cylindricity Combines roundness, straightness and parallelism, and is applied to the surface of a cylinder. To avoid problems with inspection it is often preferable to tolerance roundness, straightness and parallelism separately.	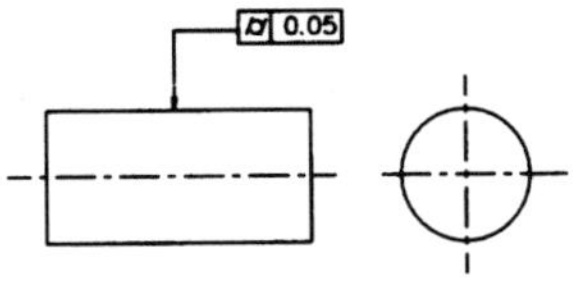 0.05	The curved surface of the cylinder must lie between the surfaces of two concentric cylinders 0.05 apart.
Profile of a line A profile tolerance controls the degree of variation from the theoretical or perfect form of a line.	⌒ 0.4	The tolerance zone is normally equally disposed about the true contour and defined as the area between two lines that encompass circles (with centres on the true contours) and of diameters equal to the tolerance. 0.4 True contour
Surface profile This controls the degree of variation from the theoretically perfect form of the surface.	⌓ 0.01	The tolerance zone is defined in the same manner as for the line profile, but using spheres and surfaces rather than circles and lines. Possible surface True contour Sphere 0.01 dia

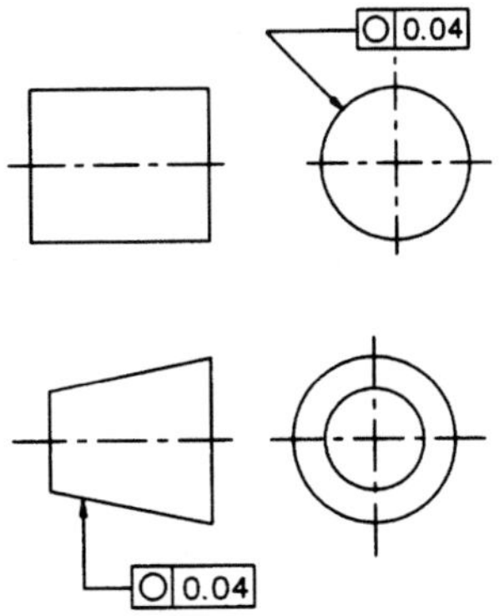

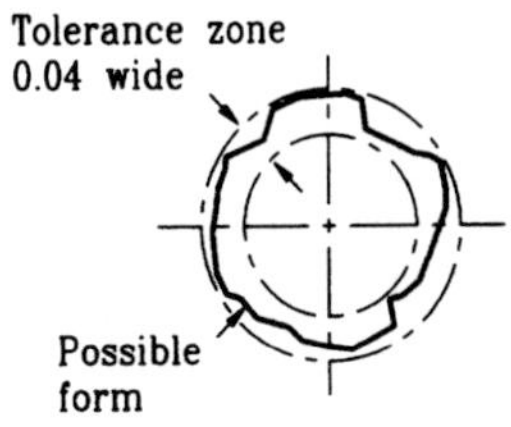

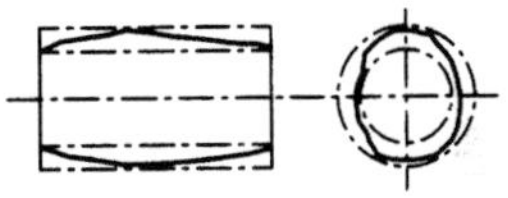

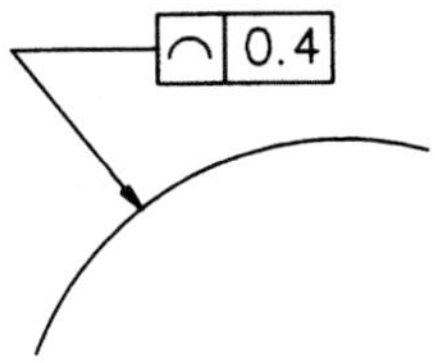

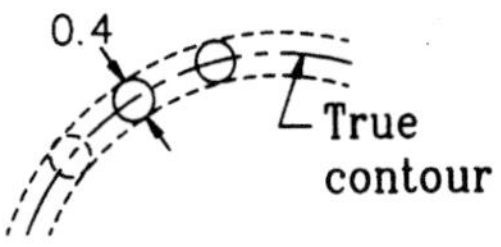

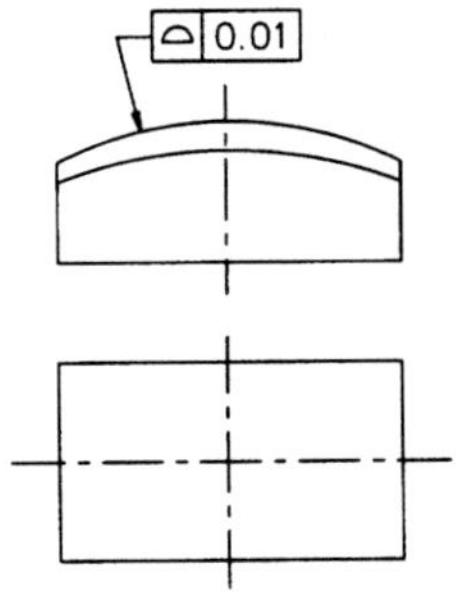

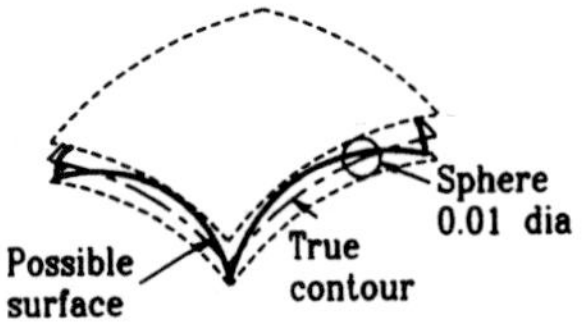

(b) Tolerances of orientation. These need to be referenced to a datum.

<table>
<tr><th>Characteristic</th><th>Example</th><th>Interpretation</th><th></th></tr>
<tr><td>Parallelism
This controls the parallelism of a feature (line, surface or axis) relative to a datum which can also be a line, a surface or an axis.</td><td>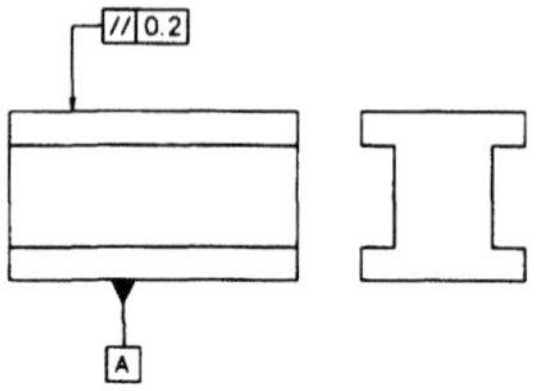
</td><td>The tolerance zone is the area between two parallel planes which are parallel to the datum plane and equally disposed about the true surface. The tolerance value is the distance between the two planes.</td><td>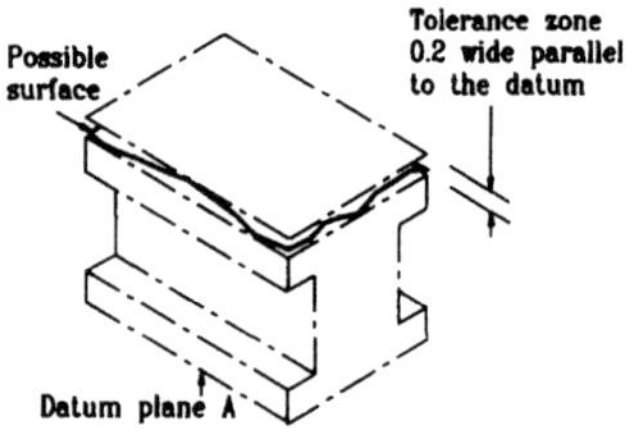
</td></tr>
<tr><td></td><td>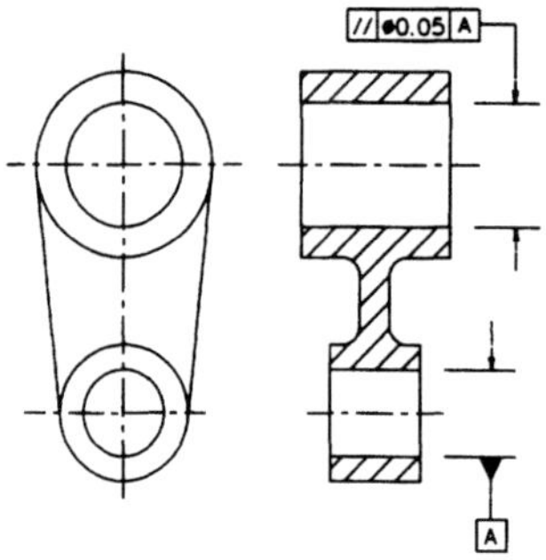
</td><td>The upper axis must lie within a cylindrical tolerance zone 0.05 diameter, with its axis parallel to the datum axis.</td><td>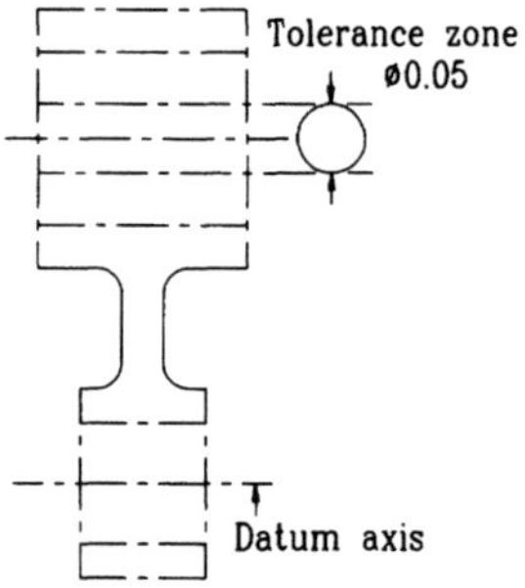
</td></tr>
<tr><td>Squareness
This controls the squareness of a line or a surface relative to a datum which can be a line, a plane or the axis of a hole.</td><td>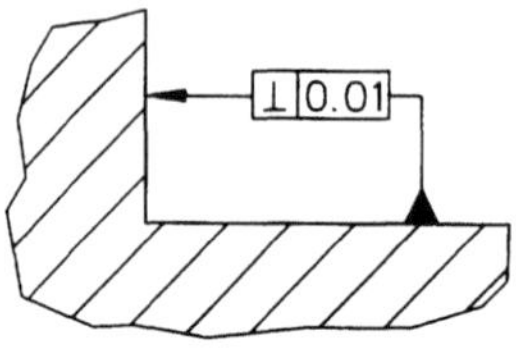
</td><td>The feature must lie between two parallel lines 0.1 apart, each of the planes being perfectly square to the datum.</td><td>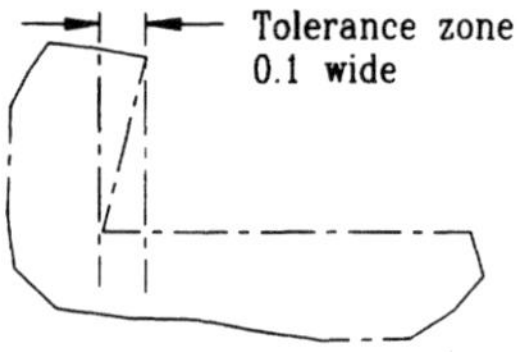
</td></tr>
<tr><td>Position
This controls the amount the actual position of a feature can deviate from its true position.</td><td>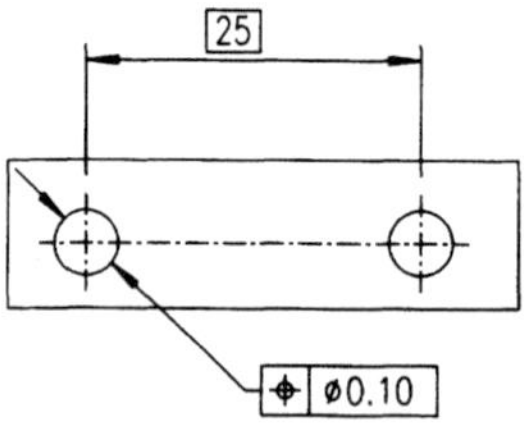
</td><td>The centre of each hole must lie within a circle of diameter 0.1 centred about its true position given by the boxed dimension.</td><td></td></tr>
</table>

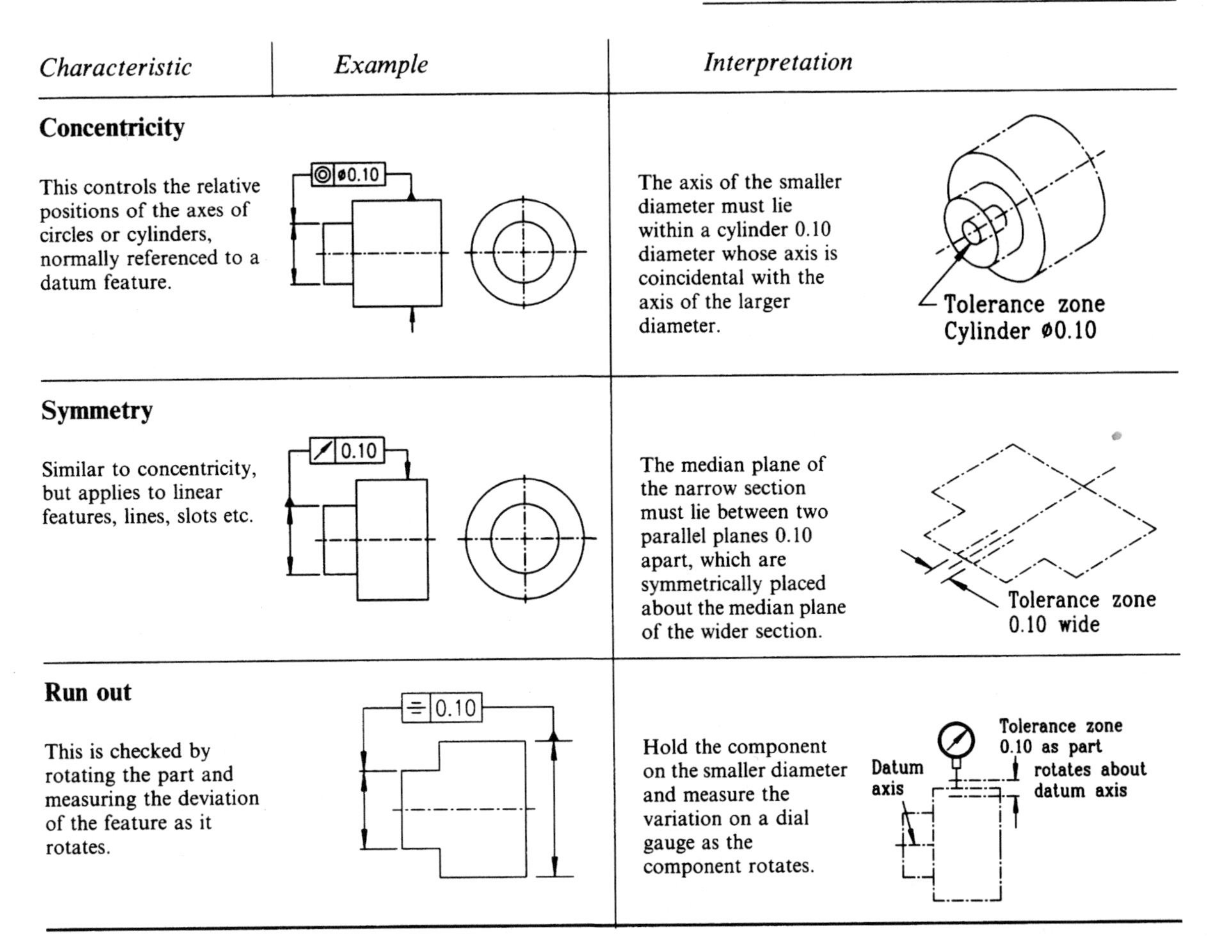

Characteristic	*Example*	*Interpretation*
Concentricity This controls the relative positions of the axes of circles or cylinders, normally referenced to a datum feature.	◎ ⌀0.10	The axis of the smaller diameter must lie within a cylinder 0.10 diameter whose axis is coincidental with the axis of the larger diameter. Tolerance zone Cylinder ⌀0.10
Symmetry Similar to concentricity, but applies to linear features, lines, slots etc.	⁄ 0.10	The median plane of the narrow section must lie between two parallel planes 0.10 apart, which are symmetrically placed about the median plane of the wider section. Tolerance zone 0.10 wide
Run out This is checked by rotating the part and measuring the deviation of the feature as it rotates.	≡ 0.10	Hold the component on the smaller diameter and measure the variation on a dial gauge as the component rotates. Datum axis Tolerance zone 0.10 as part rotates about datum axis

3.4.4 When to use geometric tolerancing

Adding geometric tolerances to a drawing is a very efficient manner of specifying closely the requirements of the designer. It would, however, be all too easy to over-tolerance a component, the result being an increase in the cost to manufacture. They should be applied for all requirements critical to the function and interchangeability of parts, except where it is known that the machinery and manufacturing techniques can be relied upon to achieve the required standards of accuracy. For example, if a component has a number of diameters that need to be concentric, provided that they are all cut during one set-up on the lathe, their concentricity will be determined by the accuracy of the machine tool. If previous similar components have performed satisfactorily when produced in this way, adding geometric tolerances will only add to the production costs, i.e. another dimension to check.

It also depends on the type of drawing. The drawing for a die for a metal pressing might need a number of geometric tolerances to be identified. The actual pressing will obviously need to be checked to ensure it is geometrically correct. However, the die and the pressing process will control the final form of the component, so its drawing for any subsequent machining operations does not need to include the geometric tolerances (unless they are directly affected by the machining process).

Only apply geometric tolerances where essential, to avoid unnecessary cost increases.

In all cases, the tolerance should be as large as possible, subject to meeting the design requirements.

3.5 Maximum Material Condition

We have already met the Maximum Material Condition (MMC) in section 3.1.4, with the unilateral method of showing dimensions. An understanding of the concept can be used to relax certain tolerances without affecting the performance of the component.

Take as an example a shaft which is required to fit into a cylindrical hole. The shaft and hole will each have size tolerances on their diameters, to ensure that the correct fit exists between the two. If either the finished shaft or the hole are at MMC, the clearance between the parts will be at its minimum. Any departure from MMC (although still within the tolerance limits) will obviously increase this clearance. However, if the shaft is curved, but still within the diametral tolerance limits, it may not be able to fit into the hole. A geometric tolerance could be used to control this condition, i.e. the out-of-straightness. The severity of control on straightness will depend on the clearance between the shaft and the hole. If we define a suitable limit for the geometric tolerance when both parts are at MMC, this will cater for the worst-case condition. However, it will give tighter control than needed if the parts are finished away from MMC (i.e. the tolerance could be increased in some cases). This opportunity to relax the tolerance of form may only be used provided the design function is not affected. The increased tolerance will allow greater freedom for manufacture.

The tightest condition of assembly between two components is when each feature is at its Maximum Material Condition *and* the maximum errors permitted by the geometric tolerancing are present. This leads to the concept of virtual size.

Figure 3.32(a) shows a shaft with tolerances of size on the diameter and form on the straightness. If this were at MMC and also at the limit allowed by the straightness tolerance, its effective or virtual diameter would be as shown in Figure 3.32 (b).

For a male part (e.g., a shaft):
Virtual size = MMC + Geometric tolerance

For a female part (e.g. a hole):
Virtual size = MMC − Geometric tolerance

The virtual size will give the minimum clearance condition for assembly. If the shaft diameter is below the virtual size, but still within its size tolerance, the geometric tolerance could be relaxed without the component exceeding its virtual size and still permit acceptable conditions for assembly.

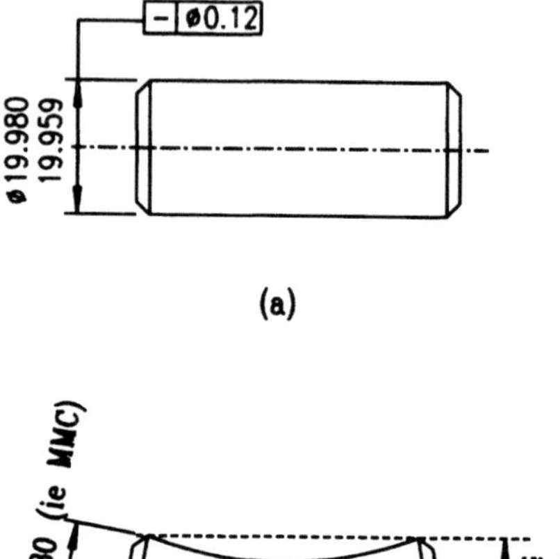

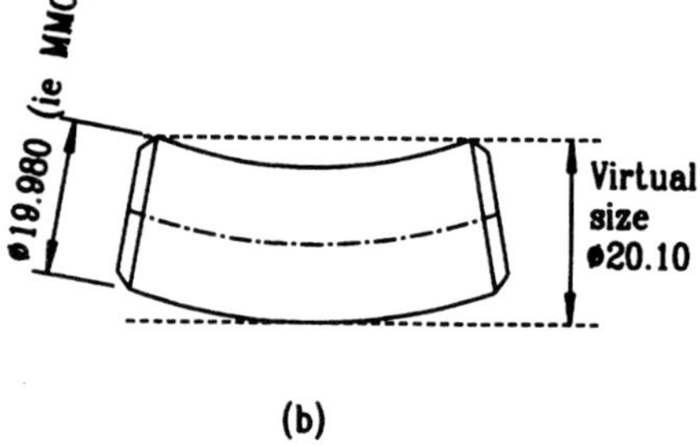

Figure 3.32 *MMC and virtual size*

If the letter M is added after the geometric tolerance value, this means that the tolerance only needs to be rigorously applied when the components are at MMC. If they are not at MMC the tolerance of form may be increased provided that

(a) the part size tolerance is not exceeded, and
(b) the virtual size is not exceeded.

This limits the maximum increase in the geometric tolerance to the difference between the actual size and MMC (Figure 3.33).

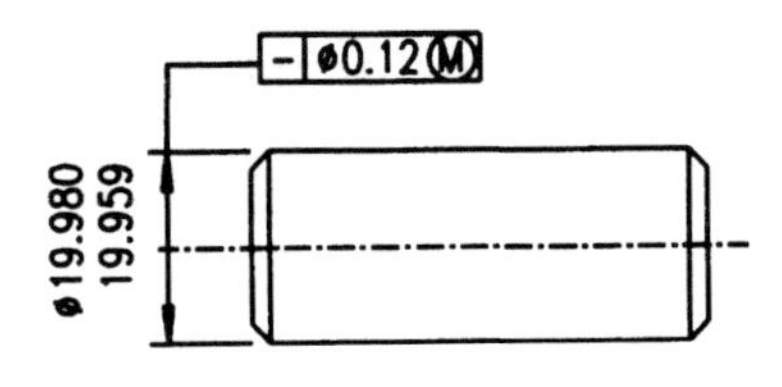

Figure 3.33 *Application of MMC*

3.6 Examples

Example 3.3

Two phosphor bronze bushes are located in the sides of a component such that their centre lines are exactly in line (Figure 3.34). A shaft with a pulley attached runs in the bushes at slow speed. If the nominal outside diameter of the bushes is 60mm and the nominal inside diameter is 35mm, what will be the actual dimensions that appear on the drawing?

Solution

The first step is to decide whether to use shaft or hole base. The outside diameter should be hole based, as there is no reason for it to be otherwise. The inside diameter, on the other hand, has a common shaft running through the bushes so should be shaft based. If you are not sure about this, refer to section 3.2.7.

From BS4500 select

OD	Light press fit	H7 p6
ID	Running fit	F8 h7

Using the tables in Appendix B, the dimensions would be as Figure 3.35. In addition to the sizes, we need to be sure that the inside and outside diameters are concentric, or we will have trouble fitting the shaft. We could do this by adding some geometric tolerances, or we could take a more practical approach and insist that they are reamed in line after assembly.

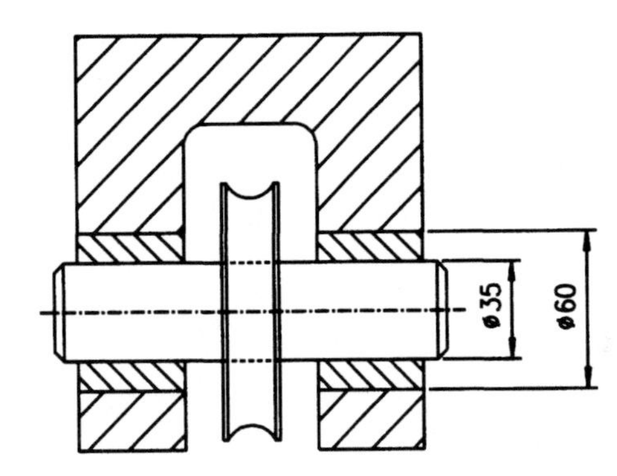

Figure 3.34 *Example 3.3*

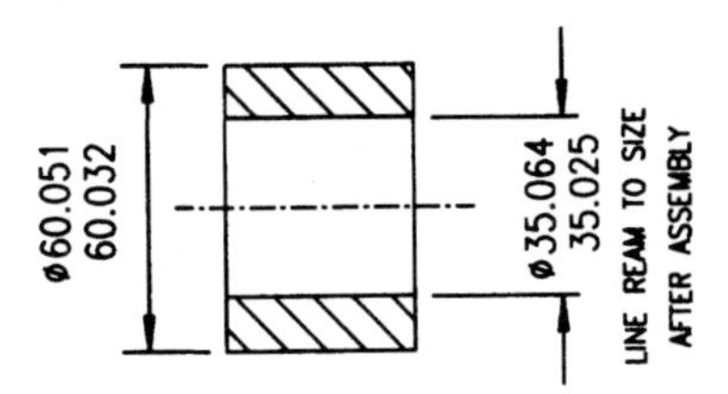

Figure 3.35

Example 3.4

Figure 3.36 shows a bracket in which holes are to be machined through and reamed in line. This figure also shows a shaft, which is to be located in the holes. The shaft requires machining and is to be a close running fit when assembled. The nominal diameters for the shaft and the holes are 25mm.

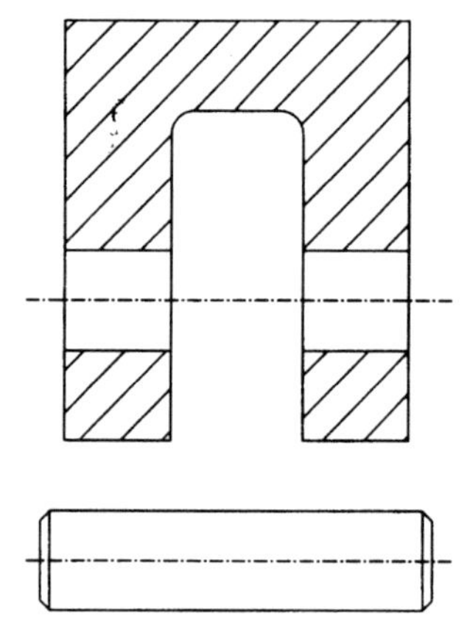

Figure 3.36

1. Dimension the shaft and hole diameters.
2. Select and add a geometric tolerance to the shaft that prevents the possibility of an interference fit. Remember that no tolerance should be any tighter than needed.
3. What is the maximum radial float of the shaft in the holes? What effect does the geometric tolerance have on the float?
4. Could MMC be used to ease the manufacturing tolerances? What would be the effect?

Solution

1. Select hole base as there is no reason to use shaft base system.

   From BS4500 a close running fit is H7 g6. Using the tables in Appendix B3, the required diameters are

Shaft	24.993	Hole	25.021
	24.980		25.000

2. The specified manufacturing process for producing the holes will allow us to assume that the tolerance on straightness for the holes will be very low: we will take it to be zero.

   The clearance will be a minimum when the shaft and the holes are both at MMC.

   i.e. Minimum float = 25.00 −24.993
   = 0.007

   This is the maximum error in straightness that can be permitted in the shaft to avoid an interference fit.

3. The maximum radial float will occur when the shaft is perfectly straight, the hole is at its maximum size and the shaft at its minimum.

   i.e. 25.021 − 24.980 = 0.041
   Maximum radial float = 0.041

   As the shaft departs from being perfectly straight the float will reduce by up to the maximum of the geometric tolerance. Figure 3.37 (a) shows how the geometric tolerance would be specified on the drawing.

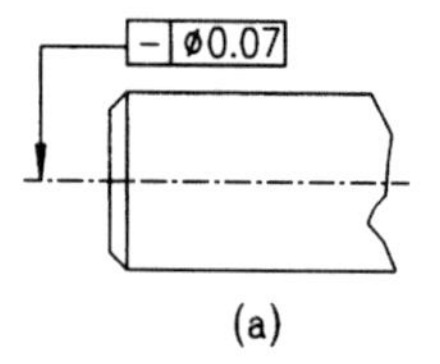

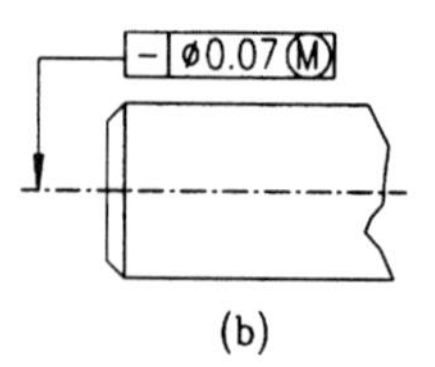

Figure 3.37

4. In this case MMC could be used to ease the straightness tolerance. It would be shown on the drawing as in Figure 3.37 (b). The effect

would be to allow the tolerance for the straightness of the shaft to increase by the difference between the actual size of the shaft and MMC.

Max increase = 24.993 − 24.980
= 0.013

The maximum departure from straightness of the shaft applies at its minimum size and is 0.007 + 0.013 = 0.20.

3.7 Summary

The success of a design can be made or lost by attention to detail. Ignore the detail and the component will almost certainly be less reliable, more difficult to manufacture or possibly have both of these handicaps. Although a wide range of detail topics has been covered in sufficient detail for undergraduate use, the serious designer will need to refer to the relevant standards for more detailed information. On completing this chapter you should have an understanding of the following points:

- The need for tolerances, and their influence on component costs.
- The ways in which they are presented on drawings.
- Functional dimensions and how they affect the selection of datums.
- Methods of allocating tolerances in such a way that they take account of the manufacturing process.
- Limits and fits, when to use hole or shaft based systems, and how to select a suitable fit.
- Surface finish, its means of measurement, selection and specification.
- The need for tolerances of form (geometric tolerances), and an understanding of how and when they should be used.
- MMC, its meaning and how it can sometimes be used to relax tolerances of form.

3.8 Questions

Now try answering these questions to check your understanding. Refer to the relevant section in this chapter (shown in brackets) if you have difficulty with any of them.

1. Name three methods of applying a tolerance to a dimension on a drawing. (3.1.4)
2. State the condition when the shaft base for limits and fits is used (3.2.7)
3. A hole and shaft have a nominal diameter of 50mm. Assuming the application requires a close running fit, what are the maximum and minimum acceptable diameters for the hole and the shaft? (3.2.6)
4. What do you understand by the term MMC? (3.5)
5. When should geometric tolerances be applied? (3.4.4)
6. Why should you avoid over-dimensioning components? (3.1.7)
7. A base plate requires one face to be machined to within 0.075mm. The current machining cost is £20. Estimate the new machining cost if the toleranced was halved. (*Hint* – see PD6470) (3.1.3)
8. A surface finish of 1.6 R_a is called for. What types of machines would you expect to be capable of producing this finish? (3.3.4)
9. What is the difference between the terms concentricity and cylindricity? (3.4.3)
10. A component 100mm long is produced from billets cut by a material stockist from 30mm bar. After machining an outside diameter, the component has a tolerance on diameter of 0.15mm. However, the designer would prefer this tolerance to be reduced to 0.04mm. The component currently costs £15 (£5 for the material and £10 for machining). Estimate the cost of the modified component. What would you judge to be the smallest realistic tolerance for this diameter? (*Hint* see – PD6470) (3.1.3)

4 Standard Components

4.1 Introduction

Today there are very few designs that do not incorporate a number of standard components. This applies to almost everything we meet in everyday life – the light-fitting overhead, the bicycle we ride, or the car we drive. All these contain a large number of standard components. Perhaps the only exception might be the free plastic toy that leapt from the cornflake packet at breakfast time!

Why is this so? If you need a new tyre for your car or bicycle, because the designer has chosen a standard product you will easily be able to locate and buy a replacement. In the same way the availability of a standard product for the car designer meant he/she was able to concentrate on his/her particular area of expertise and leave the mysteries of tyre design and manufacture to the tyre specialist.

The benefits of standardisation can be summarised under the following headings:

- *Cost:* The specialist manufacturer can produce in very high volumes and hence achieve economies of scale to give a low component cost.
- *Investment:* In many cases the levels of investment in tooling (and research) by the designer's employers is reduced. In fact the actual capital investment made by the component suppliers is often very high, the costs effectively being shared by the large number of their customers.
- *Reliability:* Standard component manufacturers are specialists in their chosen field. They will have extensive experience of the product enhanced by feedback from a wide variety of users, and will be therefore be in a strong position to predict component life.
- *Safety:* If a standard component is available but has not been selected by you, as the designer, you will need to have a very good reason particularly if the design fails in the future. The risks you take by not using a standard component may not simply relate to cost. If personal injury results from failure of a non standard item you may have to defend yourself in court.

The designer's task is to find the most cost-effective solution to a problem, so, wherever appropriate, standard components should be incorporated in the design. This chapter will introduce a number of common engineering standard components. The reader should be aware that there are many more standard items, and be prepared to search for and take council from these specialist manufacturers.

4.2 Types of standard

Standard components are usually selected in consultation with the specialist supplier, either by direct contact with their technical representative or by following the procedures in their catalogues. In most cases more than one specialist manufacturer exists for a particular range of components. In order to ensure a further level of commonality in many fields the specialist manufacturers have formed trade organisations and produced sets of standards. Their purpose is primarily to enhance trade.

For example, most cars use tyres manufactured to a national standard and hence replacements are available from a number of alternative suppliers. The resulting competition keeps prices in check, encourages efficient manufacture and is a spur to further technical advances. On the few occasions when a tyre manufacturer has tried to succeed with a product that was not accepted by his fellow manufacturers the result has been commercial failure. The Denovo tyre introduced by Dunlop in the 1970s was a technically advanced run-flat tyre fitted to a small range of vehicles. Unfortunately it was not adopted by any other tyre manufacturer. It was dropped after only a few years, largely because the replacement tyres were only available from the single manufacturer, thus limiting customer choice.

National standards: In many industrialised countries the trade organisations have worked together to create their own national standards, e.g:

BS: British Standards
JIS: Japanese Industry Standards
ANSI: American National Standards Institution
IIRS: Irish Institution of Research and Standards

Within the umbrella of, say, the British Standards, there are a series of divisions covering specialist areas, e.g. auto, aerospace, general. Similarly under the heading of ANSI there are divisions: ASME (American Society of Mechanical Engineers), ASTM (American Society for Testing Materials), SAE (Society of Automobile Engineers), etc.

International standards: Expanding world trade has generated a need for standards to cross national boundaries. The International Standards Organisation was created to address such a need. It has created a number of ISO standards, in most cases by adopting or adapting a suitable national standard.

Company standards: The greater the variety of different components that have to be handled, stored, etc. the more difficult it is for a company to keep its costs under control. Many manufacturing organisations recognise this and have produced their own systems for standardising on components. Such systems include design office manuals, and sophisticated part numbering procedures.

The cost-conscious designer must avoid 're-inventing the wheel' by checking the availability of an off-the-shelf component, and the relevant company, ISO or BS standards.

4.3 Types of transmission

4.3.1 Introduction

A transmission (see Table 4.1) is the means by which the power and torque are transferred from the source to the end user. Usually transmission systems consist of a coupling element, a unit providing a range of torque multiplication, and a means of distribution. In the case of a motor vehicle, one such transmission system consists of a clutch coupling the engine (the power source) to

Table 4.1 *Some types of transmission*

	Shaft	*Belt*	*Chain*	*Gear*	*Hydraulic*	*Electric*
Ability to change speed and torque	No	Yes	Yes	Yes	Yes	Yes
Accuracy requirements between input and output	Normally critical	Less critical than shaft	Similar to belt	Normally critical	None	None
Constraints on output shaft axis relative to input shaft	As input	Normally parallel to input	As belt	Can change angle and offset axis	None	None
Can accommodate slippage	Depends upon connections	Yes	No	No	No	Yes, depending upon design
Efficiency	High	Fairly high	Fairly high	Fairly high	Relatively poor	Moderate

the gearbox (the torque multiplier), which is in turn connected via a differential and drive shafts to the driven road wheels. This is not the only system in a motor vehicle: several others exist, e.g. one transfers energy to the battery. This normally involves a belt drive to an alternator, which outputs voltage and current to a device which converts it from alternating to direct current, and eventually connects to the battery. A number of different types of transmission exist, each having its own characteristics. Table 4.1 summarises some types of transmission.

We will now consider each of the above systems, noting some of the components required and points that the designer should note.

4.3.2 Shafts

One of the most basic forms of transmission is via a shaft. Its advantages include simplicity, compactness, and an ability to handle high powers and to operate over a wide speed range. There are, however, some fundamental limitations. There is no facility for changing speed or torque, and alignment is normally critical.

Shafts are also used as elements of other forms of transmissions. The gearbox in a motor car will, for example, use both gears and shafts to transmit the torque.

Shaft loads

In order to ensure that the shaft is strong enough we need to consider the loads to which it will be subjected. The fundamental loadings are (see Figure 4.1):

(a) Torsion
(b) Radial
(c) Axial

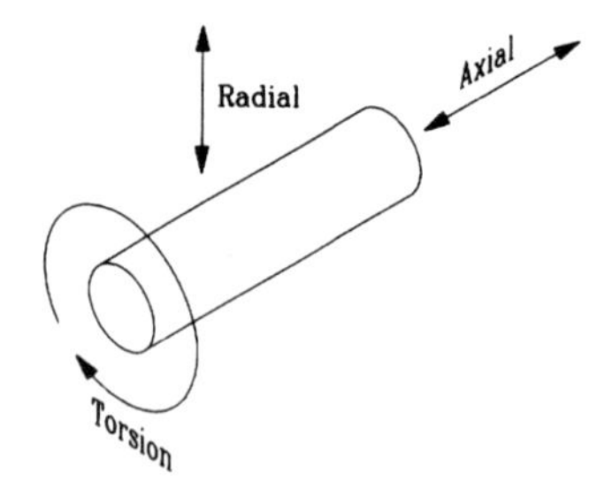

Figure 4.1

Torsion loads: A simple shaft transmission such as a vehicle drive shaft will certainly be subjected to torsional loads.

If the loading is pure torsion the basic relationship that applies is:

$$\frac{\tau}{r} = \frac{T}{J} = \frac{G.\theta}{L}$$

where

r = radius of the shaft
τ = shear stress
T = torque to be transmitted
J = polar moment of inertia of the shaft section
G = modulus of rigidity
θ = angle of twist of the shaft
L = length of the shaft

G and τ are functions of the material selected.

Rearranging the basic formula

$$\frac{T}{\tau} = \frac{J}{r_{\max}} = \text{section modulus}$$

if the shaft diameter is d

$$\frac{T}{\tau} = \frac{\pi.d^4}{32} \times \frac{2}{d}$$

$$d^3 = \frac{16.T}{\pi.\tau}$$

Hence a suitable diameter may easily be calculated if the shaft is in torsion alone.

Radial loads: Bending loads appear on shafts for a variety of reasons. For example, a pulley connected to a shaft will by virtue of the tension in the belt it carries impart a radial load to the shaft. Similarly the forces from the meshing of a pair of spur gears (see Section 4.3.5) will generate a radial force on the shafts carrying the gears. These radial forces will generally be restrained by bearings. Physical limitations ensure the radial loads and their respective restraining forces are offset. The result is a bending load on the shaft. Once the primary source of the radial load is known, it is normally a simple task to calculate the restraining forces at the bearings by taking moments. In most cases you will need to analyse the forces in two orthogonal planes, i.e. horizontally and vertically.

If the shaft is subject to both bending and torsion, then the following formula can be used to evaluate an equivalent torque which will cause the same damage to the shaft.

$$\text{Torque}_{equivalent} = T_e = \sqrt{M^2 + T^2}$$

where M = bending amount
T = applied torque

then

$$d = \sqrt[3]{\frac{16.T_e}{\pi.\tau}}$$

This will ensure that the diameter chosen does not permit the safe stress levels in the material used to be exceeded. However, in many instances shaft deflection may cause problems, even if the allowable stress levels are not exceeded. Deep groove ball-bearings, for example, will only tolerate a small degree of shaft misalignment (typically between 2 and 10 minutes of arc). The shaft designer should evaluate the radial loads on the shaft and use a technique such as *Macaulay's Method* (check your stress text) to estimate the shaft slope and hence deflections.

Axial loads: Shaft drives are often required to transmit axial loads. The shaft carrying a ship's propeller may be subjected to torsional loads transmitting the torque from the engine to the propeller, but the reaction loads from the seawater that result in the forward motion of the ship will impart an axial load on the shaft.

In practice the axial loads are normally of more concern in the provision of suitable bearings and couplings, other considerations having a greater effect on the shaft size. However, if the design includes a long thin shaft taking an axial load these effects should be checked.

Note: If a keyway is included (i.e. machined in the shaft) then increase the shaft diameter by a factor of 1.1.

Always allow for shaft expansion: NEVER axially fix both ends of the shaft. Fix one end and allow the other to float axially.

Keys and keyways

Whichever type of coupling is employed, a means of connecting the coupling to the appropriate shaft is needed (unless it forms an integral part of the shaft). The most common solution is to use a key and keyway.

The section of the key is determined mainly by the diameter of the shaft requiring the connection. BS4235 lists suitable sections, together with the necessary tolerances.

The length of the key has to be sufficient to avoid failure. Loading is a complex function of clearances and the elasticities of the shaft, hub, and key. However, some simple checks can be made by considering the loads in shear and compression.

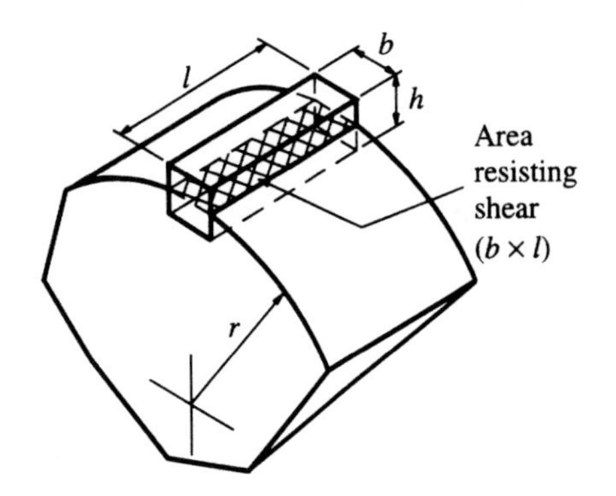

Figure 4.2 *Keyway in shear*

Shear: (see Figure 4.2)
The shearing force is:
torque/radius (T/r)

The force resisting shear is:
allowable shear stress × area ($\tau \times b \times l$)

Hence if,

$$T/r = \tau \times b \times l$$

this will give the *minimum* length to resist shear.

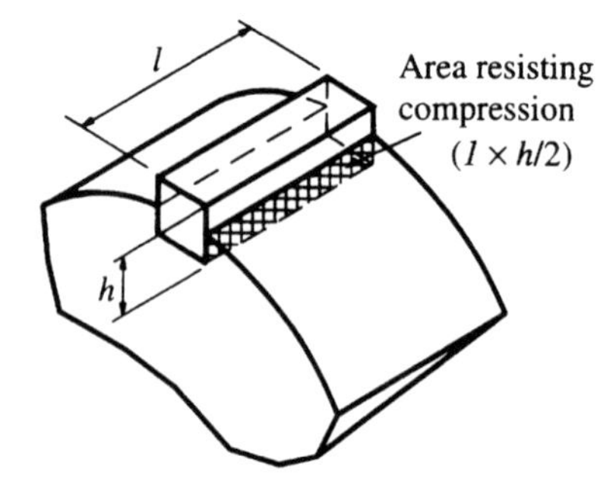

Figure 4.3 *Keyway in compression*

Compression: (see Figure 4.3.)
A check should also be made to ensure that the compressive stresses (σ) on the side do not exceed acceptable levels. That is,

$$T/r = \sigma \times h/2 \times l$$

As a general rule, make the key the same length as the boss being joined to the shaft, then check the stresses and modify if necessary.

Another means of making the connection is by the use of splines. These are often used when the connection has to accommodate some axial movement. BS2059 covers straight-sided splines, and BS3550 specifies the dimensions for involute splines.

4.3.3 Belt transmissions

Flexible connectors such as belts come in many different forms, all of which have a number of advantages over alternative systems. They absorb vibration and shock, only a minimum being transmitted to the connected shaft. They are suitable for large centre distances and are quiet. Add to this a tolerance of shaft misalignment, an ability to cope with an overload by slipping, and minimum maintenance requirements, and you have a deservedly popular form of power transmission.

Principle of belt drives

Now consider a rope or belt as in Figure 4.4 hanging over a pulley which is prevented from rotating. The tensions T_A and T_B are caused by the large and small weights respectively. Common sense tells us that the maximum differential in these tensions, assuming the belt does not slip, is determined by the coefficient of friction between the belt and the pulley.

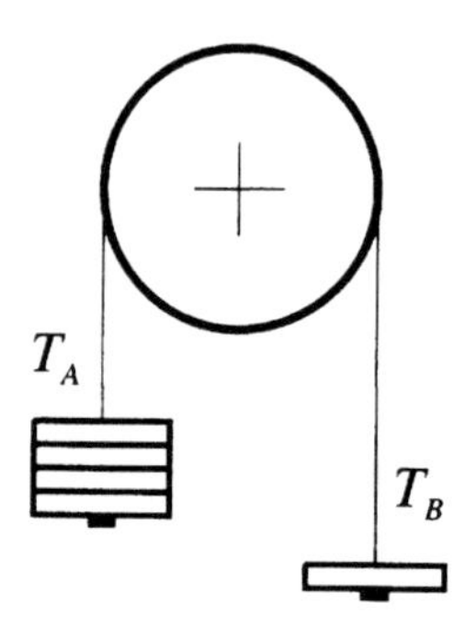

Figure 4.4

Suppose we now add an idler pulley that is able to turn freely, as in Figure 4.5. If the difference in the tensions is close to the maximum acceptable in Figure 4.5, we would expect the belt to slip on the larger pulley because the arc (or angle) of contact has been reduced. To prevent this we would need to increase tension T_B.

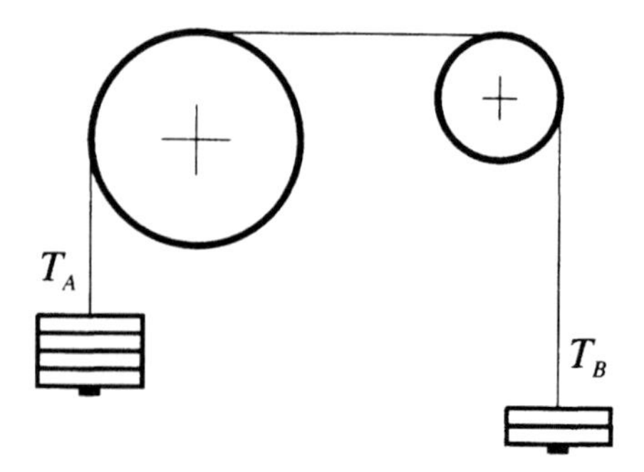

Figure 4.5

This tells us that there are at least three major interrelated factors affecting belt slippage, namely:

(a) Tension
(b) Coefficient of friction
(c) Angle or arc of contact.

If in either of the above cases the unbalanced tension $(T_A - T_B)$ is large enough to overcome the resistance, the pulley will turn, although in these examples the action will be limited by the length of the belt.

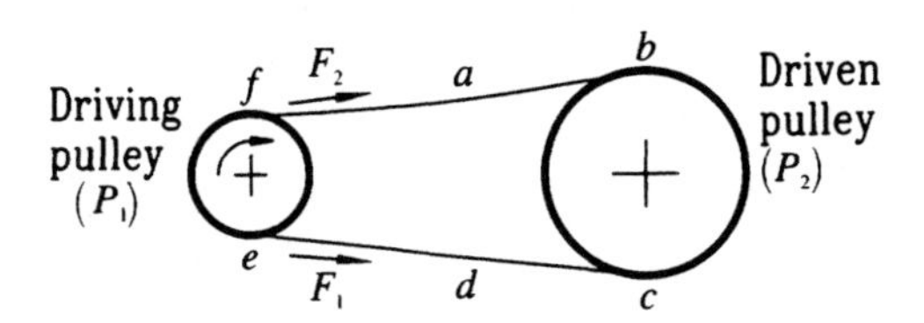

Figure 4.6

We can overcome the length problem by considering two pulleys connected by a continuous belt, as in Figure 4.6. The centre distances and the belt lengths are chosen so that no slippage occurs. Assume the smaller pulley to be driving clockwise. The force F_1 on the approaching belt is greater than F_2 on the receding (slack) side. These forces produce the driving torque: i.e. the larger (driven) pulley is subjected to a torque

Torque (driven pulley) =
$(F_1 - F_2)$ × Radius (driven pulley)

The capacity of the belt is therefore limited by:

(a) Friction between the belt and the pulley
(b) Tensile strength of the belt.

Now we will examine the stresses raised in the belt as the system rotates. Consider a point on the belt starting at a having stress s_2, i.e. relating to F_2. At b the stress increases due to bending (the level of increase depends upon the pulley radius) and becomes $(s_2 + s_{p2})$.

As the point moves round the pulley the stress will gradually increase as the tension increases from F_2 to F_1 and becomes $(s_1 + s_{p2})$, until at point d, where the belt leaves the pulley it drops to s_1 (i.e. relating to F_1) as the bending stress is no longer applied.

The stress effects as the point travels round the two pulleys are summarised in Figure 4.7. As the stress loadings on the belt are cyclic a possible mode of failure is through fatigue.

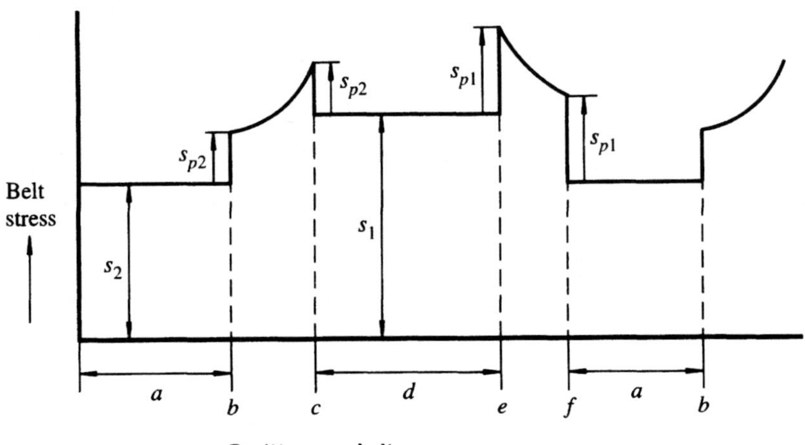

Figure 4.7

A further point to consider is that as the belt is moving on a curved path a centrifugal stress will be induced: we will take this to be a rough constant, so all the stresses in Figure 4.7 will be raised slightly.

The capacity of a belt is therefore limited by:

(a) Friction between the belt and the pulley (which depends upon tension, coefficient of friction and the angle or arc of contact)
(b) Tensile strength of the belt
(c) Fatigue effects on the belt
(d) Centrifugal effects on the belt.

Flat belts

Flat belts have been widely used in the past for power transmission, though today they have largely been replaced by more recent developments. Various materials have been used:

- Leather (oak tanned or mineral tanned)
- Solid woven (hair, cotton, or synthetic material impregnated with natural or synthetic rubber or gum)
- Nylon, faced on each side with leather (the nylon transmitting the load, and the leather providing the grip to the pulley face).

The flat profile of the belt keeps the stresses due to bending reasonably low, but some of the materials used make the predictability of their stress capability difficult.

Pulleys for flat belt drives must be cambered in order to keep the belt tracking properly. The pulley should be about 15% wider than the belt and should be cambered with a flat central portion not exceeding half the belt width. (The maximum camber should be about 1% on diameter or 5mm per 300mm pulley width, whichever is the lesser.) Efficiency tends to be high, typically 97–98%.

V-belts

A need for greater friction between the pulley and the belt in a compact design led to the development of the V-belt. The belts were manufactured with a core of high tensile cord embedded in rubber or synthetic rubber and encased in a fabric and rubber reinforcement (Figure 4.8). The core supplies the tensile strength and the casing forms the friction surface in contact with the pulleys.

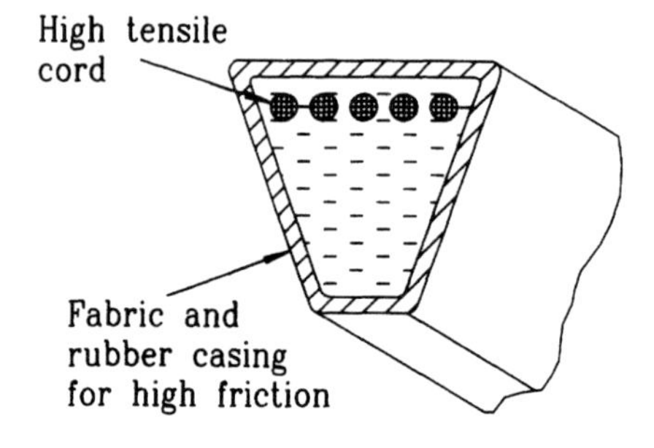

Figure 4.8

Different belt constructions are available which provide anti-static capacity, oil tolerance and flame resistant properties. All belts have a standard angle of 40°. The angle of the pulley groove varies to accommodate the slight bulging of the sides of the belt as it flexes around the pulley. BS 1440 (1971) specified six sizes of belt: Y, Z, A, B, C, and D. The height-to-width ratio of these belts is 0.73.

Wedge belts

A belt with a similar construction to the V-belt but with a deeper section giving a height-to-width ratio of 0.95 is called the *wedge belt*. This has the advantage of being able to transmit more power (greater friction between the belt and the pulley). BS 3790 specifies four sizes of wedge belt: SPZ, SPA, SPB, and SPC.

The taller section of these belts and the associated bending stresses limited the minimum size of pulley.

The next stage in belt development was due to improvements in the technology of the rubber compound, such that a better coefficient of friction was available. The *raw edge belt* did not have the fabric reinforcement on its edges, and could therefore have an extra tension member within the same outer profile (Figure 4.9).

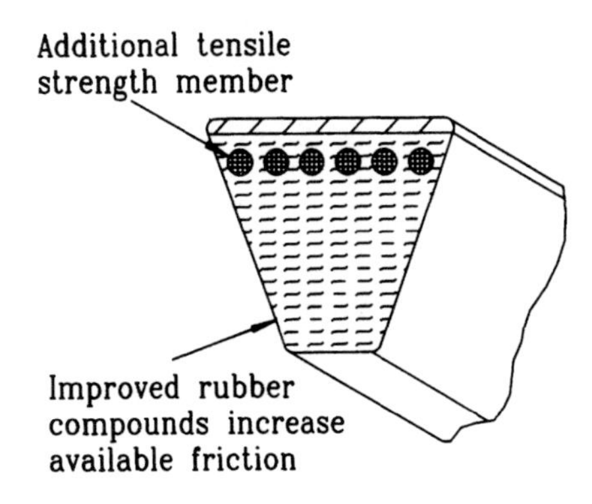

Figure 4.9

To bring us up to date, the current belts have cut-outs on the underside to allow the belts to flex more easily and therefore use smaller diameter pulleys. Note that these cut-outs do not take the drive and are intended to run on smooth pulleys (Figure 4.10). *Timing belts* are a different kind of belt; the cut-outs in this case are teeth and are used with toothed wheels where no slippage can occur.

Also note *multi vee belts*, analogous to duplex chains.

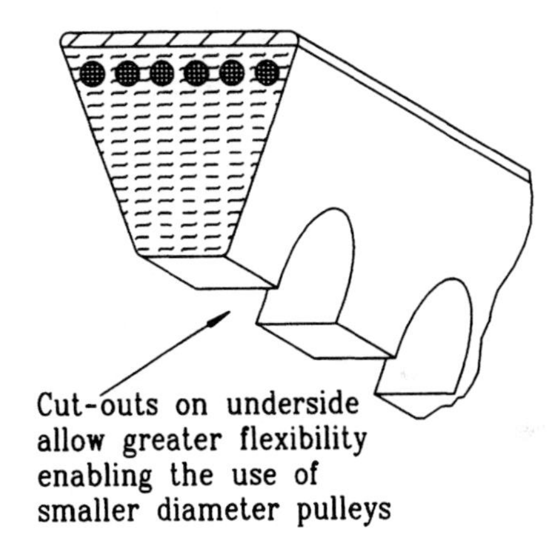

Figure 4.10

Selection of wedge belts

Although a combination of belts and pulleys are specific for a particular application, they are selected from a range of standard components which feature:

- Standard cross-sections
- Standard lengths
- Various standard pulley sizes, allowing speed ratios to be matched within 3% of the design speed.

The belt selection procedure generally starts by uprating the theoretical running power by a service factor (based on the required duty) to generate a design power. This, with the known pulley speeds, enables a suitable belt section to be chosen from an appropriate chart in the manufacturer's literature. Checks are then made with reference to a series of tables, to give the pulley sizes, the belt length, and the number of belts required. Comprehensive details of the selection process are given in suppliers' catalogues, such as the excellent guide provided by Fenner.

Suppliers' publications are based on extensive ranges of applications of their products and as such provide a valuable source of information, not just covering the product selection, but usually also giving installation details.

4.3.4 Chains and timing belts

The belt drives discussed so far have all relied on friction to transmit the power and torque. Chains, or their more modern counterpart, timing belts, do not rely on friction, but instead provide a positive drive. This is essential on a variety of installations ranging from camshaft drives on internal combustion engines to positioning devices on dot matrix printers. A further advantage of the positive drive is that the belt or chain does not need to be pre-tensioned to the same degree as the friction systems. Consequently the loads transmitted to the shafts are much lower. There is a negative consequence of this positive drive, in that under sudden overload conditions, the chain or belt will not slip and so may fail.

As with normal belt drives, chains and timing belts are efficient, relatively cheap, easy to maintain, and can accommodate large centre distances. Chains for many years have provided excellent service, but are now often being replaced by the timing (or toothed) belt which does not require lubrication, and generates less noise. It is these latter advantages that have led to the almost universal adoption of the toothed belt drive for motor vehicle engine camshaft drives. Belt life tends to be somewhat less than that for chains under these conditions, so they need to be replaced at regular intervals. Whereas chains require regular lubrication, the timing belt not only does not require lubricating, but the presence of oil or water can cause damage.

Simplex, duplex and triplex chain drives are available to transmit high levels of power and torque if needed. These are similar in principle to the multi-belt drives.

4.3.5 Gears

There are a number of good reasons for incorporating gears in a transmission system. Not only will they transmit the power and torque, but they can be used to change the speed and direction of the drive. Changing the speed is important, as this can enable the output torque (for a given input power) to be altered. We will start by considering the fundamentals of gearing.

Method of operation

Imagine a pair of discs of the same diameter fitted to parallel shafts and positioned (Figure 4.11) so that their circumferences just touch. If one shaft is turned the frictional forces between the discs will ensure the other also rotates, but in the opposite direction. Note that even though both discs are rotating, at the *point of contact there is zero relative movement.*

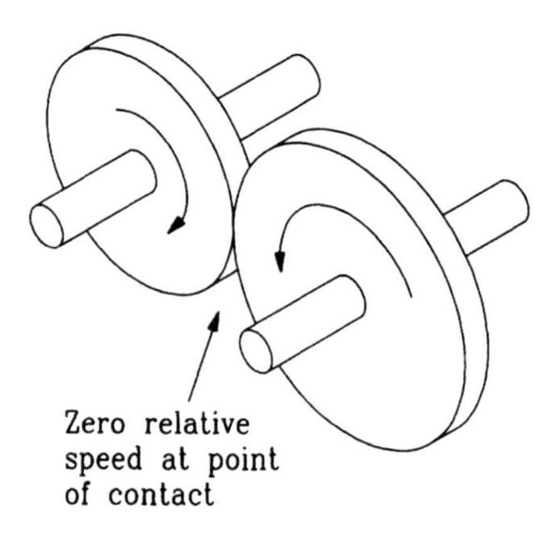

Figure 4.11

Now imagine a similar pair of discs, but this time having different diameters, D and $2D$. If the smaller disc (or gear) is rotated clockwise, the larger will turn in an anticlockwise direction. When the smaller disc has completed one revolution, a point on its circumference will have travelled πD. As there is zero relative movement at the point of contact, a point on the circumference of the larger disc must also have travelled πD, i.e. half a revolution. Hence we have a mechanism for changing speed: e.g. doubling the gear diameter halves the rotational speed. In fact

$$\frac{\text{Diameter gear A}}{\text{Diameter gear B}} = \frac{\text{Speed gear B}}{\text{Speed gear A}}$$

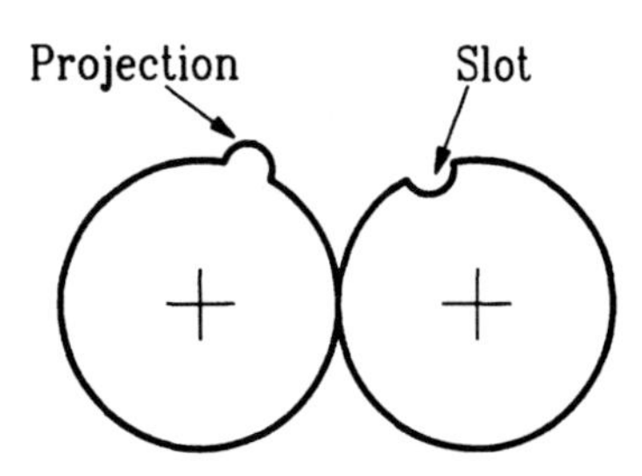

Figure 4.12

Although over the years attempts have been made at producing a geared drive with two discs, they have all been doomed to failure. There is insufficient frictional force between the discs at the point of contact. The problem can be solved by the addition of teeth, i.e. add a projection to one disc and a matching slot to the other (Figure 4.12). This naturally leads to one disc having a series of projections, and the other a series of slots (Figure 4.13). The next evolutionary step is to use the spaces between the slots to add more projections, and the spaces between the projections to add more slots. The slots and projections combine to form teeth and the result is a pair of gear wheels!

Note that during this evolutionary process both diameters of the original discs have disappeared. The datum for calculating gear ratios, the Pitch Circle Diameter (PCD), does not exist on the gear, but must still be used in the calculations.

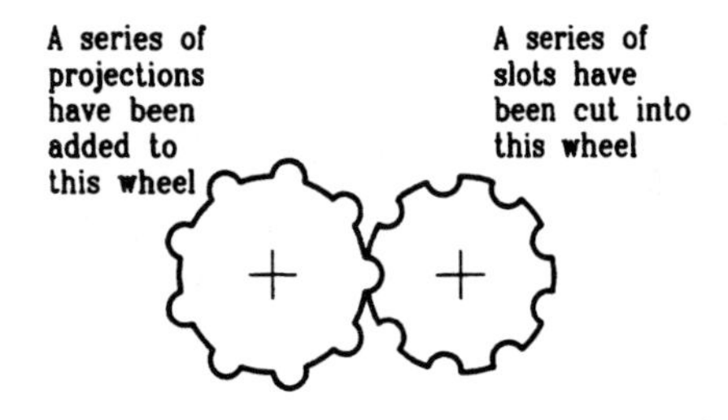

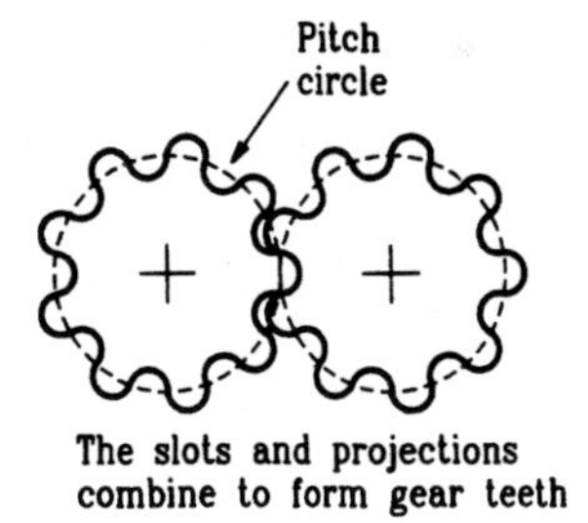

Figure 4.13

Tooth profile

Having established the need for the teeth, the next stage is to define a suitable shape. A rectangular slot and hole would, for example, simply lock the two gears together (Figure 4.14). Retaining the rectangular shape, but adding clearance, would allow the wheels to rotate. However, as well as the obvious problems of excessive wear at the points of contact, there would be a complete loss of the constant velocity relationship between the two wheels. (Remember the point of contact should be at the PCD for both wheels.) The tooth profile is critical in maintaining constant velocity between the two PCDs: it must allow the gears to rotate as if they were still discs.

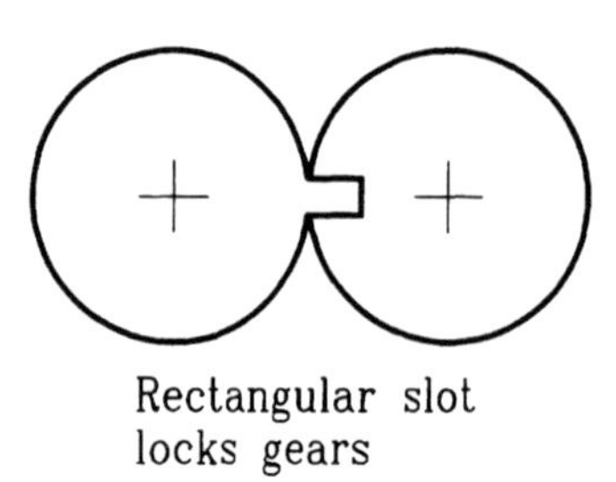

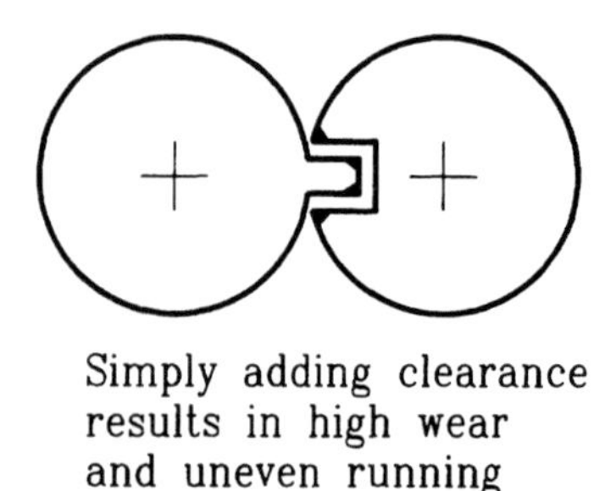

Figure 4.14

There are two geometric curves that can be employed, the *cycloid* and the *involute*.

Gears based on the cycloid do maintain the constant velocity relationship, provided the two PCDs are in contact. If for any reason (wear in the supporting bearings, manufacturing tolerances, etc.) they lose contact, the gears will not revolve evenly. The gear ratio will remain unchanged (this will be enforced by the teeth), but the gears will speed up and slow down as the engagement between the teeth varies. Vibration, noise and increased wear will result.

Modern gears are generally based on the involute profile, as it is more tolerant of any spreading between the gears. Its benefits have been understood for at least two centuries, but its general introduction had to await the development of suitable machine tools. Consider a drum with string wound around it (Figure 4.15). Slowly unwind the string, keeping it taut, and the locus of the end of the string will describe an involute curve. The diameter of the drum from which the string has been unwound is called the *base circle*. The gear tooth will only use the first part of the involute, from where it leaves the base circle. This is always smaller than the pitch circle (PCD), and must be concentric to it.

The relationship between the pitch and base circles sizes is important. The ratio of their radii is usually given as an angle – the *pressure angle*. Any pair of involute gears working together must have the same pressure angle. Altering the pressure angle of the tooth will change its shape, and also vary the length of the contact path. The greater the pressure angle, the longer will be the line of contact. Today, most gears have pressure angles of 20° or 14.5°, with the former being the most common. (Some specific applications such as gear pumps often use other values.)

In practical terms the tooth will extend below the base circle by a small amount to provide clearance between the top of one tooth and bottom of the other. This also allows room for a small blend radius to add strength to the root of

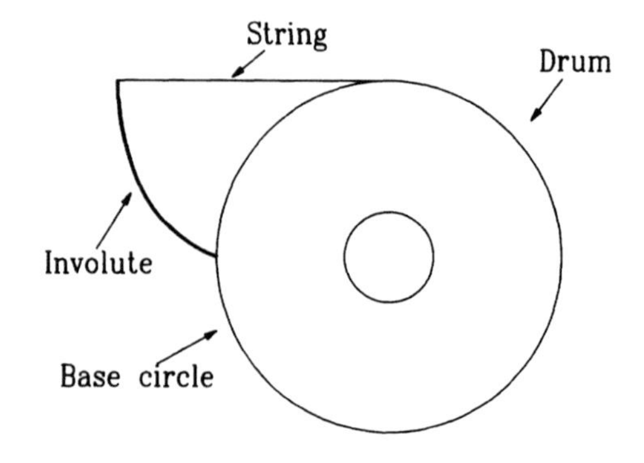

Figure 4.15

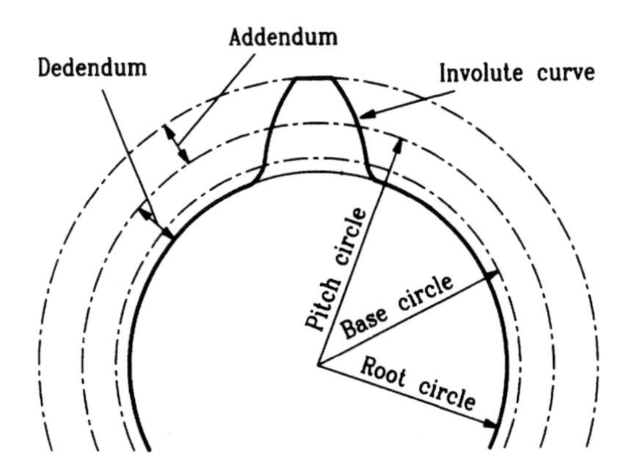

Figure 4.16

the tooth. For more details on tooth proportions refer to BS436.

Tooth size definition

So far we have only discussed the shape of the tooth, not the size. A gear of 80mm diameter could have 20 teeth, or 200 teeth. Both could be of the involute form, and even have the same pressure angle, but would not run together because their physical sizes would be incompatible.

There are three basic methods of specifying the physical size of the teeth:

1. *Circular pitch:* The distance from a point on one tooth to a corresponding point on the next tooth measured around the PCD.

   C, circular pitch = $\pi \times$PCD/number of teeth.

2. *Diametral pitch:* The number of teeth a wheel has per inch of pitch diameter, e.g. a gear with a PCD of 2″ and having 40 teeth is said to be 20DP.

3. *Module:* The pitch circle diameter in mm divided by the number of teeth is the module.

   m, module = PCD(mm)/number of teeth.

Figures 4.17/4.18
Source: K. Hurst, *Rotary Power Transmission Design* (McGraw-Hill, 1994).

Gear types

Gears are produced in a variety of types, with some common variants discussed below:

(a) *Spur gears:* These have teeth cut parallel to the shaft axis, and are therefore only suitable for transmitting power between parallel shafts. They are quite efficient, around 98–99%, and do not generate any axial forces. They are noisy at high speed (listen to a car reversing at speed) so are not normally used with peripheral speeds above about 20 m/s. The speed ratio between a pair of gears would normally be less than 10:1, with greater ratios being achieved with multiple sets of gears (Figure 4.17).

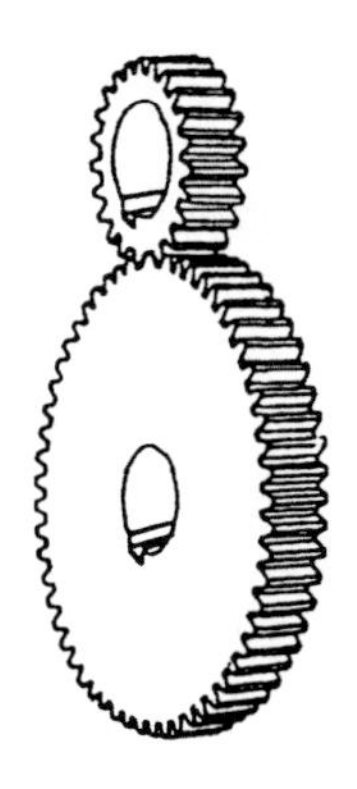

Figure 4.17 *Spur gear*

(b) *Single helical gears:* In this case the teeth are cut at an angle to the shaft axis, so the teeth follow a spiral path. This makes tooth engagement more gradual, and results in

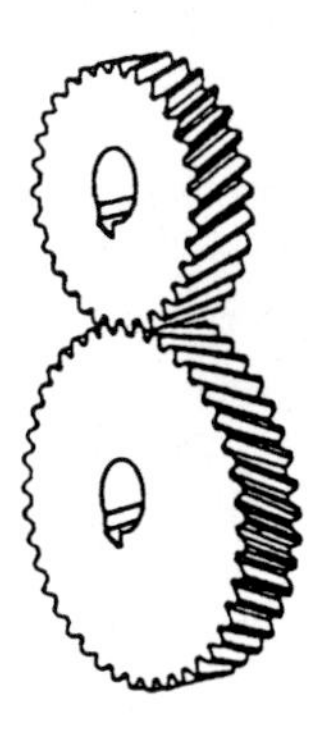

Figure 4.18 *Single helical gears*

quieter running, lower dynamic loads and higher permissible speeds. Their applications are similar to the spur gears, in that they are used to connect parallel shafts, and can accommodate ratios of up to about 10:1. Their efficiency is slightly less than that of spur gears due to the greater sliding contact at the tooth faces. However, the gearbox designer must take account of the axial loads generated. In general they can be used for speeds of up to 25 m/s in general use, with up to around 50 m/s being possible for tightly toleranced gears (Figure 4.18).

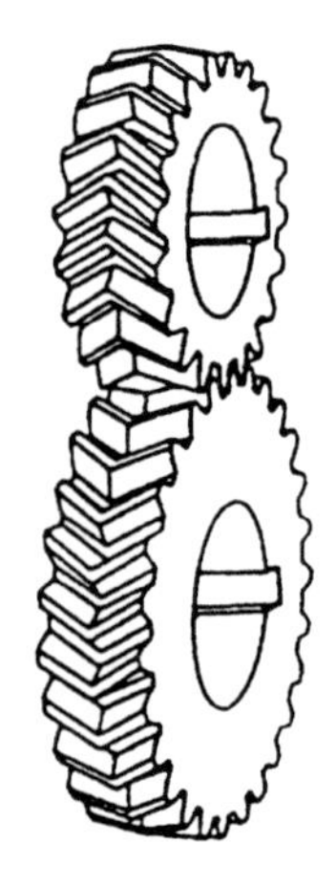

Figure 4.19 *Double helical gears*

(c) *Double helical gears:* These are a variant on the single helical gears whereby the teeth are formed on each gear by helices of identical angles but of opposite hands. As a result there is no net axial thrust. In other respects they can be considered similar to the single helical gears (Figure 4.19).

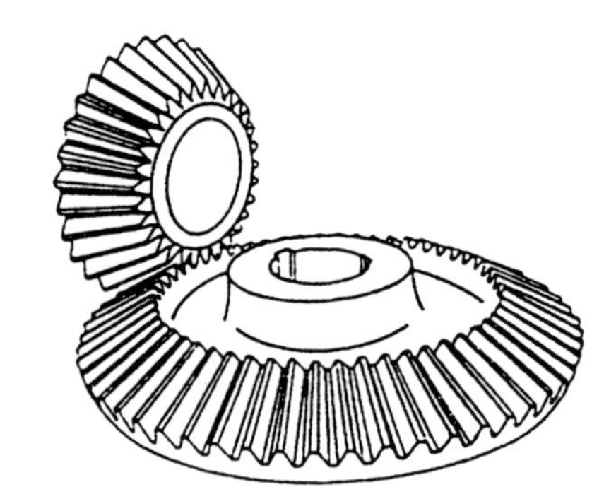

Figure 4.20 *Straight bevel gears*

(d) *Straight bevel gears:* These connect shafts whose axes intersect. The teeth of bevel gears converge on the point of intersection of the axes, called the apex. Axial loads have to be catered for. Ratios of up to 4:1 are possible, and maximum speeds are normally limited to 20 m/s. Typical efficiencies are 95–98% (Figure 4.20).

(e) *Spiral bevel gears:* These relate to bevel gears in the same way that the helical gears relate to spur gears. They are essentially quieter and can cater for higher speeds, up to about 40 m/s.

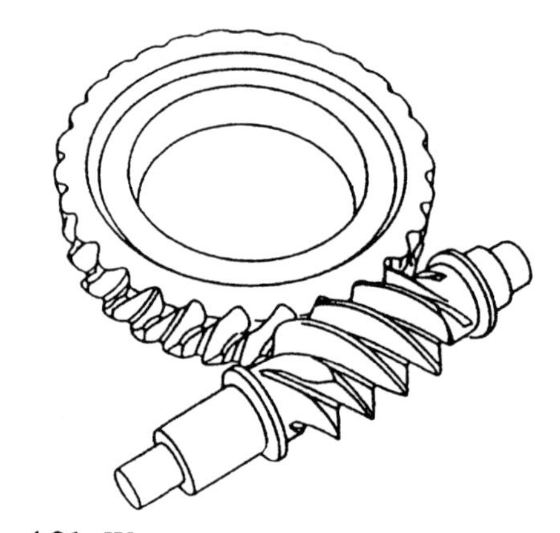

Figure 4.21 *Worm gears*

Source: K. Hurst, *Rotary Power Transmission Design* (McGraw-Hill, 1994).

(f) *Worm gears:* The worm gear is used to connect two non-intersecting shafts, usually but not always at right angles to each other. They offer a large speed reduction, ratios of around 75:1 being common. The low peripheral velocity of the wheel tends to result in quiet operation. There is a penalty in terms of efficiency which can range between 20% and 95%, depending upon the design (Figure 4.21).

Modes of failure

To help understand the application and use of gears, a good starting point is to consider how the gear is likely to fail. There are two common modes of failure: tooth wear and tooth breakage.

1. *Wear:* This occurs when pitting and scoring of tooth surface imperfections eventually form cracks which spread, and cause small pieces to break away from the tooth surface. Tooth wear is a complex subject, but relevant factors will include the hardness of the material surfaces, the quality of the surface finish, and the lubrication conditions.

2. *Breakage:* To avoid tooth breakage we need to calculate the gear load capacity. This is affected by many variables, so unfortunately no single formula exists to select a suitable gear. Over the years much empirical data has been collected and a number of standards exist to enable the selection of a suitable gear via the use of extensive charts and tables. BS 436 (spur and helical gears) was introduced in 1946, along with more recent standards such as ANSI and DIN.

Gear forces

When selecting gears, some basic checks need to be made to ensure that appropriate sizes for the teeth and the gear shaft are chosen. As a simplified introduction we will consider spur gears, and assume that the teeth have been adequately hardened, have a suitable surface finish and are well lubricated so that the strength of the tooth is more critical than the wear considerations.

Tooth loads: First we will examine the loads on the tooth. As in Figure 4.22 the geometry is such that the tooth load will be tangential to the base circle and will pass through the point of contact between the teeth on the pitch circle. Knowing the pressure angle this can be resolved into two forces, one tangential to the pitch circle and one

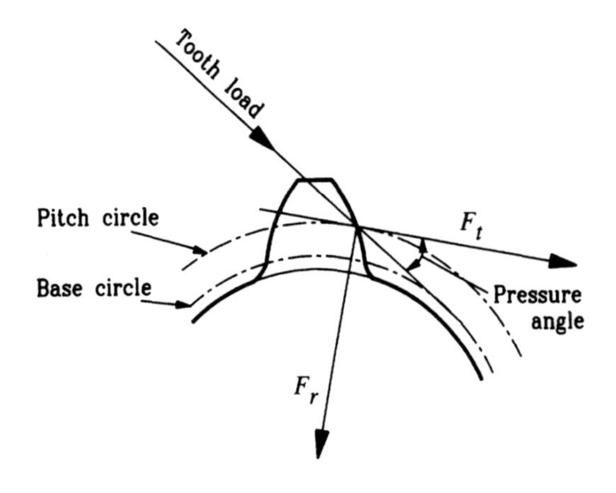

Figure 4.22 *Gear forces*

normal to it. We should already know the power to be transmitted and the speed of the gear, hence the appropriate level of torque can be determined (Power = Torque × Speed). As we know the diameter at which the torque is transmitted (PCD), the tangential force F_t can be found from

$$F_t = \frac{2T}{PCD}$$

The total tooth load, F, is given by

$$F = \frac{F_t}{\cos\psi}$$

where ψ is the pressure angle, and T is the torque transmitted.

Shaft size: These loads will be transmitted to the gear shaft, and thence to its supporting bearings. Checks will need to be made to ensure that the shaft diameter is sufficient to both transmit the torque loads and avoid excessive deflection of the shaft (see section 4.3.2). The allowable shaft deflection is likely to be affected by the choice of support bearings. For spur gears as considered here, there will be no axial loads, but for other types of gears these may need to be considered.

Tooth size: We can consider the tooth as a cantilever beam subjected to a force F_t, the tangential component of the total tooth load. The greatest moment will be at the root of the tooth, so this will be the area of likely breakage (see Figure 4.23).

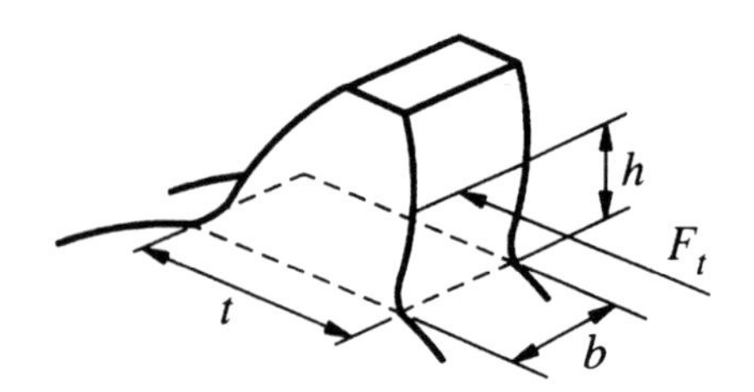

Figure 4.23

From engineer's theory of bending

$$\frac{M}{I} = \frac{\sigma}{y}$$

where

M = applied moment
I = second moment of area of beam section
σ = direct stress generated
y = distance from the neutral axis

so,

$$M = \frac{\sigma I}{y}$$

$$F_t h = \sigma \left(\frac{bt\,3}{12}\right)\left(\frac{2}{t}\right)$$

$$F_t = \sigma b \left(\frac{t^2}{6h}\right)$$

$$F_t = \sigma\, b\, m\left(\frac{t^2}{m6h}\right)$$

$$F_t = \sigma\, b\, m\, Y$$

where m = module, and Y = Lewis form factor.

This equation is called the Lewis equation after Wilfred Lewis who first derived it in 1893. The form factor, which effectively incorporates tooth geometric information and numeric constants, can be obtained by reference to tables. From the formula it is apparent that increased tangential forces can be accommodated by increasing the allowable bending stress of the material, the gear width, or the module.

Standard sizes

Suppliers of standard gears often offer a very wide range, but many sizes are unlikely to be available direct from stock. Where possible it is best to use preferred sizes. Preferred standard modules are:

0.5, 0.8, 1, 1.25, 1.5, 2, 2.5, 3, 4, 5, 6

Preferred numbers of teeth are:

12, 13, 14, 15, 16, 18, 28, 22, 24, 25, 28, 30
32, 34, 38, 40, 45, 50, 54, 60, 64, 70, 72, 75
80, 84, 90, 96, 100, 120, 140, 150, 180, 200,
220, 250

4.3.6 Hydraulic transmissions

A hydraulic transmission employs a positive displacement pump (usually a gear pump) to generate a supply of high pressure oil. This is then piped to a positive displacement hydraulic motor which converts the energy from the high pressure oil back into a rotational format. The hydraulic motor is effectively a gear pump running in reverse.

This system offers advantages in being very flexible; the distance between the input and output points can be large, and can handle high values of power and torque. However, to set against these points, it is not very efficient ($< 80\%$), it is expensive, and will require good oil filtration.

4.3.7 Electric transmission

Electricity is probably the most widely used form of energy transmission. A prime mover at a power station will generate energy in the form of a rotating shaft, which is then converted into electricity simply because this is a highly convenient method of transmitting the energy. In many cases this energy is then used to impart some rotation to a shaft in a range of implements such as a washing machine, a freezer compressor, a fork-lift-truck motor and so on. It may also be used for other purposes, such as heating, or powering computers, but we will concentrate on generating a rotational motion.

In some ways this form of transmission is similar to the hydraulic transmission, the generator or alternator being the equivalent to the gear pump, and the electric motor fulfilling a similar role to that of the hydraulic motor. As the subject is so wide we will concentrate on points to consider when selecting an electric motor.

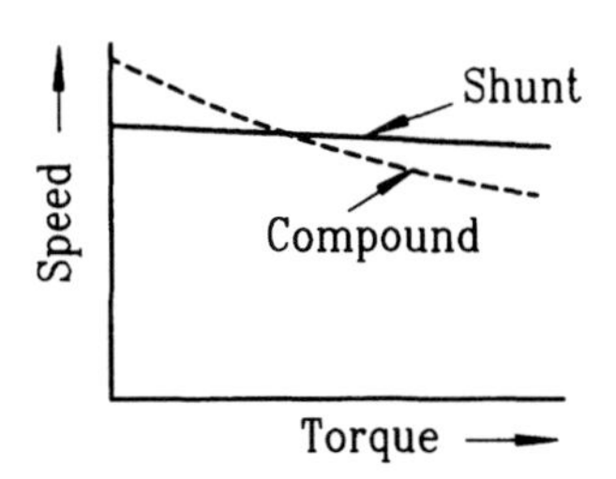

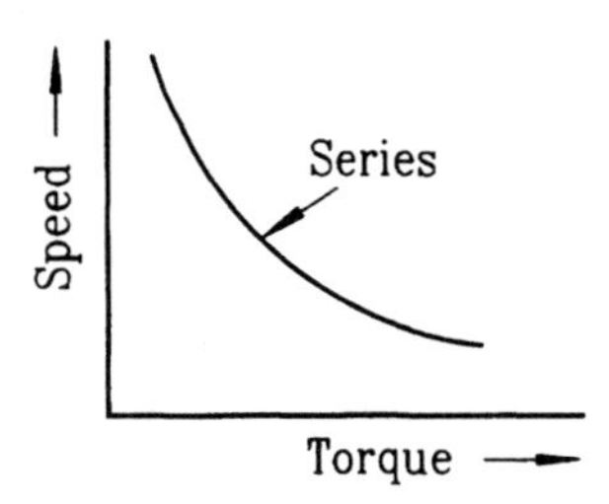

Figure 4.24 *DC shunt/series characteristics*

Motor types

Different types of electric motors have vastly different torque speed characteristics, so the first step in any application is to consider the type of load that the motor will be required to provide. Once that has been decided we then need to consider the alternative types of motors.

DC motor: This design consists of an armature with a conventional commutator rotating within an electromagnetic field. The field can either be connected in series, or in parallel with the armature, resulting in what is called a series wound or shunt wound motor. The torque speed characteristics are as shown in Figure 4.24. As shown, the DC motor can be designed to generate either constant torque or constant power. A series motor generates greater starting torque, but as the load increases this will lose more speed than an equivalent shunt wound motor. The reverse is also true in that if the current is maintained and the load is reduced, the motor speed will increase. They should never therefore be employed with belt drives, where a broken belt could result in the motor running to a dangerous speed. Series motors are used for applications where a great starting torque is required, as for example in a crane or train motor. Shunt motors are used where a steady speed that can be controlled largely independent of load is required, such as driving machine tools. Compound motors are a variant that attempts to combine the characteristics of these two types.

Speed control of the shunt wound motors can be achieved simply by varying the input voltage. Reversing this voltage will, of course, reverse the direction of rotation. The presence of a commutator means brushes will be employed, which will require replacing at some point in the motors' life.

AC motor: Induction motors run on alternating current, do not employ a commutator and therefore have no need for brushes. They are widely used in industry, having a reputation for simplicity and reliability. They are normally constant speed devices, the speed being a function of the number of poles and the frequency of the alternating current. Speed control is possible by varying the frequency of the input current, although this is much more difficult (i.e. more expensive) than simply varying the voltage on a DC motor. These motors are suitable for running on either three-phase or single-phase supplies. The best general purpose machines are the three-phase variety as these are self-starting, and will run at a constant speed regardless of load. The high starting current means the switchgear should be sized accordingly (see Figure 4.25).

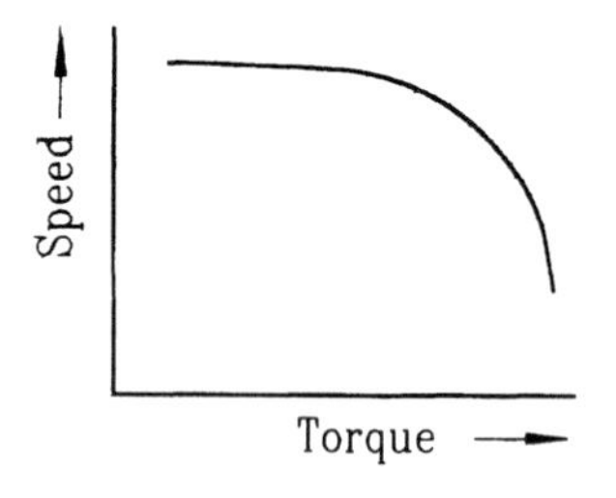

Figure 4.25 *AC motor characteristics*

Single phase motors are not normally self-starting, so a split phase is created artificially, normally by placing a reactance or resistance in series.

AC/DC motors: These universal motors are self-starting and can run on either AC or DC supplies. They are often found in power tools and domestic appliances.

DC servo motors: This type is used when accurate control of speed or position is called for. A tachometer is used to provide feedback to an electronic control system. As the motors are DC, most use brushes, although brushless variants are available but with a cost penalty. The rotor position is detected by Hall effect sensors, so that the electronics can perform the switching effects of the commutator.

Stepper motors: These are normally DC permanent magnet motors that operate in steps. These steps are usually 200 per revolution (1.8° per step) or 400 per revolution (0.9° per step). This type is often found in low-power devices where it is important to index the motor by a known amount, such as the paper feed on a printer. They are inherently brushless, and as they do not require feedback, can use simpler, lower cost control electronics. Provided the inertias and torques encountered are not excessive enough to result in overshoot, there should be an exact correlation between the number of pulses sent to the motor and the change in rotor position.

Operating environment

The operating environment should always be considered when specifying a standard motor. It can affect the choice of casing design as well as the materials used in the construction of the motor. One of the most important is the maximum ambient temperature, as this will affect the temperature of the cooling air. Temperatures above 40°C will derate the motor, by approximately 5% for every 5°C rise above 40°.

In a similar fashion, altitude will affect the cooling as the air density will vary. For example, a derating of about 8% will apply to a motor at 2000m compared with its capability at sea level.

The choice of insulation materials used will depend on the available cooling. The most common classes of insulation are as follows:

Max. temp. rise (°C)	Insulation class (BS4999)
75	E
90	B
140	F

Temperature is not the only factor. For example, the presence of some liquids could affect the performance adversely. Various alternative enclosures are available. The three most common types are:

- *Drip-proof* (DP): These have ventilation holes in the end covering, for cooling, and are protected against minor dampness such as condensation. They are generally one of the cheaper types of enclosure.
- *Totally enclosed/fan cooled* (TEFC): In this case all the working parts are fully enclosed. Cooling is provided by air forced over the casing by a fan mounted on the non-driving end of the motor shaft. These are suitable for contaminated environments, provided flammable substances are not present.
- *Flameproof* (FP): These are similar to the TEFC but are more robust, the design being capable of withstanding an explosion of defined flammable gas within the motor case.

Other types of enclosure are available and are fully described in the relevant BS.

4.4 General standard components

So far we have discussed several alternative systems of power transmission, mainly by looking at the major components of those systems. Most, although not all, of these have been specific to a particular transmission system. We will now expand the field of standard components and look at some common components that are applicable to a wide variety of applications.

4.4.1 Couplings

Shafts can form an integral part of many types of transmission systems and, as such, they need to be connected in some way to the other elements involved in the power transmission process. The connection is made via some form of coupling which can either be permanent, or capable of being disabled in some way. If the coupling is capable of this type of adjustment it is normally called a clutch.

In theory, shafts being connected should be in perfect alignment. Practice is, however, somewhat different, with potential misalignment arising from a variety of sources, including:

- shaft deflections
- bearing wear
- insufficiently rigid mountings
- tolerance build-up
- temperature changes

Your choice of coupling design will be influenced by the level and type of misalignment that it can accommodate, as well as the normal performance criteria (duty, power, torque etc.).

A wide range of standard couplings is available. Any selection should be made in accordance with the manufacturer's literature. As an introduction we will now describe some of the more common types of coupling.

Rigid couplings

These are the simplest form of coupling (see Figure 4.26). They are generally keyed to the shaft or forged as part of the shaft. An integral forging has the advantage simplifying the coupling to shaft joint, but may require split seals and bearings to be used to support the shaft.

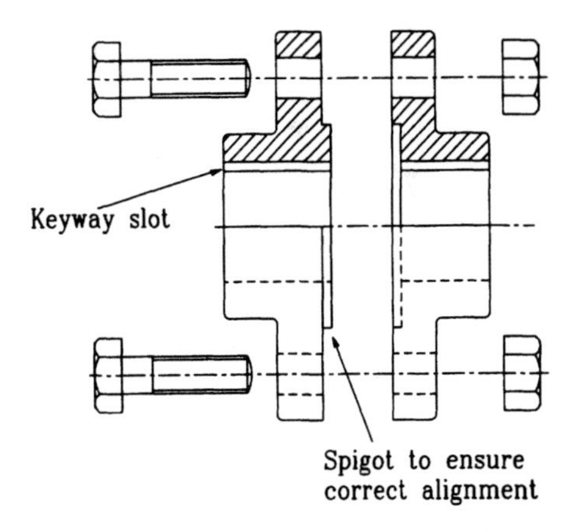

Figure 4.26 *Rigid coupling*

The number of bolts can be calculated taking into account the allowable shear stress. Note that in theory this is an over-design, as bolts should normally only take loads in tension. It allows for the bolts to be incorrectly tensioned. The correct method for sizing the bolts will be covered later in Chapter 13.

A major disadvantage of a rigid coupling is that it will not cater for any misalignment.

The flange thickness of the coupling is generally 0.25 to 0.3d, where d is the shaft diameter.

Flexible rubber disc

The simplest type of flexible coupling evolved from the ordinary rigid or flange coupling with some kind of resilient material such as rubber forming bushes in the bolt holes, thus allowing a small degree of flexibility in two or three directions. The next stage in development was to insert a third element, a rubber disc between the two flanges (Figure 4.27). The rubber disc is normally bonded to two steel plates which are then bolted to the flanges, the bolts being applied alternately to each flange. This design has the ability to cope with misalignments of up to 5° and can absorb occasional shock loads of up to 50% above the normal loading.

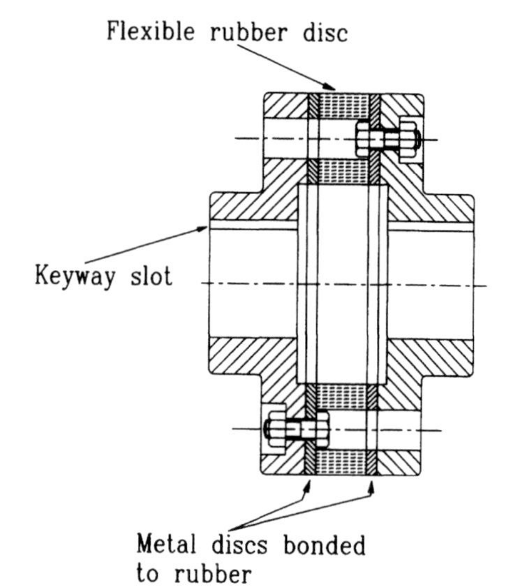

Figure 4.27 *Disc coupling*

Spider coupling

This design also incorporates a third element made of rubber, but in the form of a spider, as shown in Figure 4.28. It will cater for similar misalignments to the disc type, and can also accommodate end floats up to about 0.8mm.

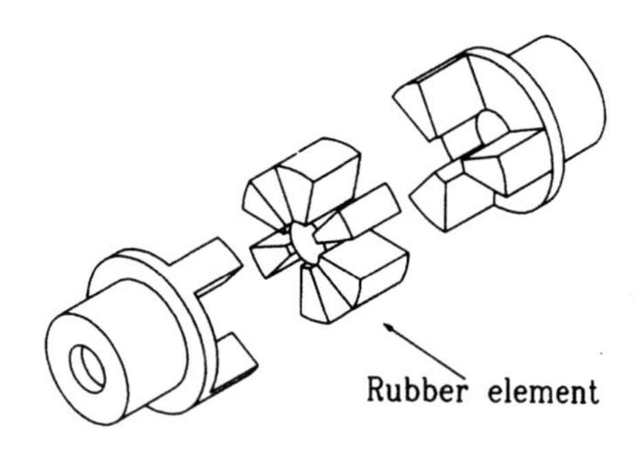

Figure 4.28 *Spider coupling*

Oldham coupling

In this design the third element is not made from flexible rubber; instead, the two orthogonal slots allow some misalignment (Figure 4.29). Although it cannot cater for end float, it will permit parallel misalignments of the shafts of up to about 3mm.

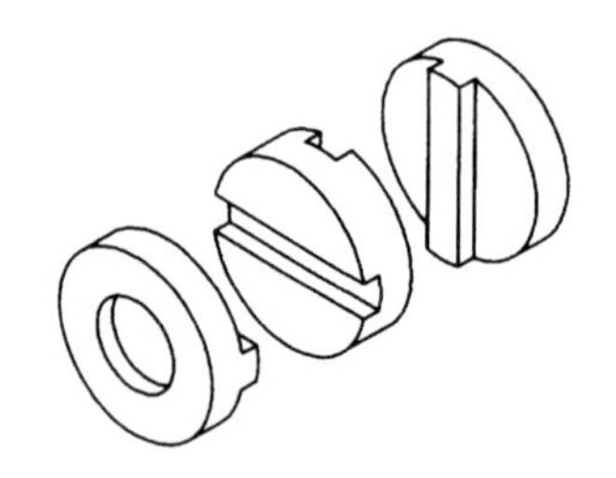

Figure 4.29 *Oldham coupling*

Chain coupling

This design consists of two similar chain sprocketed wheels, keyed to their respective shafts and connected together via a suitable length of duplex roller chain. The torque is transmitted through the chain, which can be removed to disconnect the shafts. It is generally enclosed so that lubricant can be retained. It is suitable for shafts of up to 150mm diameter, the largest being capable of transmitting 250 to 1500 HP according to speed when used for steady drives.

4.4.2 Clutches

In some cases a means of connecting and disconnecting the coupling is required. This function is normally performed by a device called a clutch. There are a wide number of variants. We will introduce some of the common types.

Dog Clutch

This is the simplest form of clutch. It provides positive engagement, but does not allow any slippage. It can normally only be engaged at rest (see Figure 4.30).

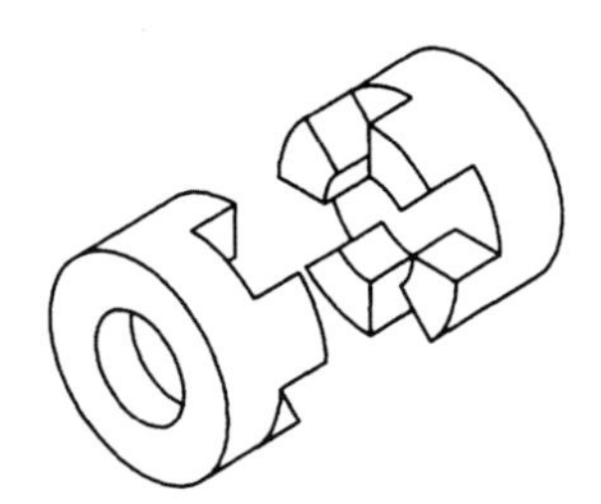

Figure 4.30 *Dog clutch*

Friction plate clutch

There are a number of variants on this design, the most common being that used in motor vehicles. A plate is attached to the input shaft (normally the engine flywheel). A second plate or disc is attached to the output shaft (normally the vehicle gearbox input shaft) via splines so that this disc has a small degree of axial movement. Between these two discs is yet another disc covered on both sides with a friction material. The drive is made by sliding the 'gearbox' disc towards the 'flywheel' disc, trapping the friction disc and thus transmitting the drive. In a car a diaphragm spring is used to hold the three plates together. The clutch pedal is usually linked to the disc at the gearbox end, and when depressed disengages the clutch by separating the plates. The smooth take-up is achieved by slippage between the discs (see Figure 4.31).

Centrifugal clutch

This is another type of friction clutch commonly found on mopeds, lawn mowers and electric motor drives that need to be started on low loads. It has the advantage of being automatic in action. The driving shaft is connected to an assembly consisting of a number of shoes which have friction material attached to their outer surfaces. The shoes are supported by pivots and springs so that the assembly can fit inside a drum attached to the driven shaft. There will be a small clearance between the outside diameter of the shoes and the inside diameter of the drum at rest. As the driven shaft rotates, centrifugal force overcomes the spring loading and the shoes contact the drum, thus transmitting the drive (see Figure 4.32).

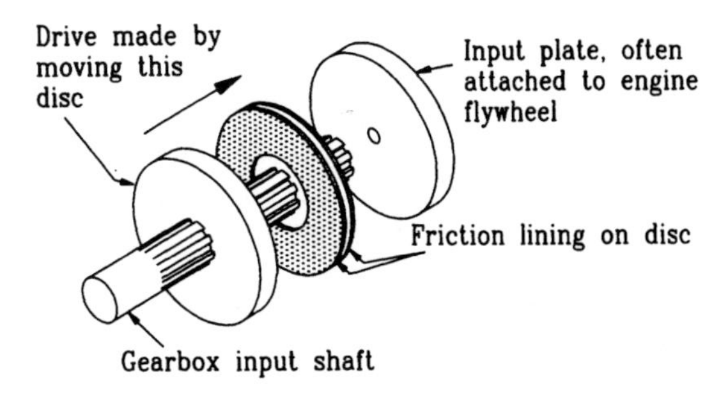

Figure 4.31 *Friction plate clutch*

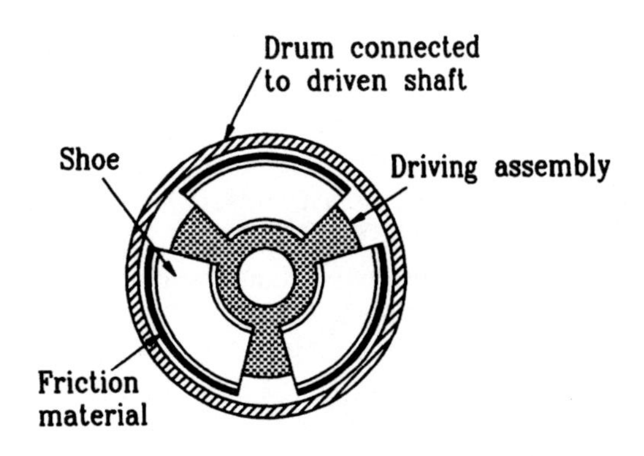

Figure 4.32 *Centrifugal clutch*

Over-running clutches

There are many instances when the designer wishes the drive to only operate in one direction, such as the freewheeling action on a bicycle. Figure 4.33 shows one design using balls or rollers that run on an inner race profiled such that one direction of rotation causes them to wedge the inner and outer races together, transmitting the drive. If rotated in the opposite direction, the outer race is allowed to rotate freely.

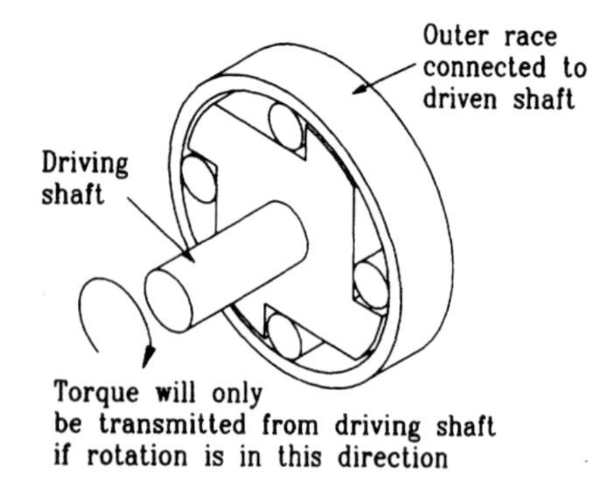

Figure 4.33 *Over-running clutch*

Sprag clutch

This is an over-running clutch, but instead of employing balls or rollers it uses a specially shaped device called a sprag. Its special shape generates a similar action to the above example, but does not need specially profiled races. There is a further advantage in that the sprags can be packed more closely than the balls and are able to transmit greater levels of torque.

Other types include fluid couplings (often found on motor vehicle automatic transmissions), powder clutches, where the application of a magnetic field allows the drive to be made, and many more which are beyond the available space in this text. By now you should understand the principles behind clutches and be capable of extracting suitable information from the relevant specialist suppliers to meet the needs of your design problem.

4.4.3 Bearings

Whenever surfaces in contact rotate or slide, the resulting friction generates heat. Energy is wasted, and the parts wear, all of which leads to loss of product efficiency and a reduced component life.

Bearings are designed to minimise these effects. They do this by allowing relative movement between the components of machines whilst carrying a load and providing some type of location between the machine components.

Direction and nature of load

The design or choice of a bearing starts with an examination of the type of load to which the bearing will be subjected. The list below summarises the different types of load:

- Radial load
- Axial load
- Combined radial and axial load
- Steady state applied load
- Dynamic or rotating load

Bearings fall into one of two general categories, depending on the type of relative motion employed. They can be either sliding contact, or rolling contact. For each type we will introduce a number of the possible variants, and consider methods of lubrication. Then we will discuss the merits of each type, finishing with the procedure for selecting standard items.

Sliding contact

Sliding contact bearings can be subdivided into three classes:

1. *Plane slide:* This type offers limited relative motion for non-rotating elements such as a splined joint or a machine slide.
2. *Thrust:* These are designed to withstand force along the axis of a shaft, the relative motion being primarily one of rotation. They are often combined with journal bearings.
3. *Journal:* The most common type of sliding contact bearing is the journal bearing, which

is used to support an axle or shaft in a radial direction. Typical examples are ships' propeller shaft bearings, reciprocating engine crankshafts, big ends, gearboxes. The motion can either be a regular rotation or, as in the case of gudgeon pin bearings, intermittent rocking. (A gudgeon pin connects a piston with its connecting rod.) Care must be taken (in both the design and assembly) to minimise misalignment, as this can result in localised metal-to-metal contact, leading to a breakdown of the lubrication film and subsequent failure of the bearing. Self-aligning bearings can be used to overcome this potential problem, but are expensive.

Journal bearings come in a variety of types, some of which are discussed below.

Direct lined housing: The housing is lined directly with the bearing material. The choice of material is limited by the practicality of keying or bonding the bearing material to the housing surface. It is generally limited to ferrous housings with low-melting-point, white-metal bearing surfaces. (Light alloy and zinc base housings are difficult to line directly with white metal.) In the early days of the internal combustion engine this type of bearing was often used for the main and big end bearings in motor vehicle engines. Replacement of worn bearings was a skilled and expensive task, so in this application they were superseded by the lined insert for almost all post-war vehicles.

Liners: The need for simplified servicing led to the development of this variant, which consists of a liner inserted into a previously machined housing. There are a number of variants.

Solid insert: The insert is manufactured wholly from suitable bearing materials such as aluminium alloy, copper alloy, or white metal, and is available as machined bushes (plain or with flange), half bearings, and thrust washers. The housings normally need to be machined to relatively tight tolerances. An insert may be finish machined after assembly, or a prefinished precision liner added as a final operation. Typical applications are small diesel engine bores, crankshaft main bearings, bushes for gearboxes, steering gear and vehicle suspensions.

Lined insert: These are similar to the solid insert, but have a backing material such as cast iron, steel, or a copper alloy which has been lined with a suitable bearing surface of aluminium, copper alloy, or of white metal. They can be supplied as a one-piece insert, a split bush, half bearing, or as a thrust washer, e.g. petrol engine big-end bearings. They perform a similar function to the directly lined type, but offer the advantage of much simplified spares replacement.

Wrapped bushes: These are rolled from a flat strip of bronze or steel lined with white metal, lead bronze, copper lead, or aluminium alloys. A wrapped bush is cylindrical in shape, having a continuous split from one end to the other which closes when the bush is fitted into the housing. One advantage of the wrapped bush is that its construction allows it to be self-retaining, requiring no additional location. It is therefore suitable for high-volume, cost-effective applications such as motor vehicle controls, electric motor spindles, and numerous consumer goods applications. They can be supplied either as finish machined or to be finish machined on assembly. They are suitable for bushing applications in which the tolerable wear will not exceed the thickness of the lining.

Sintered bronze bushes: Typically these are formed from a tin/copper/graphite composition which has been compacted and sintered. This provides a porous structure with some 10–35% of its volume capable of retaining lubricating oil. Typical applications include gearbox shafts, electric motor and dynamo shafts, instrument spindles etc.

Non-metallic bushes: These bushes can be manufactured from a wide variety of materials such as nylon, ptfe, and carbon. Potential advantages over metal bearings include corrosion resistance, electrical insulation, and good damping. Potential disadvantages include poor heat dissipation, and the fact that some plastics (particularly nylon) swell in the presence of water. Typical uses are hinges, catches, low-speed parts in domestic appliances, cameras, copying equipment.

Lubrication of sliding contact bearings

The correct functioning of sliding contact (or plain bearings) is absolutely dependent on satisfactory lubrication. There are three basic systems:

1. Hydrodynamic
2. Hydrostatic
3. Semi-lubricated and non-lubricated

Hydrodynamic lubrication: The lubrication system is called hydrodynamic when the relative motion between the surfaces is such that it causes the lubricating oil pressure to build up enough to support the load without metal-to-metal contact. As in Figure 4.34, at rest there is metal-to-metal contact. When the journal starts to rotate clockwise there is initially some rubbing of metal on metal and the journal climbs upward to the right. As the oil adheres to the journal's surface, an oil film is drawn between the surfaces with the rotation and the journal moves to the left of the bearing centre. This is the equilibrium position. If correctly designed the wedge-shaped chamber will ensure that the pressure holds up until there is no metal-to-metal contact.

With hydrodynamic bearings the metal-to-metal contact existing at rest continues during start-up. Hence to avoid excessive wear it is wise to start machines under light or no load, especially for machines that are started frequently. However, in many cases, as with turbine rotors, this is not possible. Another limitation of hydrodynamic bearings is in those cases where the bearing rotates too slowly for the hydrodynamic action to build up a separating film. Finally, thrust bearings are not inherently designed for producing the hydrodynamic action.

Hydrostatic lubrication: To reduce the high friction that would otherwise exist in these circumstances, oil is pumped into the load-carrying area at sufficient pressure to support the load. Where the film is maintained by the oil flow due to external pressure, the lubrication is said to be hydrostatic.

Sometimes gas (usually air) can be used in a hydrostatic bearing, although its load-carrying capacity is quite low, typically 10 psi (0.7 bar). Coordinate Measuring machines often employ this type of bearing.

In some turbines the journal may be floated in oil at start-up, but once up to speed the subsequent hydrodynamic action may be permitted to take over.

Semi and non-lubricated: To cater for those circumstances where lubrication and maintenance would prove difficult, impossible or costly, some bearings have been developed to contain sufficient lubrication for their lifetime.

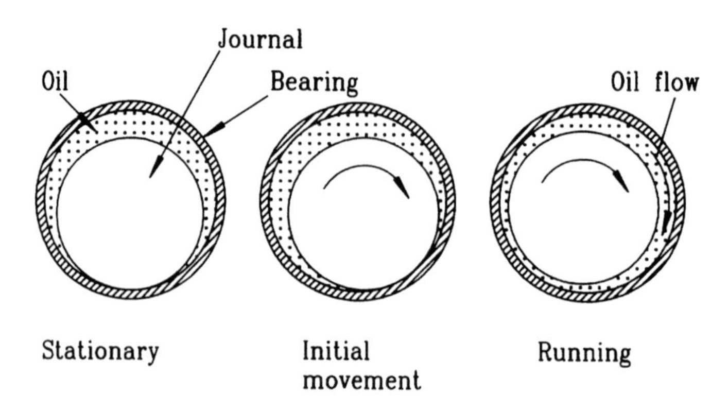

Figure 4.34 *Hydrodynamic lubrication*

Sintered bearings as already described are pre-impregnated with oil which comes to the surface when the bearing is subjected to a higher temperature or pressure.

Non-metallic bushes often require no lubricant, particularly if, as in some cases, a solid lubricant filler is added to the plastic compound during manufacture to enhance the bearings' performance.

Carbon graphite inserts are another form of dry bearing, the mixture of carbon and graphite acting as its own lubricant. They can tolerate temperatures of 750°F.

Rolling element bearing types

A rolling element bearing consists of four main components: an inner race, an outer race, the rolling elements (balls or rollers), and a cage used to separate the rolling elements. The inner and outer races contain hardened tracks in which the balls or rollers run.

Consider an installation where the inner race rotates with the shaft, and the outer race remains static within its housing. Any radial load from the shaft will be transferred to the inner race, and then, via the rolling element, to the outer race, which in turn is supported by its housing. It should be evident that the load will at any given time be supported by only one or two of the rolling elements. If these are balls the load will be transmitted via a point contact between the ball and the races. As this would theoretically result in an infinitely high pressure point, some distortion of the components must occur so that the contact area can increase to reduce the localised pressure to an acceptable level. As the inner race rotates, this localised point of distortion will move around the race. If we examine one point on the race, each time it completes a revolution it will suffer high load, distort, then the load will be removed. In other words, the load will be cyclic, and fatigue will be the most likely failure mode of the race. This, as we will see later, is used to predict the life of the bearing.

Rolling element bearings are available in a number of variants, the more common ones being described below.

Deep groove ball: This is probably the most commonly used type, having the capability to take both radial and axial loads. The load capability is, however, much greater for radial loads. They are normally equipped with a single row of balls, but are also available in a double row variant. Tolerance of any misalignment is poor. Under normal service condition a single row bearing can tolerate between 2 and 10 minutes of arc, but the double row version can only accommodate up to 2 minutes of arc misalignment. Many sizes can be supplied with integral seals (see Figure 4.35).

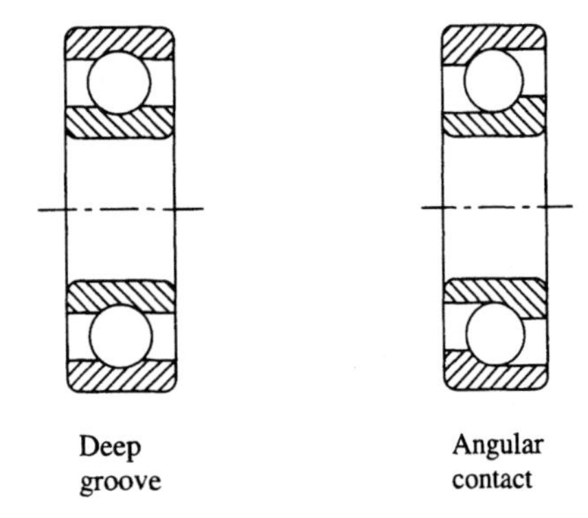

Figure 4.35 *Radial ball bearings*

Angular contact (see Figure 4.35): An angular contact bearing will take both radial load and axial loads, but the axial load can be in one direction only. They are mounted in matched pairs, back to back, so that together they can cater for axial loads in both directions. They are suitable for applications generating greater axial loads than could be covered by deep groove ball-bearings, such as front wheel bearings of motor vehicles.

Cylindrical roller (see Figure 4.36): Size for size, a cylindrical roller bearing will be able to take a greater radial load than its deep groove ball counterpart. However it cannot take any axial load. (Special variants are now available that will take some axial loads.) Misalignment must be kept below 3 to 4 minutes of arc. Integral seals are not available as an option for single row cylindrical row bearings.

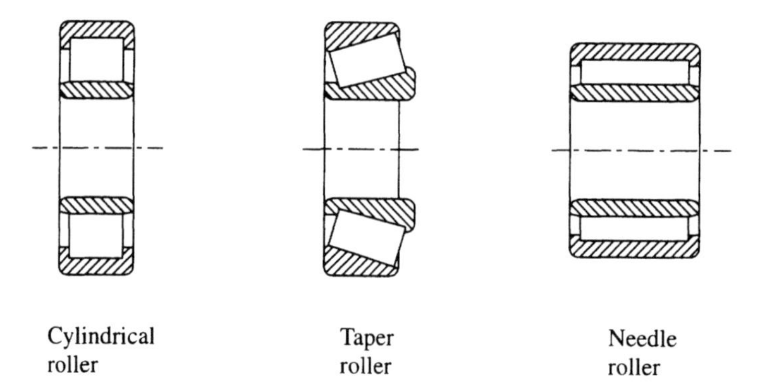

Figure 4.36 *Radial roller bearings*

Taper roller: These can be considered as the roller bearings version of the angular contact bearing. They are installed in pairs and are able to take significant axial and radial loads.

Needle roller: If radial space is very limited the needle roller bearing may offer a solution for radial loads. The rollers are long, but small in diameter. The outer race is produced from hardened sheet steel. Although inner rings are available, if the shaft can be hardened and ground they are not needed.

Thrust: Thrust bearings are designed to take axial loads and are available with ball, roller, or needle roller elements (see Figure 4.37).

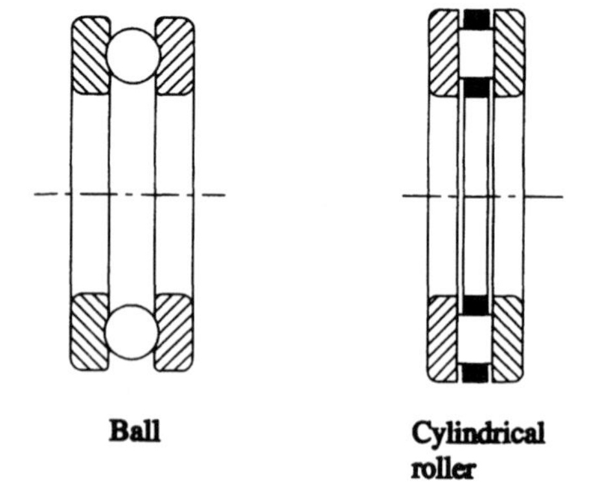

Figure 4.37 *Thrust bearings*

The types of bearings discussed here summarise most of the commonly used variants. Other types, such as self aligning and full complement cylindrical roller bearings, are available and are fully detailed in the appropriate literature available from the major bearing suppliers.

Lubrication of rolling element bearings

If rolling element bearings are to operate reliably they have to be adequately lubricated to prevent direct metallic contact between the rolling elements, raceways and cages, to protect against wear and to protect the bearing surfaces against corrosion.

Those bearings supplied with seals at both sides are filled by the supplier with (normally) a lithium-based grease which is suitable for operating at temperatures within the range −30 to +110° C. These are lubricated for life and should be considered maintenance free.

Bearings without such seals can be lubricated with either grease or oil. We need to compare the pros and cons of each, as follows:

(a) Grease

Advantages:

- It is more convenient to apply and retain, and serves as a seal against intrusion of foreign matter.
- The grease is retained in the bearing when not in use thereby preventing rusting during idle periods.

- Minor assemblies packed with grease can operate over extended periods without replenishment.
- Suitable for food and textile machinery where the lubricant must be sealed effectively.

Disadvantages:

- Very high speeds require special consideration, such as a grease relief system where grease is fed in at one side of the bearing and at the opposite side a rotating flinger expels excess or used grease through a wide cavity into a collecting tray or container. Consequently the efficiency of lubrication is dependent on frequent grease additions.
- Bearings fitted in close proximity may have problems due to differential churning.
- Not suitable for bearings from which heat must be dissipated.

(b) Oil

Advantages:

- Less tractive resistance and therefore suitable for light precision machinery.
- Suitable for speeds beyond the limits of grease lubrication.
- Can be used as a coolant to dissipate heat from a bearing.

Disadvantages:

- The installation is more complicated than that for grease.
- Efficient sealing is required if leakage must be eliminated.
- Drains away from the bearing while at rest.
- Does not protect the bearing from intrusion of foreign matter.

Overall, grease is normally the first choice of lubricant for rolling element bearings as it is generally cheaper to install and maintain. Oil is usually the second choice, because of the complications of feeding and sealing, but would be used when other components in the unit require oil lubrication, high speeds are required, or high temperatures are involved.

Bearing selection

The first stage in the selection process must be to choose between rolling element and sliding contact element bearings. Unless the duty is very light and can be accommodated with a simple bush, the designer would select a rolling element type for the following reasons:

1. *Easier lubrication:* Plain bearings require specialised lubrication, often needing a supply of pressurised oil. Rolling element bearings normally can be supplied as greased for life.
2. *Simpler selection:* Plain bearings tend to be designed for a particular application, i.e. the designer has to design the actual bearing he is going to use. A wide range of rolling element bearings are available as standard products. The selection of a suitable standard bearing is made relatively simple by the availability of extensive historical information from the suppliers.
3. *Starting friction is similar to the operating friction:* During start-up the metal-to-metal contact in most plain bearings results in high friction (and the associated wear) until the lubricating film becomes fully effective. The rolling action of the rolling element bearings means that they do not suffer from this problem.
4. *Require less axial space, but more diametral space:* Although this is true, the wide range of different types of rolling element bearings means that this should not be a problem. For example, if a deep groove ball-bearing is too 'tall' perhaps a needle roller could solve the problem.
5. *Preloaded bearings allow stiffer designs:* This is a positive plus for the rolling element type.
6. *Noisier:* As the bearing wears a change in the noise emitted is often an early indication of imminent failure.
7. *More expensive:* True, when comparing the individual bearings, but do not forget that the plain bearings are likely to need a more expensive lubrication system.

8. *Life is usually limited by eventual fatigue failure:* The theoretical point load between the ball and the race results in a cyclic deformation of each as the bearing rotates, thus leading to fatigue failure. However, sufficient historical data is available from the manufacturers to enable the bearing life to be predicted so that this does not become a problem.

Rolling element bearing selection

Once the decision to use rolling element bearings has been made, the first stage is to choose the type of bearing, based on the loading and the physical space available.

The next stage is to select a standard bearing that will be capable of taking the required load for an acceptable life. At this point some definitions of bearing life and loads are appropriate.

- *Life:* If all normally avoidable causes of failure (incorrect mounting, inadequate lubrication, etc.) are eliminated, then the only way in which a rolling element bearing can fail is by fatigue of the material. The life of an individual bearing is defined as the number of revolutions made before evidence of fatigue develops in the bearing.
- The *basic life* rating is that associated with 90% reliability: i.e. in a sufficiently large group 90% of the bearings will exceed the basic life. In fact, the median life will be about five times greater than the basic life.
- The *basic dynamic load* rating C is used for calculations involving dynamically stressed bearings, i.e. when selecting a bearing which is to rotate under load. It expresses the bearing load which will give a basic rating life of 1 000 000 revolutions. It applies to loads that are constant in both magnitude and direction.
- *Basic static load* rating C_o is used in calculations when the bearings are to rotate at very slow speeds, are to be subjected to very slow oscillating movements, or are to be stationary under load during certain periods. It must also be taken into account when heavy shock loads of short duration act on a rotating (dynamically stressed) bearing. It is defined as that load which will produce a total permanent deformation of the rolling element and raceway on the most heavily stressed rolling element/raceway contact of 0.0001 of the rolling element diameter.

Life and load are related by the ISO equation for basic rating life which is:

$$L_{10} = \left[\frac{C}{P}\right]^p$$

where

L_{10} = basic rating life (millions of revolutions)
C = basic dynamic bearing load (N)
P = equivalent dynamic bearing load (N)
p = exponent of the life equation
p = 3 for ball bearings
p = 10/3 for roller bearings

The selection process is as follows:

1. From your knowledge of the application calculate the load (P) the bearings are required to take. If the bearing is to take both radial and axial loads, the equivalent dynamic load P should be calculated using

$$P = XF_r + YF_a$$

where

F_r = actual radial bearing load (N)
F_a = actual axial bearing load (N)
X = radial load factor for the bearing
Y = axial load factor for the bearing

Values for X and Y will be found in the bearing supplier's catalogue.

When calculating the loads the bearing will be subjected to, account should be taken of the type of load. Gears, for example, may require an additional factor to be added to the load, depending on the quality of the gears. Loads imposed by belt drives may need to be uprated by factors depending on the type of belt employed. Loads imposed by

Vee belts may need to be uprated by a factor varying from 1.2 to 2.5. The supplier's literature should be examined, and if any doubt exists the supplier should be contacted.

2. Decide on the life you require, normally expressed as a number of hours. Your PDS (Chapter 5, section 5.4) should have the information in terms of a service life and duty cycle.
3. Your knowledge of the application should also give you the required speed of the bearing in rpm. On the life calculation chart (Figure 4.38) identify the speed (rpm) and life (hours) points and by joining them with a straight edge read off the value for C/P.
4. In step 1 we calculated P so you now have C, the basic dynamic load rating.
5. A combination of physical constraints, and possibly some stressing calculations, will have determined a suitable size for the shaft diameter. This may need to be increased to the nearest suitable standard bearing inside diameter. Within the bearing catalogue, for a given type of bearing for each ID there will be listed several alternatives. Use the calculated value for C in the previous step to make your selection.

If you have any doubts about the particular application you should always contact the bearing supplier.

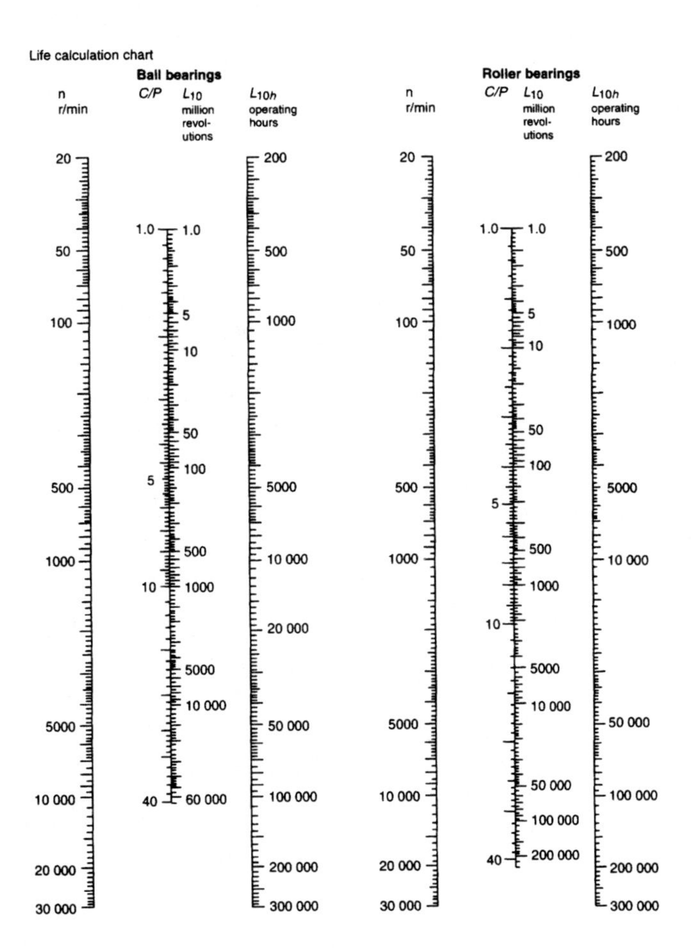

Figure 4.38 *Bearing life calculation chart*
Source: SKF (UK) Ltd.

4.4.4 Seals

The primary purpose of a seal is to separate two fluid media, usually to prevent the egress of fluid from machinery and the ingress of dirt from the working environment. There are two main groupings of seal types:

1. *Static:* the seal is between stationary surfaces.
2. *Dynamic*: the seal is between surfaces with a relative motion.

Many of the seal types require the bounding surfaces to have high dimensional accuracy and a good surface finish. For example, 'O' rings require a surface finish better than 0.6 μm R_a.

Static seals

Static seals can be formed in a number of ways using a sealing medium. A gasket or an 'O' ring (see later for details) could be used to seal an end cap. The gasket material chosen depends on a number of factors including the pressures, medium being sealed, temperatures, and types of, and surface finishes of, the materials involved. The gasket can be a single material, a laminate (such as the head gasket on a car engine), or can even be applied in liquid form from a tube to the surfaces being connected.

Some static seals can be achieved without the use of a gasket. If two components are joined by means of a shrink fit a good seal is generally formed between the components. In some cases, provided the surface finish of the adjoining surfaces is high enough, and sufficient pressure is exerted between the components, a satisfactory seal can be achieved without an intermediate gasket. This technique has been used on some internal combustion engines, but is not common.

Dynamic seals

Dynamic seals can be achieved with non-rubbing seals. They depend for their effectiveness on the sealing efficiency of narrow gaps, which may be arranged axially, radially, or combined to form a labyrinth. This type has the advantage of almost zero friction and wear. It is not easily damaged, and is particularly suited to applications involving high temperatures and high speeds.

In general, however, rubbing seals are more common. They rely for their effectiveness on the elasticity of the material exerting and maintaining a certain pressure at the sealing surface. The peripheral speeds and the quality of the sealing surface will affect the choice of seal.

Felt washers can be used, mainly with grease lubrication, providing a seal suitable for speeds of up to 4 m/s and temperatures of up to about 100°C. The felt washers are normally soaked in oil prior to assembly.

Lipseals

Probably the most common form of standard dynamic seal is the lipseal. This design consists of a rubber casing with a metal insert to give rigidity fitted into a housing, as shown in Figure 4.39. The seal is formed by a lip in the rubber moulding which just touches the surface of the shaft being sealed. The circular format of this lip is maintained by a small diameter garter spring. This design of seal works well provided the seal housing is maintained within a fairly close tolerance on concentricity with the shaft. These seals are more suitable for higher peripheral speeds than are felt washers. As a general guide,

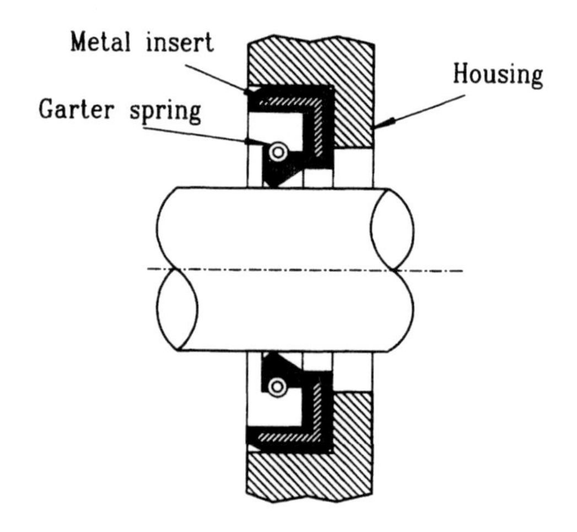

Figure 4.39 *Lipseal*

at peripheral speeds of over 4 m/s the sealing surfaces should be ground, and over 8 m/s hardened, or hard chrome plated and fine ground or polished if possible. The maximum rubbing speed is about 15 m/s. The surface finish on the shaft should be in the range 0.2 to 0.6 μm R_a.

The orientation of the seal is important. If the main requirement is to prevent leakage of lubricant from the bearing, the lip should face inwards. However, if the main purpose is to prevent the entry of dirt, then the lip should face outwards.

A wide range of standard lipseals are available from a number of manufacturers, who should be contacted for installation details. The following are typical of the recommendations made by seal manufacturers:

- In general the most suitable shaft materials are either stainless steel or hard chrome plated. Malleable cast iron can also be used, but care should be taken to ensure the surface is free of imperfections at the sealing point. Pores in the material should be less than 0.03mm in diameter.

- Tolerance on the shaft diameter should be to BS4500 h11.

- Shaft eccentricity should be kept to a minimum, normally less than 0.4mm.

- Shaft and housing should be concentric, any offset being less than 0.25mm.

- Suitable operating temperature range depends on the choice of seal material. Typical figures are:

Nitrile	−40°C to 120°C
Polyacrylic	−20°C to 170°C
Silicone	−60°C to 200°C
Fluorocarbon	−40°C to 250°C

O-ring seals

Another popular type of standard seal is the O-ring. This is normally moulded in nitrile rubber (although is also available in a variety of alternative materials) to a toroidal shape. The ring is contained within a housing which elastically deforms the ring cross-section, thus creating an initial seal. Any further pressure applied by the fluid being retained deforms the ring against its housing, increasing the seal efficiency. Although the seal itself is a fairly low cost item, care must be taken with the design of the housing in terms of both size and surface finish for an efficient seal.

They can be used for both static and dynamic applications. A standard O-ring in a static application will handle up to 100 bar; this compares with a maximum for the lipseal of only a few bar. However, the O-ring can only handle a fairly low rubbing speed in dynamic situations: 0.2 m/s compared with about 15 m/s for the lipseal. A common dynamic use of the O-ring is as a seal on a reciprocating piston.

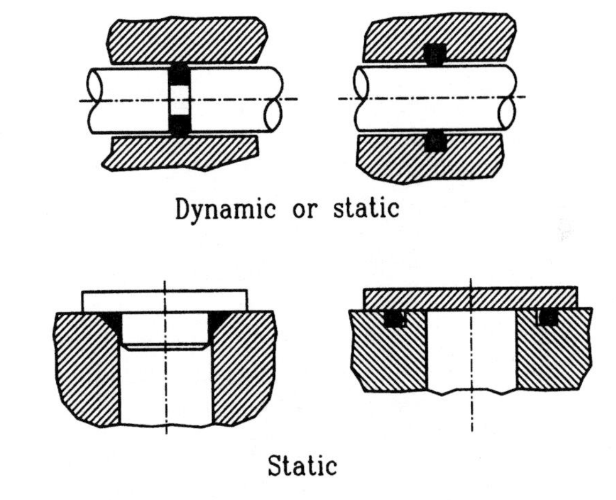

Figure 4.40 *O-ring seal applications*

Figure 4.40 shows some of the possible installations. The finish on the critical surfaces is important, normally 0.4–0.8 μm is required. For further details the reader should refer to the supplier's literature or BS 4518: 'Specifications for metric dimensions of toroidal sealing rings (O-rings) and their housings'.

4.5 Joining devices

Many different techniques are used to join engineering components. These include bolting, welding, adhesives, and appropriate fits, to name but a few. The method chosen depends on a wide variety of factors such as: the rigidity requirements of the joint, type of load to be carried, need to dismantle, frequency of dismantling, tools required, cleanliness of environment, time to make the joint, skill available, appearance, cost, etc.

This section will concentrate on the use of standard devices for making joints, but the reader should also refer to Chapters 9 and 13 for further information on joining techniques.

4.5.1 Threaded fasteners

Thread form

The screw fastening is one of the most popular methods of joining components using standard items. Its success as a standard component is dependent on the acceptance of a common thread form. The main characteristics (Figure 4.41) of a thread form are:

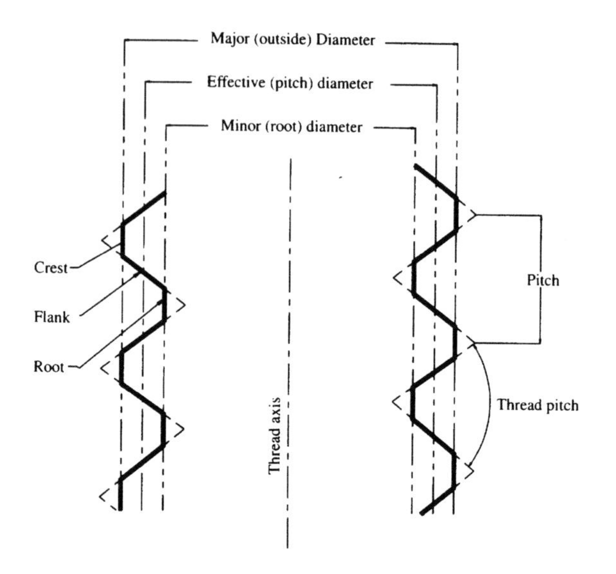

Figure 4.41 *Thread details*

- *Effective diameter*: the nominal thread size.
- *Major diameter:* the outside diameter of the threaded portion of an external thread, or maximum diameter of an internal thread.
- *Minor diameter:* the diameter of an imaginary cylinder that just touches the root of an internal thread or the crest of an external thread.
- *Pitch:* the distance, measured parallel to the thread's axis, between corresponding points on adjacent surfaces.
- *Flank:* the straight sides which connect the crest and the root.

The simplest form of thread (Figure 4.42(a)) is a poor design. Its sharp crest is likely to cause injury, and the sharp root will generate a stress concentration point. In the USA the Sellers thread (introduced in 1861) with a flat crest and root (Figure 4.42(b)) limited these problems and was the standard there for many years.

In the UK the standard was the Whitworth thread (introduced in 1841) with rounded crests and roots (Figure 4.42(c)), having better fatigue resistance than the Sellers thread. As time passed various other thread forms evolved such as BSF (a thread with a finer pitch than BSW), and BA (a range of threads for sizes below 6mm diameter). Although a wide choice of threads may be beneficial for meeting engineering functionality, in practice the benefits of the specialist threads are minimal and offset by the problems of interchangeability.

In Europe, since 1965, the recommended first choice for a thread form has been the ISO metric thread. This uses the 60° angle of the old American standard and some of the rounded features of the Whitworth thread. The proportions are as shown in Figure 4.43. In the USA the unified threads UNC and UNF are still in common use.

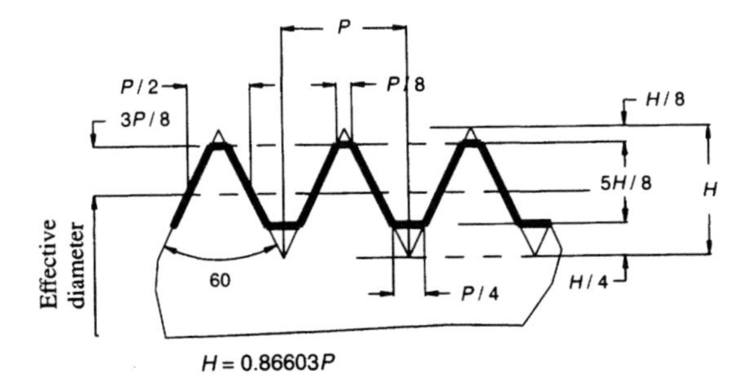

Figure 4.43 *Isometric thread*

Isometric thread

From these proportions you will note that there are two variables, diameter and pitch. A metric thread is designated by the letter M followed by the diameter and pitch in mm, e.g. M6 × 0.5. The relevant standard BS3643 allows two series of pitches:

(a) *Coarse*: the pitch varies according to diameter.
(b) *Fine*: the pitch must be specified from a limited range available.

If a fine pitch is required, *both* diameter and pitch must be specified (e.g. M6 × 0.5). However, a coarse pitch only needs the diameter to be specified (e.g. M6).

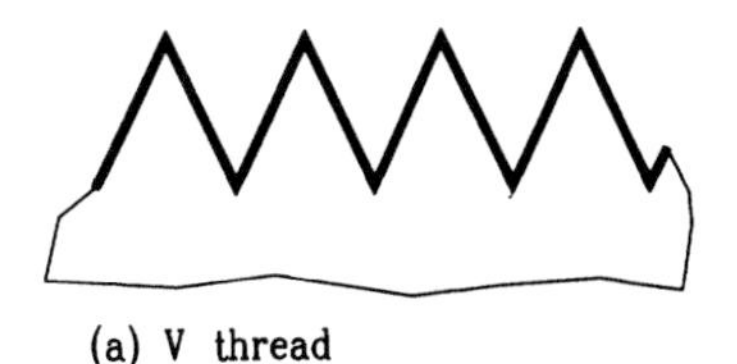

(a) V thread

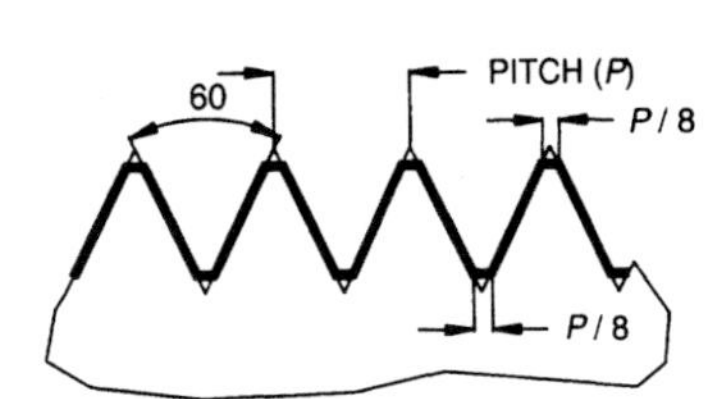

(b) Sellers thread

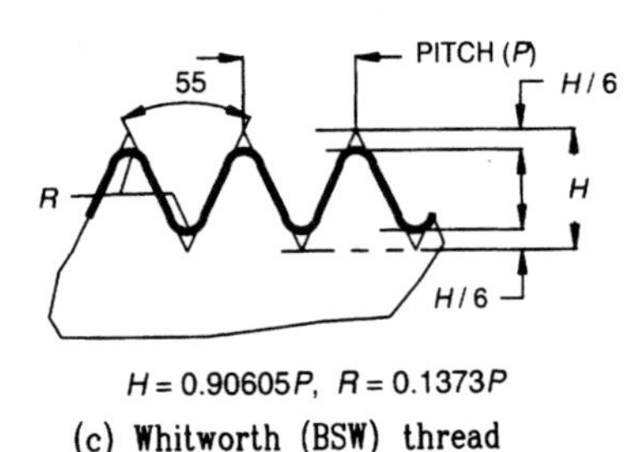

(c) Whitworth (BSW) thread

Figure 4.42 *Thread developments*

BS3643 also specifies three grades of fit:

1. *Medium fit (H6/g6)* This is suitable for most general engineering situations.
2. *Close fit (H5/h4)* Expensive to produce, so only used when close tolerances are essential.
3. *Free fit (H7/g8)* Use when quick assembly is required even if the threads are damaged or dirty.

It is normally assumed that a medium fit would be appropriate so it is only necessary to specify a fit if this is not the case.

There is another thread form created by self-tapping screws. These, as their name suggests, cut their own threads when driven into a pre-drilled hole of the appropriate size.

Bolts, screws, studs

- *Bolt:* A fastener with a hexagonal head and a thread for part of the shank length. The simplest application of threaded fasteners is probably by using the combination of a nut and bolt to clamp components together. The designer only has to provide two relatively low cost clearance holes, surfaces for the nut and the head of the bolt to bear upon, and clearance for the head of the bolt and the nut.

- *Screw:* A fastener threaded along the length of its shank and available in a variety of head styles. Socket headed screws are often used where space for spanner access is limited.

- *Stud:* A threaded fastener without a head, either threaded along its length or with the central portion without a thread. They are particularly useful if the hole is in a weak or brittle material, as the studs can remain in place during subsequent dismantling, thus reducing the wear and tear on the hole. A stud can provide an initial location for the components being joined. For example, studs help support a car road-wheel as the nuts are tightened. Note that they do not provide the final location; the holes in the wheel are clearance holes.

Strength grades

Standard bolts (BS3643) are available in a number of material strength grades, each designated by a two-digit code visible on the head of the bolt.

The first digit when multiplied by 10 gives the bolt materials UTS in kgf/mm^2.

The product of the two digits gives the yield strength in kgf/mm^2. (Remember kgf/mm^2 $\times$ 9.81 gives N/mm^2.)

For example, a 6.8 bolt has a UTS of 60 kgf/mm^2 and a yield strength of 48 kgf/mm^2.

Locking devices

Although, as we have seen, fasteners can come loose for a number of reasons, the greatest danger lies in the unscrewing of the fastener. A wide variety of locking systems has evolved over the years: the following is a limited selection.

Locknuts: In essence these increase the friction between the threads of the fastener. There are two basic types:

(a) Two nuts are used and tightened against each other so that the nuts press on the threads in opposite directions.

(b) The second type uses a taller nut with a modification to the threads at one end. This can be in the form of slight thread deformation, or a plastic insert in which a thread is cut as the fastener is tightened. Many of these designs should be used only once.

Lock washer: The prime purpose of a plain washer is as a bearing surface for the head or nut and to distribute the load over a greater area. Lock washers are often a coil or spring design with sharp points that become embedded in the underside of the nut, and the component thereby increasing its reluctance to unscrew.

A second type is the tab washer, where part of the washer locates in the shaft and part on the nut by being distorted against one of its flats.

Other techniques include:

- Adhesives.

- Split pin with castellated nut.

- Wire locking, etc.

4.5.2 Non-threaded fasteners

Threaded fasteners are particularly useful on those components that are likely to need to be taken apart during the product's life. However, there is a penalty for this flexibility. They tend to be time-consuming (i.e. expensive) to assemble, assembly is difficult to automate, and machining tapped threads can be expensive, to name but three disadvantages. We will mention briefly a few of the vast range of alternatives.

Rivets

Rivets are used to make permanent fixings between components. Their popularity is due to their inherent simplicity, low production costs and good reliability. The rivet is passed through aligned holes in the components to be joined, and the end formed over to complete the joint (Figure 4.44).

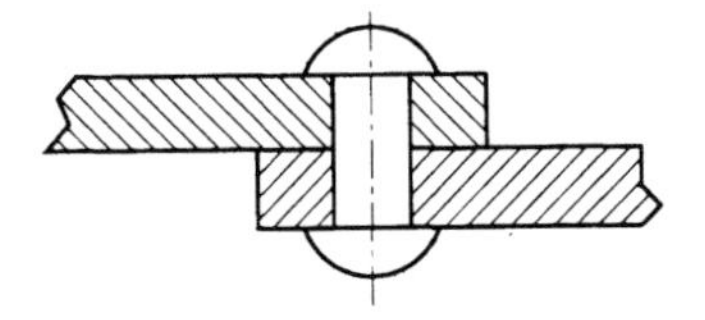

Figure 4.44 *Typical riveted joint*

Unlike bolts, which should normally only take a tensile load, rivets are designed to resist shear, and so should not be subjected to tensile loads.

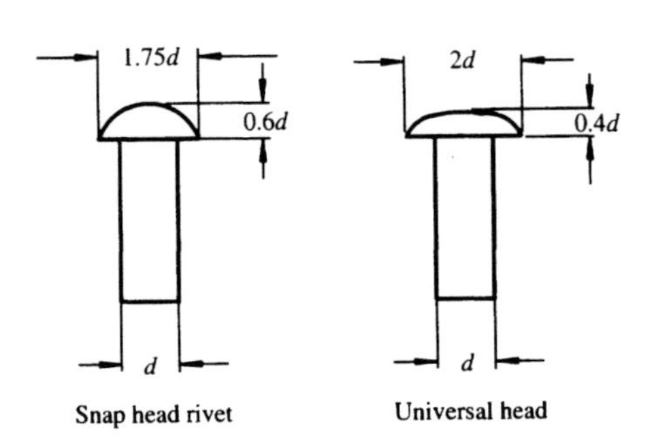

Figure 4.45 *Solid rivets*

Solid rivets (Figure 4.45) are available in a variety of sizes and head shapes. BS 4620 specifies a range of sizes from 1 to 39mm diameter.

Hollow rivets (pop rivets) are also available. These have the advantage of only requiring access to one side of the joint and are often used for fastening sheet metal components. The hollow rivet has a mandrel fitted in its bore. This assembly is passed through aligned holes in the components, the mandrel being gripped by a special riveting tool. This pulls the mandrel, distorting the rivet to form a head, as shown in Figure 4.46, eventually causing it to snap. The remains of the mandrel are discarded. These rivets are available in a wide variety of styles, often in steel and aluminium.

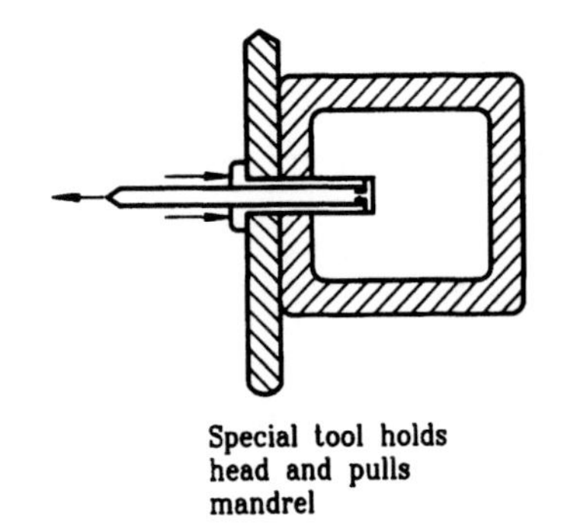

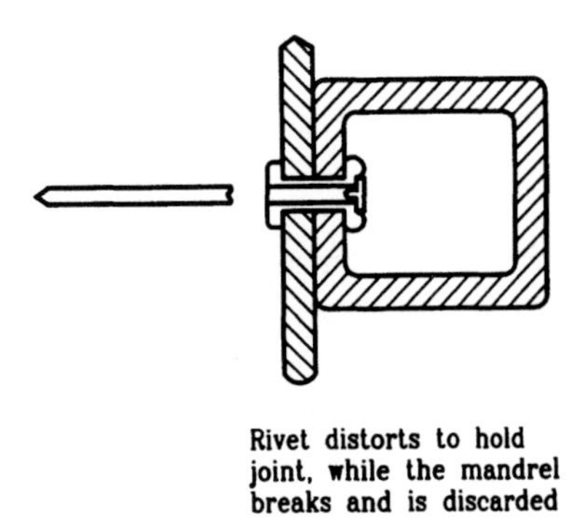

Figure 4.46 *Pop rivet*

Circlips

These items (see Figure 4.47) are designed to fit within a groove on a shaft, or in a hole, and at the same time stand proud of the shaft or hole, thereby effectively creating a removable shoulder. The ends of the clips usually incorporate small holes for a special tool to fit into in order that they may be distorted for easy assembly or removal. They are manufactured from spring sheet steel or wire, which is then hardened and tempered. Sizes of standard circlips, together with the appropriate groove dimensions, are detailed in BS 3673.

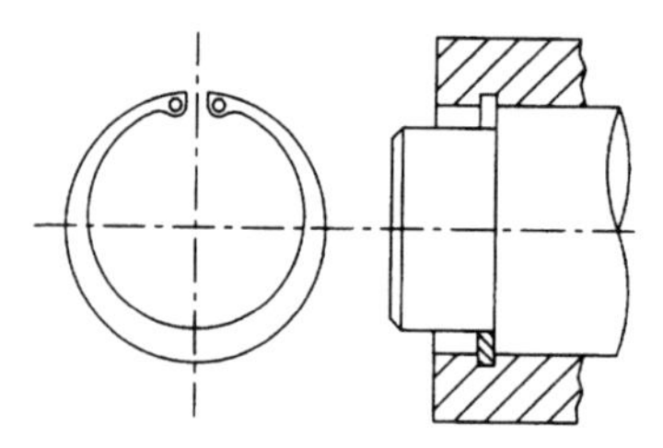

(a) Internal

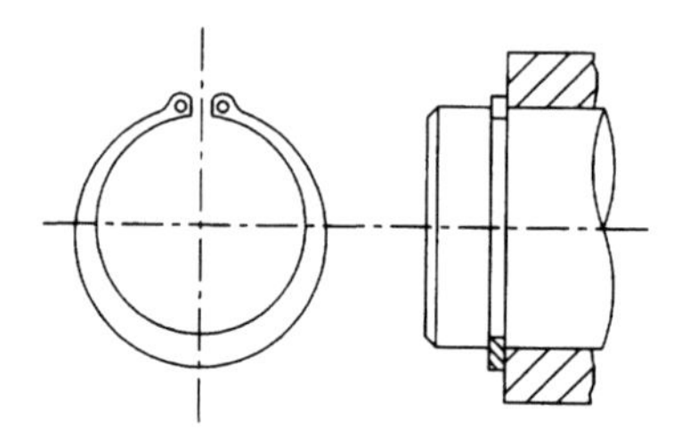

(b) External

Figure 4.47 *Circlips*

Split cotter pins

These pins are designed to fit into clearance holes in a shaft, the ends being splayed to prevent their removal (see Figure 4.48). They are often used to retain clevis pins on small linkages such as the handbrake systems on motor cars. They can also be used to provide a locking system for a suitable design of nut, such as is used to retain the wheel bearings on motor vehicles. Details of standard sizes can be found in BS 1574, or ISO/R 1234.

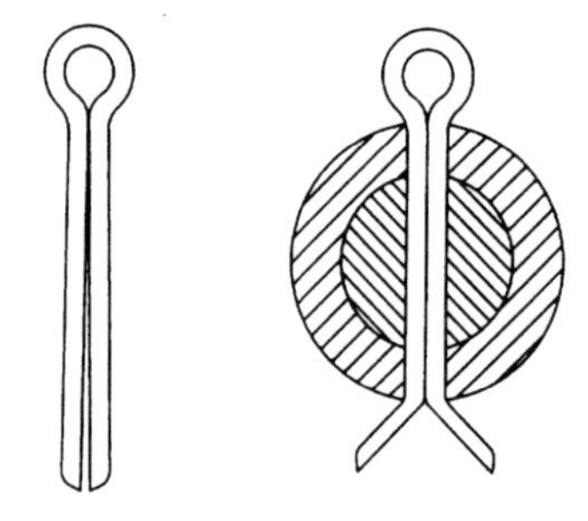

Figure 4.48 *Split pin*

Dowels and rollpins

Dowels are generally parallel-sided steel pins that are a good fit in aligned holes in the two components being joined. The purpose of the dowel is to ensure that the components are replaced in the same relative positions after disassembly. Obviously, the relative positioning of the two holes is important, and will require some careful tolerancing. This tolerancing can be avoided by drilling the holes after initial assembly, but the penalty is that the components will then not be interchangeable. Details of standard dowel sizes are given in BS 1804.

A variant on the standard dowel is the rollpin (Figure 4.49). This is a pin produced by rolling spring steel, and can almost be thought of as a small diameter circlip that is very thick. This pin can be inserted in a hole slightly smaller than its outside diameter. The spring effect of the pin ensures that it is retained in the hole.

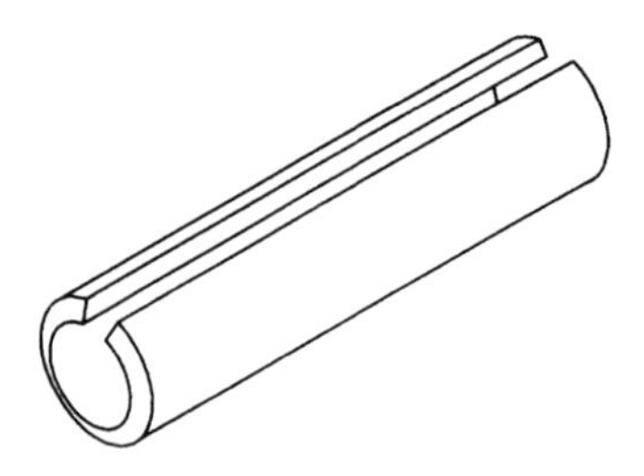

Figure 4.49 *Rollpin*

4.6 Summary

Although a few new products introduce totally new principles and concepts, the majority of new designs are rearrangements of not only existing principles and concepts, but also existing standard components. The principle of standardisation offers potential benefits in terms of cost, reduced tooling investment, increased reliability, and better safety standards. In addition, it allows the designer to concentrate on his or her particular application without having to reinvent the nut, bolt, bearing, or wheel.

This chapter started with a brief summary of the types of standards organisations, and then moved to introduce a variety of standard components. As such components form an important part of the staple diet of the engineering industry it is not possible in a text of this size to do more than introduce a small selection of components. For example, the current SKF catalogue on rolling element bearings includes 185 pages of information regarding the selection and application of bearings. With the data tables on standard bearings the catalogue runs to some 975 pages!

On completing this chapter the reader should appreciate the benefits of using standard components and that the good designer will always check:

- the availability of standard components
- the relevant standard (international, national, or company)
- the component manufacturer for selection and applications data.

4.7 Questions

The following questions are designed to allow you to check your understanding of the areas covered in this chapter, but should be supplemented with some practice selection of standard components using suppliers' catalogues.

1. Name three of the major benefits available through standardisation. (4.1)

2. A gear is attached to a shaft via a keyway. The shaft is supported between two rolling element bearings such that the gear meshes correctly with another gear which is driven. One end of the first shaft is fitted with a pulley and used to power a small machine tool.

   (a) If the single helical gears are used, name the different forces that will be applied to the shaft. Will there be any change in these forces if spur gears are used? (4.3.5)

   (b) When selecting the keyway, part of the information will come from a standard. What is the information and which is the BS standard? A key will normally fail in one of two modes, what are they, and how does the designer check for them? (4.3.2)

   (c) When selecting suitable rolling element bearings for this application, what criteria would affect your choice. Would a pair of cylindrical roller bearings, for example, be suitable? (4.4.3)

3. When would you use a sprag clutch? (4.4.2)

4. If a standard bolt has the number 6.8 embossed on its head, what does this mean? (4.5.1)

5. Discuss the differences between

   (a) the pop rivet and the solid rivet

   (b) the roll pin and the dowel. (4.5.2)

6. Explain the process by which most rolling element bearings eventually fail? (4.4.3)

7. Discuss the pros and cons of using oil or grease as a lubricating mechanism. (4.4.3)

8. If you select an electric motor as TEFC, what does this mean? (4.3.7)

9. What do the terms 'pressure angle', 'involute' and 'base circle' mean? (4.3.5)

10. Discuss the difference between lip and O-ring seals, giving some examples of their application. (4.4.4)

5 The Design Process

5.1 Introduction

As its title suggests, this chapter will explore the design process. This may conjure thoughts of a specialised, highly efficient process that designers use to create new products. It is, in fact, a methodical approach that leads designers in an efficient way towards producing new products. However, it is also much more than that.

Although engineers do design new products, much of their work covers modifications or improvements to existing products. These tend to be prompted by a wide range of problems, such as difficulty with the method of manufacture, change of component supplier, premature failure of the component, a need to reduce the manufacturing cost, and so on. Design of this type starts with the identification of a problem and is essentially the process of generating a suitable solution. It can be therefore be considered to be a form of problem solving.

Designing a new product is, in fact, very similar, except that it usually involves a wider range of components. In essence, it also is a type of problem solving. The designer's approach to finding solutions can be applied to almost all forms of problem; it is much more than just a tool for designers.

5.2 The process

To demonstrate the process we will consider a simple problem: a request to provide a support for a television set.

The first step is to consider why a support is needed. The answer may be that the available horizontal surfaces are either too small, at the wrong height, or are unable to support the weight. If this is the case, we have completed the first step by establishing that a true need does exist.

The next logical step is to quantify the need, i.e. establish factors such as the weight to be supported, the dimensions of the set, the required height, etc. These factors are all directly related to the need. In practice, the design will also be constrained by a number of other parameters. These may include cost, time, type of floor or wall construction, proximity of power or signal sockets, and so on.

Once all this information has been collated, we have effectively defined the problem, and can move to the initial steps of generating a solution. We need to consider a number of potential solutions – e.g. shelf, small table, ceiling fixing – and how these are to be manufactured and finished, e.g. materials to be used, painted, plated, natural finish, etc.

By comparing the potential solutions we will eventually arrive at the most suitable, and move towards the detail steps of deciding component sizes, obtaining the materials and completing the job.

Although this is a trivial example, it serves to illustrate the basic stages in the design process, which are:

1. Identify the need
2. Define the problem to be solved
3. State the limiting parameters
4. Generate ideas
5. Evaluate ideas
6. Preliminary design
7. Detail design
8. Implementation.

The first three steps can be summarised as problem finding, with the remaining stages covering problem solving.

There is a natural tendency when faced with a design problem to attempt to generate a solution before properly considering the problem. This tendency must be resisted by the designer, as a poor solution to a problem is much more useful than an elegant solution to the wrong problem.

In fact, as it is pointless to solve the wrong problem, we could state that *problem definition is more important than problem solving.*

There are many everyday examples that prove this point. The Sinclair C5 was an elegant alternative to the bicycle or moped, but as it did not satisfy a need it was doomed to be a failure. On the other hand, the Sony Walkman provided an elegant solution to the problem of playing tapes while performing another activity. It fulfilled a need, i.e. solved a problem, and has proved an enormous success.

A colleague once proposed what he called the 'Birmingham theory'. If you are planning a journey, your first step must be to define the end point of the journey. Unless you first decide that your destination is Birmingham (for example) it will be impossible to select the most suitable route to take. Design is no different. Unless you first decide upon the need for your product, service or modification, it is most unlikely to succeed.

Let us apply this to the design process itself. Why is such a process needed? Is it simply to ease the job of the designer who is working to satisfy the specific need for a particular product, or are there other needs?

To answer this question we must start with the product and consider whom it can affect. Two groups come into this category. The customer (or end user) for whom the item was nominally designed is obviously very important. Unless the product meets his/her expectations in terms of function (performance, reliability, aesthetics, etc.), cost, and availability, high sales are unlikely. The second group is the producing organisation. (In its simplest form this would be a company designing, manufacturing and selling products.) Here the product has to fulfil a different set of criteria to be considered a success. It must be capable of generating a cash return for the company, sufficiently in excess of the cash input required to develop and produce the product, to satisfy the owners of the producing organisation. If an organisation has too many products that do not succeed in these terms, the organisation itself will not be able to survive, let alone prosper.

We now have the basis for the definition of the design process.

5.2.1 Definition

The detailed process which leads to the generation of a product or service that meets the needs of both the consumer and the producing organisation. The outcome of the process is normally a set of detailed instructions that enable the product or service to be supplied regularly in a manner that satisfies both consumer and producer.

The starting point of this process will be the creation of a detailed product design specification.

5.3 Product design specification

The Product Design Specification (PDS) is the foundation of the design process. Unless sufficient care is taken in its generation, the route to a successful product will be very difficult.

The PDS is effectively a list of all the criteria that the product must meet to be successful. It must both meet the functional requirements of the customer, and generate adequate profit for the producing organisation. It has to be systematic and thorough, but must avoid predicting the outcome. If it makes assumptions about the solution it will reduce the potential for innovation. For example, if the specification for the Mini motor car had contained details of the size of the differential in the rear axle (a feature common to almost all cars at that time), the possibility of driving the front wheels would not have arisen. Before considering the detail of its contents, we should spend a few moments on the production and maintenance of the PDS.

The producing organisation will almost always contain a wide variety of departments and associated skills, such as research, marketing, sales, design, production, purchasing, and testing. As each of these will play a part in the evolution of any new product, they must be involved in the generation of the PDS. In this way the PDS does more than simply define the product requirements: it provides a goal upon which the organisation can focus, and thereby help the various departments to work together. It also keeps a sense of realism to the specification by involving a range of expertise. For example, the marketing people could easily identify a market need for a motor car that caused no pollution, carried six passengers, and was capable of achieving 100 mph and 100 mpg at the same time, but the input from the research and design areas would no doubt temper this dream with a touch of reality.

The time taken to develop a new product can vary from a few months to a number of years depending upon the type of product. During this development phase various influences (internal or external) may require changes to the specification. Changes could be needed because your own research department has developed a new material, your manufacturing facility may have developed an alternative manufacturing process, a competitor may have introduced a product with a feature that marketing may now consider to be essential in your product, or a court case may result in legal implications for your product. Unless you respond to these influences your PDS could result in a product that is obsolete before it even reaches the production phase. However, if the need does arise for modifications to the specification, any changes can only be implemented with the full agreement of all those involved in creating the PDS. Remember, the PDS is evolutionary – it is not a static document.

As the PDS is effectively the foundation stone for the design process it must obviously be a written document. This may seem obvious, but the author has experience of a number of organisations where a formal document has not been produced. All sorts of problems can arise where the designer has, for example, included an additional feature, possibly as the result of an informal conversation with, say, the sales manager, only to find him/herself severely reprimanded because the target cost has been exceeded. A formal document, agreed, dated and with an issue number is the best defence against this sort of problem.

At this point we can summarise the requirements for the PDS as:

1. Will contain all the facts relevant to the product outcome.
2. It must avoid predicting the outcome.
3. The restraints must be realistic.
4. All areas involved in the product introduction process must participate in its generation.
5. It is an evolutionary document, but can only be amended with the agreement of all involved.
6. It must be a written document.

Now we need to consider the detail of the content of the PDS. The detail will obviously vary according to the type of end product. However, it is possible to use a generic check list of headings that should cover most products. Using such a list should ensure that no major areas are missed and will lead to a systematic and thorough design.

It is important at this stage that the reader appreciates that the PDS provides a standard by which the design can be judged. If it meets or exceeds all the elements of the specification, it can be considered a success (subsequent success of the product in the market will, of course, depend on the suitability of the PDS). It follows that if the PDS is to be used as a benchmark, its elements must be quantifiable. The importance of this aspect cannot be over-stressed.

As an example, consider a simple piece of equipment such as a bench top laboratory centrifuge, where the level of noise it produces is one of the PDS criteria. It would be easy to include a statement such as 'This centrifuge should be as quiet and vibration free as possible'. As a target for the designer this statement is meaningless. How quiet is quiet! By all means include the statement as a general comment, but it must be followed with a definition of an acceptable noise level, and how this is to be measured. For example:

'This centrifuge should be as quiet and vibration free as possible. The noise level is not to exceed 64 dBA when measured 1 metre in front of the machine with the XWZ rotor running at a speed of 4000 rpm.'

This quantifiability applies to all elements of the PDS. In some areas it will be reasonably easy to quantify the requirements. In others it may prove more difficult. Much data exists in various standards, or even specifications for previous products, and can be used in the PDS. If this is done, the source should be acknowledged simply to help producers of subsequent specifications. In other cases the information may be gained from some simple practical tests. The noise example was based on a real case where feedback from customers indicated that the current model was too noisy when compared with the competition. Some competitors' machines were purchased and noise measurements made. These formed the basis of the figures in that element of the PDS. The final product that was the result sold well and had a reputation for being a quiet machine. It may prove almost impossible to provide numbers for some situations. In these cases assumptions may be made, provided they are logical and are recorded.

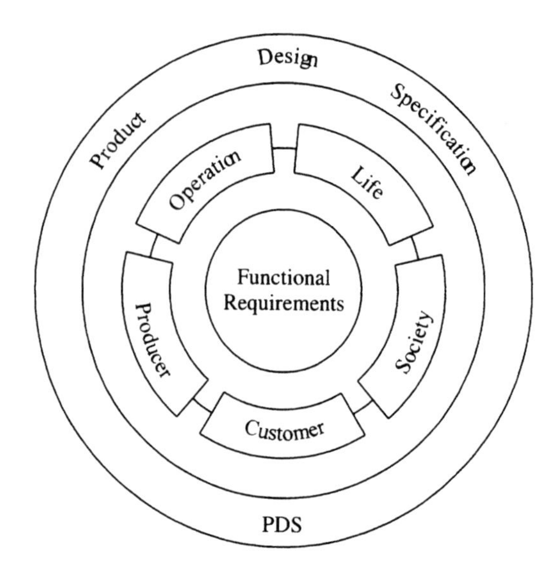

Figure 5.1 *Product design specification boundary*

5.4 Elements of the PDS

The PDS can be thought of as a set of functional requirements, born of a need and limited by five major groupings of constraints as illustrated in Figure 5.1. The manner in which the product meets the requirements will determine its success. It must operate to meet the need, for an adequate length of time (life), while satisfying the needs of the customer, producer and society alike. (Remember our definition of the design process in section 5.2.1.)

A good PDS will start with a simple statement of the origin of the need and then move to each of the five constraint areas with quantifiable statements defining each of the criteria the design must meet.

We will now examine some of the more commonly used elements of the PDS under each of these headings. The reader should understand that, although this section will cover the type of information that should be included under the various headings, not all headings will be applicable to all products.

5.4.1 Origin of need

A PDS should always start with a brief introduction giving the reason for the product. Taking our small centrifuge as an example, the introduction might read:

'*This centrifuge is to be a small bench-top machine to replace the Model ABC. It will meet the latest BS4402 safety requirements and, in addition, have an increased carrying capacity, lower noise level, and lower sample temperature rise.*'

Note the lack of quantification. This is the *only* area of the specification where this is permissible.

5.4.2 Operation

By operation we mean those aspects that are likely to have a direct physical bearing on the product. A typical PDS will include the following.

Performance

Here the performance likely to be demanded must be defined as precisely as possible. Ask a series of questions:

- How fast?
- How slowly?
- How often?
- How closely does the speed need to be controlled?
- What loads are anticipated?
- What fuel consumption is expected?

The answers to these and similar questions need to be defined to specify the required performance. Wherever possible, use numbers to avoid ambiguity. It may also be useful to include the source of your numbers. There are sometimes non-numerical methods of quantifying performance requirements. In the early 1970s when Ford were introducing their own design of diesel engine to the Transit van, the specification called for the acceleration performance to be no less than that of the then current Volkswagen van. To validate this aspect of the design, much fun was had by the engineers comparing the performance of the two vehicles by means of a series of 'drag races' on the Ford test track!

Care must be taken to ensure that only realistic requirements are included. Asking for the ultimate in performance may result in a product that the customer is unable to afford. Another potential danger exists with low volume products, where inadequate information is available from the customer to fully define his/her requirements. In these cases it is particularly important to make sure that the source of any data is identified.

Environment

In this instance we are concerned with the physical environment(s) that the product is likely to encounter. The effects on our environment will be covered elsewhere. All possible aspects should be included:

- *Temperature range*: For example, standard electric motors are designed to operate with cooling air at a maximum of 40°C, giving a maximum temperature for the windings of 140°C. If the ambient exceeds 40°C, the motor must be derated to avoid overheating the windings. At the other end of the temperature scale there may be problems with wax forming in diesel fuel, clogging the filters.

- *Pressure range*: At high altitudes, the lower pressure and hence less dense air will reduce the output from internal combustion engines.

- *Humidity*: High levels will promote corrosion or fungal growth. Even if humidity is not anticipated in the operating environment, does it need to be considered when the product is shipped to markets elsewhere in the world? Remember the small packet of silica gel that is normally packed with electronic equipment shipped from the Far East.

- *Corrosion*: Here we need to define the likely source of corrosion, e.g. hydraulic fluid, salt spray, etc.

- *Shock loading*: This will include any instance of a sudden increased loading: e.g. dropping the item, the effect of an emergency stop on the various components of a car, the effect of a car tyre hitting a pothole at speed, etc.

- *Dirt or dust*: The first British-built cars sold in Australia performed poorly when driven on the many dirt roads. Although the cars were reasonably watertight, they failed dismally in preventing dirt entering the vehicles. Presumably the specification did not call for this characteristic.

- *Degree of abuse*: The type of people who may use the equipment must be considered. Public telephones are attractive to vandals so must be designed to offer some form of resistance to abuse.

There are numerous further headings that may be used in the general category of environment, including noise levels, vibration, insects, electromagnetic radiation, and so on. It should also be noted that a number of environmental changes may be experienced prior to the product reaching the end user. These may occur on the shop floor, during storage at the works, on site, during transportation, etc.

Size

Any size constraints should be specified. These will have a direct effect on the product and may arise for a variety of reasons. There may be a functional requirement: for example, a portable CD player must be large enough to take the CD, but small enough to still be portable. There is a legal requirement for the overall length of an articulated lorry not to exceed a maximum value. Shipping costs may be greatly increased if the product cannot fit within an ISO container, so this may impose a size limit.

Weight

Often for a given type of product there is a close correlation between the weight of the product and its cost. The tendency is, therefore, to design for minimum weight. This is not always the case. For a number of years the car industry has reduced the weight of its products, with consequent gains in performance, economy and presumably savings in material costs. The Citroen AX of the late 1980s is an excellent example of this philosophy. In the minds of some customers, vehicles of this type lacked an obvious solidity which might affect both longevity and safety, so the next generation of cars from this company, the Peugeot 106, actually put on some weight.

Whatever the reasons, the PDS should define some weight targets.

Ergonomics

As the interface between the operator and the product is obviously important, any critical points should be defined in the specification. For a motor vehicle these will be the positions of the controls and instruments relative to the driver, and the forces needed to operate them. It may also include access to and easy identification of those areas under the bonnet requiring routine maintenance.

Aesthetics

In almost all cases the customer sees the product before trying or buying it. If this initial sighting is unfavourable, the product may not even have the opportunity to display its functional prowess.

Colour, shape, form and texture should be considered from the outset. This is very important for consumer products, but is becoming increasingly important with engineering products.

As an example, in the food processing field almost all items that come into contact with the food products are manufactured from stainless steel. A simple item such as the body halves of a butterfly valve would normally be produced from forgings or castings. As stainless steel is a relatively expensive material, one company produced a valve with the bodies made from stainless steel pressings, thereby reducing their material costs. Functionally these performed as well as the more sturdy looking valves made from forgings or castings. However, because their aesthetic appeal was diminished, the pressed valve did not sell well and was only able to command a lower market price.

5.4.3 Life

There are a number of different aspects to the life of a product that need to be defined in the PDS.

Product life

How long do we expect to be able to sell the product for? The product life-cycle graph will always be the same general shape (Figure 5.2), and only the proportions will vary. An estimate of the expected proportions of this graph will help to assess the sensible levels of investment in generating and producing the new product.

Errors in this area can prove expensive. After the Second World War the British army declined to accept the design for the Volkswagen Beetle,

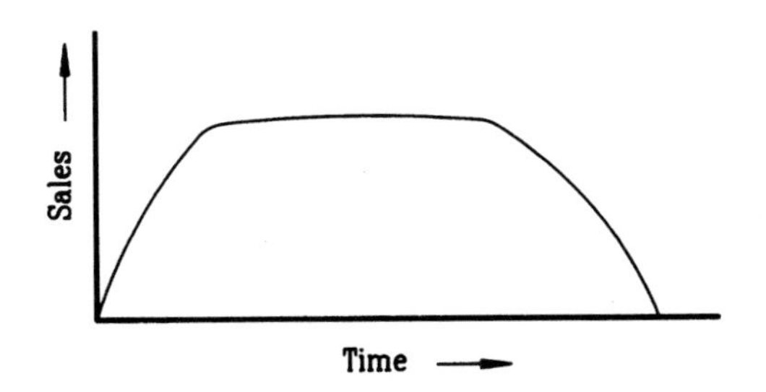

Figure 5.2 *Product life graph*

presumably taking a view that the potential for a long production run was not high. How wrong they were! However, the German VW company made an equally serious mistake in the late 1960s, by assuming that production could continue almost indefinitely. They spent vast sums of money on highly specialised automated production equipment. When sales eventually started to fall the company almost went bankrupt!

The product life can vary enormously for different products. Some typical examples are:

Motor car engine	10 years
Computer	18 months
Camera	12 months
Mars bar	50 years

Service life

How long do we expect the product to last in the hands of the end user? This will obviously depend on how frequently it is used, so we must consider the *duty cycle*. The service life is usually expressed as a number of hours based on a duty cycle of x hours per day, y days per week, z weeks per year.

Take an electric lawnmower as an example. This might be used for, say, half an hour per week for 25 weeks of the year. The owner might reasonably expect it to last for 5 years. This would give a service life of $\frac{1}{2} \times 25 \times 5 = 62.5$ hours.

Note that the service life of individual components of a product may differ from the service life of the whole product. The life will obviously depend on the duty cycle for that individual component. Some typical examples might be:

Motor car	3000 hours
Electric motor on cooling fan	150 hours
First gear on a car	4 hours

Shelf life

This is the length of time that the product might reasonably be expected to survive without deteriorating when stored. For many items it will be important to define the storage conditions. For example, rubber seals may degrade if subjected to sunlight. The shelf life might be 3 years provided the seals are stored in black plastic bags. Both the required length of storage time and the means of combating decay or corrosion should be specified.

Reliability

How long does the customer expect the product to last before failing? This can be defined as the number of hours to first failure, and the mean time between failures. The normal way of proving that the design has met these criteria is by having a statistically valid number of tests performed on a valid number of prototypes. The results of the tests can be analysed to give a defined level of confidence that the reliability requirements have been met.

Testing

Specific tests may need defining. These may meet in-house test procedures, or may be defined in an external standard. Will a performance test certificate be required by the end user? If so, the test operations need to be costed as part of the manufacturing process.

Maintenance

This will have a bearing on a number of aspects of the product. The service life will, of course, assume that any specified routine maintenance has occurred. It may be possible to design out much of the routine maintenance, but this may incur too heavy a cost penalty. A note on the customer's normal practice should be included.

Ease of access to parts likely to require regular attention may be important. At least one car company has colour-coded such items in the under-bonnet area.

The spares policy of both customers and the producing company need to be considered. As an example, consider a belt drive. If this uses a standard, readily available size of belt, this may please your customer, as a spare will be available from a number of sources. However, if a non-standard size is chosen, any spares will have to be supplied from your company, which will no doubt charge a premium. A compromise needs to be reached to satisfy both needs (see the definition of the design process).

The desirability of special tools should be considered. Again they may tend to favour the producer's service organisation, in that in the short term more in-house service business may be generated.

Operating costs

Actual money values would not normally be included in the specification, but the service intervals and the time required to complete the servicing might be.

Operating efficiency of the product may also have a significant influence on the running costs. It is important to understand the requirements as perceived by the customer. As an example, a company supplying pumps for use in the food process industry introduced a new range of pumps that offered as an option a high efficiency version for a small extra charge. It assumed that as the pumps often ran continuously for 24 hours a day, a reduced energy consumption rate would be important. Unfortunately, those targeted with buying the pumps were more interested in the initial cost of the pumps than the running costs. Consequently very few high efficiency pumps were sold.

Note that the maintenance and operating cost sections could just as easily have been included under the operation set of constraints. The name of the category is far less important than the need for the criteria.

5.4.4 Producer

Here we need to consider any constraints that relate to the producing organisation. These could be many and diverse. This can be illustrated by an example from the motoring world. A sports car produced by the Rootes Group in the 1960s (the Sunbeam Tiger) used a Ford V8 engine on the grounds that the car was a limited production volume vehicle, and an engine of that type did not exist within the producing organisation. The Rootes Group was subsequently sold to Chrysler, a strong competitor to Ford, and almost instantly production of the Sunbeam Tiger ceased, simply because the producing organisation did not want to be seen selling a product so obviously using one of its competitor's engines. Today, the car companies take a more enlightened approach, but at the time pride within the producing organisation could have a direct effect on its products. More common elements to include in the PDS are:

Quantity

This is of vital importance in the specification as it will affect the levels of investment, and the choice of manufacturing processes and materials. The product life cycle graph (Figure 5.2) is important as it will indicate not only the estimated annual sales volumes, but also the rate of growth in the early months of sales.

Product cost

Every PDS should indicate a maximum product cost. This is normally an estimate of the actual cost of production, and not the sales value. At the start of the project this cost estimate may be somewhat inaccurate, but as the project proceeds the cost should be refined gradually until in the final issues of the PDS it will be an accurate reflection of the cost of production.

Company constraints

The developing and producing organisation may have internal policies that may affect some of the design decisions. If this is the case they should be included. For example, if the company has a foundry it may prefer the designer to use in-house-produced castings rather buy in forgings. The internal costing system may automatically take care of this. On the other hand, the company may have a policy involving only in-house assembly, all of the component manufacture being subcontracted. If the company has any policy that is likely to affect the engineering decisions, it should be included as part of the PDS.

5.4.5 Society

Today, the impact that products can have on society in general has generated a need for these to be considered at the earliest possible stage of the process. Such factors will include ensuring competence in the design for safe operation, both during production and in subsequent use, minimising environmental damage, etc. Typical headings in this area of the PDS would be:

Standards

Any relevant standards should be identified. (See Chapter 4, section 4.2, for the types of standards.)

Codes of practice

These provide guidelines for the design of commonly used components. They are written in such a way that their correct application will normally lead to a satisfactory and safe result. They tend to be conservative, but are updated frequently to maintain a balance between past success and present technological advances. The available range deals with a wide variety of procedures including design of pressure vessels, gears, welded joints, application of preservatives, etc.

Although there is no legal obligation to comply with standards and codes of practice, the potential threat via product liability legislation should be enough to encourage compliance. A good defence against any legal action implying inadequate design could be the use of standards, etc.

Regulations and laws

There are a variety of regulations that must be complied with. These can arise from sources such as:

The factory inspectorate
Health and safety executive
Environmental health officers
Nuclear inspectorate
Local by-laws
Customs regulations
EEC regulations
Product liability laws

Environmental effects

Here we are concerned with the effect that the product or its manufacture could have on the environment. In the past this may not have been considered important, but now, particularly with the possibility of retrospective legislation applying, it is prudent to consider this aspect at the earliest possible stage. For example, the type of materials selected and the possibility of subsequent recycling may be important; the manufacturing processes chosen may use scarce world resources, or result in harmful by-products that require careful disposal.

5.4.6 Customer

The whole purpose of designing, developing and producing the product is to meet the needs of the customer. It is vitally important that these needs are understood clearly when both generating the specification and developing solutions. Discussions with customers, customer clinics and customer surveys of competitors' products are all ways in which the designer needs to build up the background information necessary for a

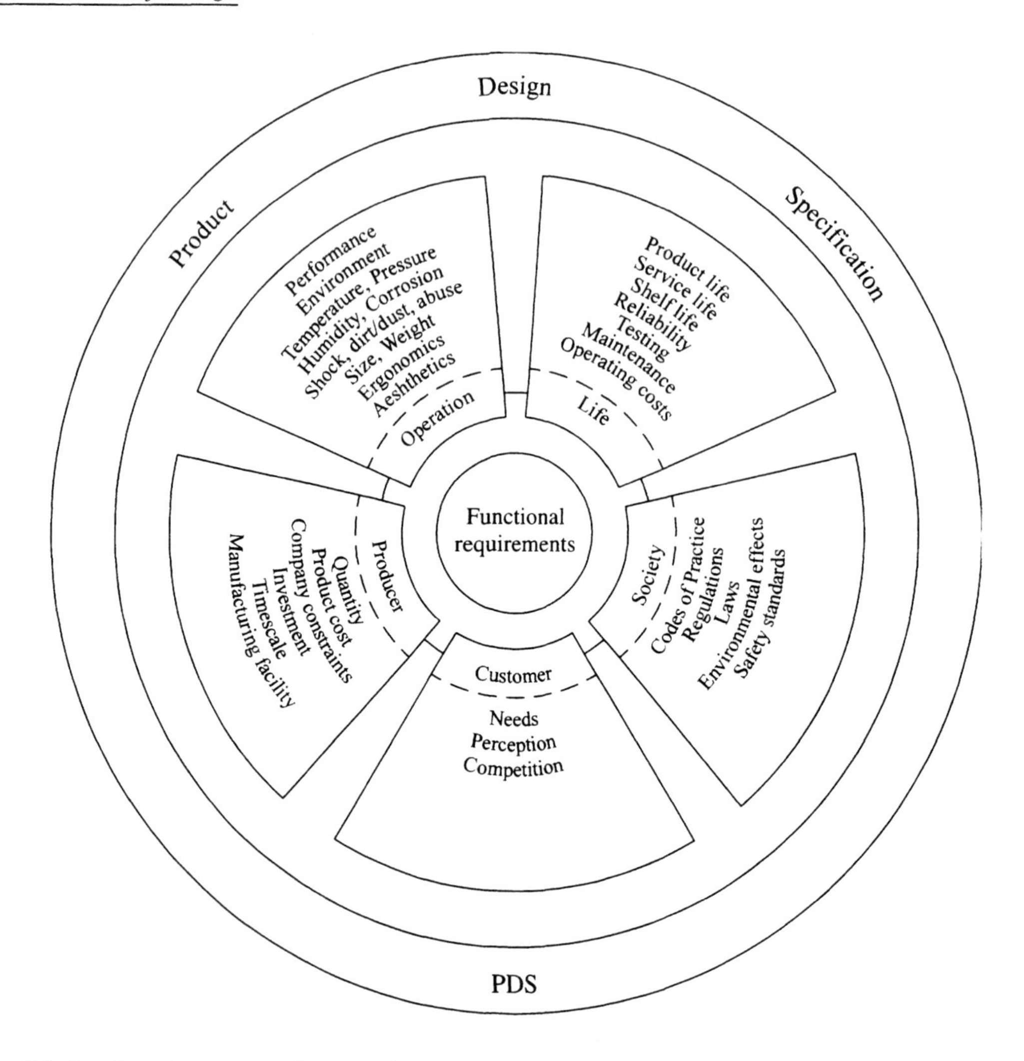

Figure 5.3 *Product design specification elements*

successful product. It may be that most customers are within a certain age range, and if so, this should be included. There may only be a short time window when the customer will want to buy the product – Easter eggs will not sell well in the summer. This window need not only be seasonal – the first thin flat screen television could take the lion's share of the available market.

Different products will require emphasis on widely differing criteria. However, if each of the above areas has been considered (and included as appropriate) the resulting PDS should be a comprehensive definition of the problem and its constraints. The basic elements of the PDS forming the design boundary are summarised in Figure 5.3. The next section introduces a number of sources of information for the designer. Some of these sources (e.g. standards) are likely to be used in the generation of the PDS.

5.5 Information sources

Although not impossible, it is very rare for a completely new design or concept to emerge. Most new designs incorporate already-known features or concepts; their novelty is usually in the application or combination of ideas and concepts. A vital first step in the design process is to increase one's store of relevant information.

5.5.1 Standards

Types of standards are covered in Chapter 4. Here we simply remind the reader of their existence and note that they may act as another source of information. Even if the PDS does not specify compliance with a particular standard the prudent designer should always check for the presence of a relevant standard or code of practice. A PDS has a habit of changing, particularly when a competitor introduces a feature or claims compliance with a particular standard.

5.5.2 Component manufacturers

Component manufacturers (particularly the large manufacturers) tend to be experts in their field, and can be a valuable source of information, both from their published literature and from their applications engineers. They are normally in an excellent position to provide advice, which is usually free. Any large car manufacturer will, for example, take the advice of a recognised adhesive's supplier when selecting the method of retention for the windscreen.

5.5.3 Competitors' products

The good designer should always be aware of his/her competitors' products. Many large organisations include product planning departments, where a very close check is kept on the opposition; however, such a department does not absolve the designer of his/her responsibility. In fact, such a department is likely to be a prime mover in creating a change to the PDS if a competitor adds a surprise feature that appears attractive to customers. The rapid addition of strengthening bars in car doors and seat belt tensioning devices are two such features that were added rapidly to a number of motor industry specifications.

The best way to gather information on a competitor's product is to buy an example, so that it may be tested and examined in great detail. The practicality of this obviously depends on the product and the industry: while it may be common practice in the car industry, the aircraft industry will be a different matter. If this option is not possible, existing users of competitors' products can be canvassed, exhibitions visited, and competitors' literature consulted.

5.5.4 Library searches

Libraries represent a vast store of information, particularly with the ILL (Inter-Library Loan) service where texts may be obtained even if physically not available within your local library.

The search procedure employed may well depend on your existing level of knowledge on the particular subject. If it is a new technical area to you, a good starting point may be via encyclopedias or dictionaries. These will provide a broad view of the subject matter.

The most up-to-date technical information is likely to be found in one of the vast number of specialist technical journals. Your search for relevant information will start with the appropriate abstracts. Many libraries are supplementing or replacing the conventional hard copy abstracts with computerised systems. The abstract can be supplied on a CD ROM which, with suitable software, provides a much more efficient method of information searching. For example, Compendex, supplied by Engineering Index, has in excess of 2 million abstracts on its database.

5.5.5 Patents

Yet another information source is available from the Patent Office. In case the budding designer should invent the perfect mousetrap, we should dwell for a few moments on the operation of patents.

Principle

When patents are raised with new designers they often imagine a system that will provide a restriction to their creativity, because an existing patent may limit the use of an idea. In fact, quite the reverse is normally true. The patent system was devised to encourage disclosure of information. It recognises that it is in the public interest that new manufacturing techniques and novel products are introduced, so that all may benefit from the inventor's expertise.

In fact, the only way an inventor can profit from his/her idea is by putting it into practice. This could be done by the inventor him/herself, whereby he or she could gain a short-term advantage over his/her competitors, at least until they managed to copy the idea. However, if this short-term advantage is insufficient for the inventor to recover the costs of developing the idea and make a profit, he or she may be discouraged from making further advances.

The principle of the patent system is that the originator (or owner) of the idea is granted a temporary monopoly, in return for disclosure of the idea. The inventor now has a clear period of time during which he/she may profit from it by either using it him/herself or by allowing others to use it in return for a fee (royalties).

When the Patent Office grants a patent it does not provide any automatic enforcement of the monopoly protection. It simply gives the owner of the patent (the patentee) the right to take legal action if he/she considers that this is in his/her best interests. The fact that a patent exists may be sufficient to deter some potential infringers, and may enable the patentee to gain some financial recompense without resorting to legal action.

Legal action should not be undertaken lightly. It can only be taken in the civil courts and may result in considerable costs. The infringer may not only deny the infringement, but may bring a counter-claim that challenges the validity of the plaintiff's patent. The success of such cases can often depend on how skilfully the patent specification was drawn up. The costs of drafting a patent can be high. The lone inventor should also consider the costs of defending the patent in court; a large organisation may decide that its greater spending power may win the day!

The point to remember is that possession of a patent only offers the right to defend the temporary monopoly in court.

How long does it last?

The maximum period is 20 years from the date when the application was filed. To keep it in force an annual fee has to be paid from the end of the fourth year.

What can be patented?

The Patent Act 1977 lays down certain conditions that must be fulfilled for an idea to be patentable.

- *Be new*: The patent application must be the first public disclosure of the invention. Any prior disclosure would prevent the granting of a patent. This covers publications, exhibitions, common usage, etc.
- *Innovative step*: If the potential invention would be obvious to someone with good knowledge of the subject, it would not be acceptable. It must involve a degree of innovation.
- *Industrial application*: The invention must be of a practical form, i.e. a device, material, industrial process, agricultural product, etc. It would not include an impractical process such as one that contravened known laws – a perpetual motion machine, for example.
- *Excluded inventions*: There are a number of areas that are not patentable, including intellectual ideas, aesthetic concepts, a discovery, a mathematical model, scientific theory, a computer program, methods of diagnosis, treatment or surgery.

How are patents obtained?

Although most industrialised countries operate patent schemes, at present it is not possible to obtain a world patent. The UK resident has three possible alternatives:

1. UK patent: This will provide protection in the UK only.
2. European Patent Convention (EPC): In general, to obtain protection abroad a separate application is required for each foreign country. If the appropriate fees are paid, a European application will provide protection in up to 12 European states.
3. Patent Cooperation Treaty (PCT): This enables protection to be granted in the UK and 39 other countries.

The application procedure is similar in all three cases and involves defining the invention carefully, publication of the patent application, and various searches to ensure that it qualifies to be patentable.

Pitfalls and benefits

Starting with the negative elements, as the patentee's monopoly can only be enforced in the civil courts, it can often prove difficult to counter the spending power of a wealthy company. By going for a patent you have to publish your invention, i.e. explain it to the world. In some cases better protection may be obtained by keeping the process secret.

On the positive side, patent documents are a large source of technical information. They started in 1449. Between 1916 and 1977 some 1.5 million specifications were published. These, along with those published by EPC and under PCT, are comprehensively classified and indexed, thus providing a readily accessible information source for the designer. The first page of the patent specification is an abstract. These abstracts are published weekly. Twenty-one UK libraries hold these pamphlets.

5.6 Idea generation

Now that we are clear in our minds about the problem that requires solving, all that remains is to generate a suitable solution. History has over the years produced a number of eminent solution generators such as Leonardo da Vinci, Alec Issigonis, Frank Whittle, etc. It may however be difficult to emulate the thought processes of these individuals as they are very exceptional people. What is needed is a system, or at least guidelines to help us innovate.

5.6.1 Common obstacles

We are all capable of original thought, but are also handicapped by a number of obstacles. By identifying them we can at least attempt to avoid them.

Tradition

Continuing with a well-proven method or design philosophy is a safe way forward in the short term. However, the innovator must look for alternative solutions; this is the only way to progress. As an example, children's push-chairs for many years were heavy, cumbersome, and could only be folded to a very limited degree. The McLaren buggy introduced in the early 1970s was a dramatic step away from tradition. Aluminium alloy was used for the frame, the wheels were much smaller than had been used before, and a deckchair principle was used for the seat. A conscious effort had been made to ignore tradition, or current practice, and to concentrate on the best way to solve the identified problems.

NIH syndrome

The Not Invented Here syndrome. Often individuals can be reluctant to adopt ideas from competing individuals or departments. Care must be taken to recognise this as a human failing and hence avoid the problem. Amazing though it may seem, some organisations also suffer from this syndrome and can be reluctant to adopt competitors' solutions. The author has worked with one company that put a great deal of effort into finding alternative solutions to problems simply to avoid 'losing face' by adopting someone else's solution.

Most successful organisations are now more open to using ideas, regardless of their source. Ford, along with other car manufacturers, buy competitors' products and analyse them in great detail, in the search for engineering solutions. Subsequent models often incorporate significant lessons learnt in this way. It was no coincidence that Ford's Mark 3 Cortina sported an overhead cam engine, wishbone front suspension, and four bar link rear suspension, despite the fact that none of these followed current Ford practice. These features, along with similar styling, all appeared some three years earlier on a General Motors product. None of this is to detract from Ford's actions: they produced a very successful product.

Standard practices in industry

This is similar to the problems of tradition. Prior to free-thinking Alec Issigonis, almost all cars had an in-line engine mounted at the front of the vehicle, driving the rear wheels via an in-line gearbox, prop shaft and differential. His solution for the Mini was to turn the engine through 90°, fit the gearbox and differential in the sump and drive the front wheels.

The above potential obstacles to innovation apply equally well to manufacturing processes as they do to the final product. For example, the methods used to separate the metal and rubber elements of used car tyres must have required some lateral thinking. The tyres are cooled, so that the rubber becomes brittle, vibrated to break the rubber into pieces, and then passed through a magnetic field to extract the metal components.

5.6.2 Improving creativity

Our creative ability depends largely on two properties: the data that is available and the methods used to access the data. The process is summarised in Figure 5.4.

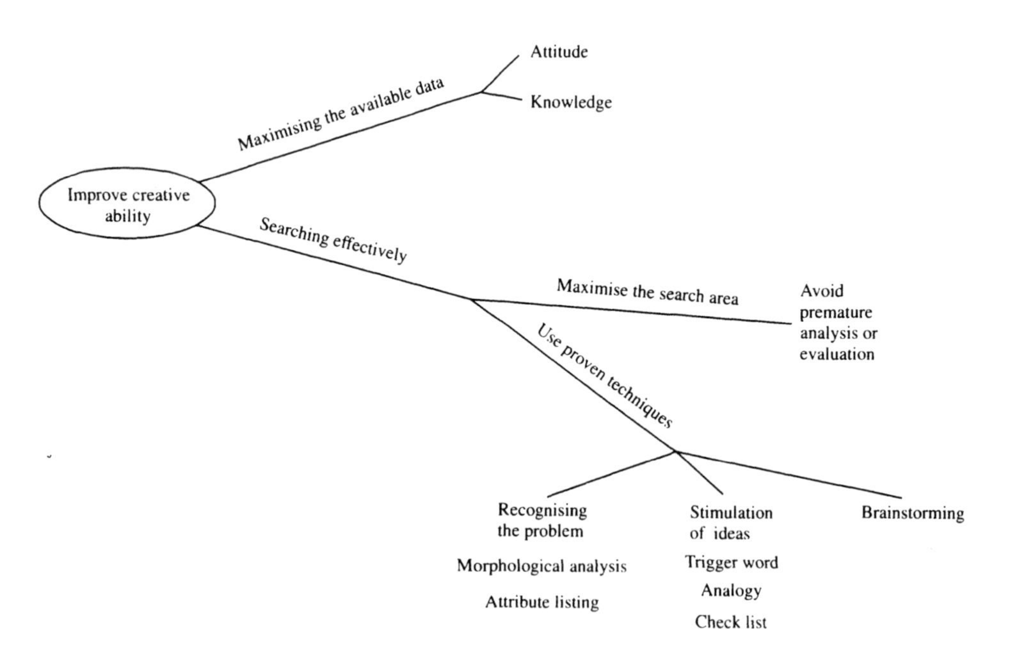

Figure 5.4 *The creative process*

Data store

As the designer attempts to create, the data we are concerned with is that which is in his or her mind at the time, rather than the contents of a filing cabinet or a computer database. This information will be a mix of the details gathered as the PDS was created, the general information available within the organisation, and the past experiences of the designer. The richness of the material will in part depend on the designers' general attitude: the more open and enquiring the individual, the more natural will be the creative process. In general the prospective innovator forever needs to increase his/her store of knowledge, and for a specific project needs to gather as much background data as possible.

Search techniques

Having the data is only of use if it can be accessed when needed. The techniques used need to maximise the search area. As Figure 5.4 shows, for this to happen, care must be taken to avoid inadvertently raising any barriers that might limit the scope of the search. Such barriers can occur by the premature rejection of possibilities, or by sidetracking the search in attempting to solve the detail rather than continuing to search for further potential solutions.

Having opened as many doors as possible in our storage area, we now need to sample as effectively as possible. We will now discuss some suitable techniques.

Recognising the problem

Although it may seem obvious, the problem needs to be recognised before solutions can be generated. There are a number of proven techniques for helping this recognition by breaking down, redefining, or rearranging the problem so that the number of potential solutions is maximised. We will discuss two such techniques.

- *Attribute listing*: This is a technique more appropriate for existing products. The concept of the solution is known already, but there is a need for some modification or improvement. The approach is to list the characteristics of the device or problem. The object is to focus the designer's attention on the relevant aspects. For example, consider a motor lawnmower:

  1. Cutting blade
  2. Power unit
  3. Roller
  4. Energy source
  5. Control system

Each of these characteristics can be further subdivided:

Control system:
(a) Accelerator
(b) Clutch
(c) Blade only control
(d) Control location

The subdivision continues as far as the complexity of the product requires. The process will help clarify the essential characteristics of the problem.

- *Morphological analysis*: This a procedure which involves analysing the problem by identifying the primary parameters that are involved and then considering alternative methods of satisfying them. The results are presented in a matrix form. The available combinations in effect represent a large number of possible solutions. Although many will be unacceptable, remember that the name of the game at this stage is to maximise the number of possibilities.

  Take as an example the lawnmower, assuming this time that we are able to start our design from scratch. Table 5.1 allows a large number of potential solutions to be generated by combining possible concepts with each of the features. Of course, many of these solutions will be impractical, but remember that the object at this stage is to maximise the number of ideas. The morphological analysis prompts us to consider one aspect of the problem at a time and generate ideas for each aspect. This leads to the question of idea generation.

Table 5.1 *Potential solutions*

Feature	*Possible concepts*
Power source	Mains electric; manual; diesel; petrol; coiled spring; hydraulic; battery
Ground support	Wheels; rollers; air cushion; skids; legs
Cutting mechanism	Scissor; rotary blade; cylinder; flame thrower
Propulsion	Manual; wheel; tracks; jet; propeller
Direction control	Steering; adjustable track speed; side thrust
Structure	Space frame; composite monocoque; steel chassis; integral engine chassis

Stimulation of ideas

Many people mistakenly believe that only the talented few are able to generate new ideas. While accepting that some are exceptionally gifted, a combination of the correct frame of mind and the application of some of the following aids should enable most people to be able to generate some novel solutions.

- *Trigger word*: Concepts or ideas can be thought of as being stored in a number of separate compartments in the brain. The problem is in accessing these areas. This technique relies on a variety of different concepts that can spring to mind when a series of words having similar meanings are recalled. This is analogous to using keys to similar compartments in the brain. A thesaurus can prove useful.

  For example, consider methods of propulsion for the lawnmower. Taking move/motion from the thesaurus we get: stir, jerk, pluck, budge, shift, manhandle, trundle, roll, wheel, shove, tug, pull, draw, rail, wave, dart, hover, fly, glide, ski, slide.

- *Analogy*: This operates in a similar way to the trigger word, the difference being that similar situations, rather than words, are sought. A good example is portrayed in the film *The Dambusters*, where Barnes Wallace is searching for a way of controlling accurately the height above ground of a very low flying aircraft. As he relaxes one evening at the theatre the sight of the spotlights playing on the actors triggers the possibility that focusing two lights from the aircraft together on the ground would be an accurate but practical method of indicating the plane's height to the pilot.

- *Checklist*: Sometimes a checklist can help, particularly in the area of problem identification, where working through a defined pattern can help with the listing of attributes.

These aids are all techniques that can be used by an individual. Often a group activity is of greater benefit because there is a greater available memory store and the search techniques are helped by the interaction between the individuals:

Brainstorming

The idea of brainstorming is to have a group of people 'throwing ideas into the pot' with the basic rule that no idea is rejected (or ridiculed) no matter how obscure. The aim is to maximise the quantity of ideas generated. It can help if the ideas are recorded in a way that allows them to be visible to the participants (e.g. on a wall

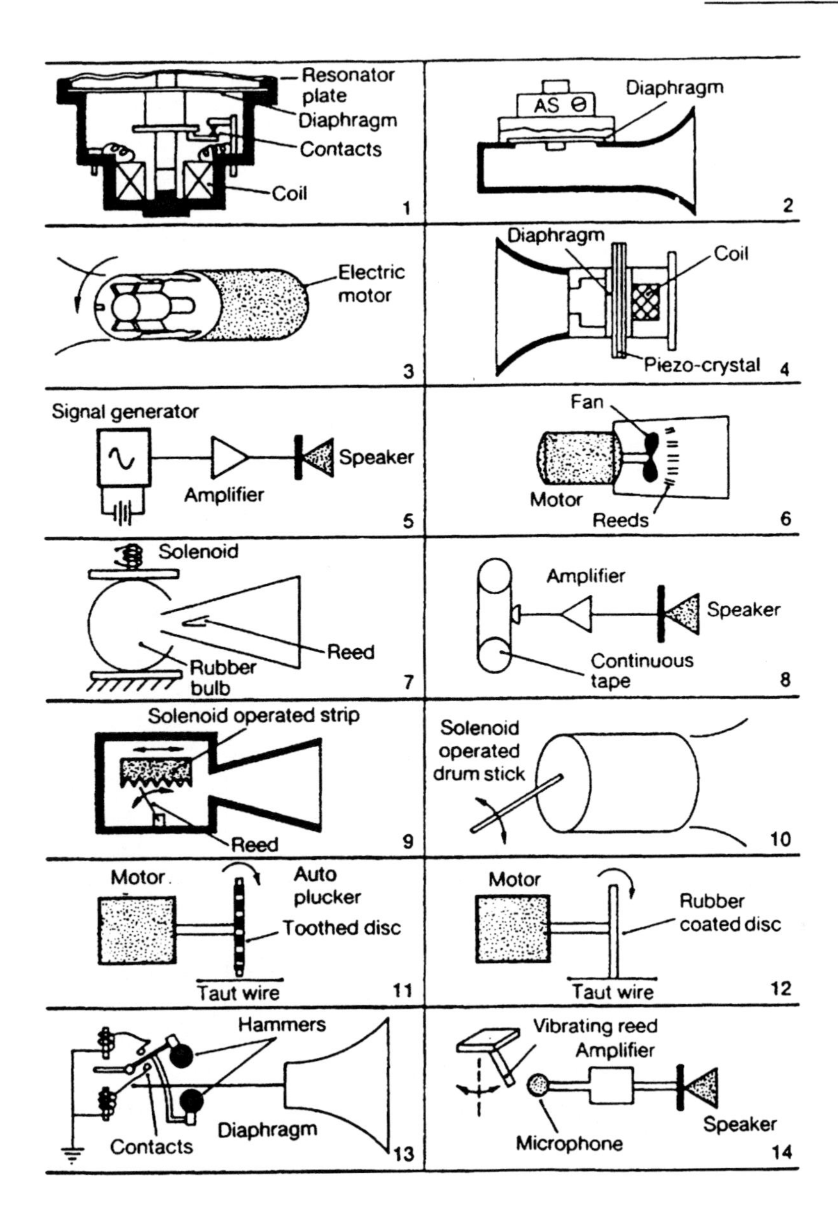

Figure 5.5 *Horn concepts*
Source: Institution of Production Engineers, *A Guide to Design for Production* (London, 1984).

board) as this will help the suggestion of various combinations of ideas. This technique is called brainstorming.

No judgements are made until after session. Most of the ideas will end up being rejected, but often the final solution can be a surprise.

The output from this solution generation activity should be a number of potential concepts, as shown in Figure 5.5. These are in response to the task of generating embryonic solutions for an audible means of warning of approach of a motor car.

5.7 Idea evaluation

So far we have concentrated on generating the maximum number of potential solutions; quantity has been more important than quality. The aim now is to separate the good ideas from the not so good. The target is to do this as objectively as possible. A number of stages may be required.

5.7.1 Initial screening

The first step must be to eliminate those ideas that stand no chance at all. A brainstorming session, for example, may have resulted in quite a large number of ideas that are doubtful. This is to be expected. Once the session has been completed, the obviously unsuitable ideas can be discarded, using common sense. (These concepts were not a waste of time; they have served their purpose in helping the free flow of ideas.)

The remaining concepts now need to be put through a more formal screening process to identify those that are worth developing further. Each concept must pass the critical TEST to proceed further. A check is made against each of the following headings:

- **T**echnology
- **E**conomic
- **S**afety
- **T**ime

To qualify for further work a concept must have a positive reply to each of the TEST headings. A negative reply to one or more headings will disqualify the concept from proceeding further.

Technology: Is the solution likely to work with today's technology? Does it obey well-established laws of physics?

Economic: Is the cost of development and associated tooling likely to be affordable? Is the final cost of the end product likely to be acceptable to the customer?

Safety: Will the solution satisfy existing safety standards? Is it a socially acceptable concept? Are we confident that any potential legal problems can be overcome?

Time: Is the estimate of development time compatible with the market needs? (This is particularly important for either seasonal, or rapidly developing markets.)

A chart listing the concepts against these factors is a good way of performing this TEST.

The potential solutions that survive this stage now require more in-depth analysis. This involves developing the concept further by examining its practicality, complexity, potential cost, and reliability.

5.7.2 Concept comparison

Continual checks must made be to ensure that the concept meets the requirements of the PDS. Remember that the PDS is not carved in tablets of stone: it is a dynamic document that can be, and often is, updated. The updating must, of course, involve consultation and agreement of all the parties involved in the generation of the PDS. Following this deeper investigation, we now need to decide which of the concepts should proceed further.

The simplest way to do this is by comparison on a chart; this time the criteria listed vertically are based on the design requirements. The PDS can provide the basis for this list. However, if we use the GO/NO GO system all the ideas are likely to pass equally, since they have been developed in line with or resulted in a modification to the PDS.

One method is to take one of the concepts as a datum and then compare each of the others to it by placing in each box an S, + or –, representing:

S same or equally as good as the datum

\+ better than the datum

– worse than the datum

If each column in the chart represents a concept, the number of Ss, +s and –s can be totalled.

Table 5.2 shows how the horn concepts in Figure 5.5 could be evaluated with this method. In this instance, Concept number 5 appears to offer considerable benefits over the original design.

An alternative would be to draw up a similar chart, but instead of comparing each concept to that of a datum, to give a score (out of 10) for each criterion, or attribute.

Table 5.2 *Concept evaluation*

Criteria	*Concept*													
Concept No.	1	2	3	4	5	6	7	8	9	10	11	12	13	14
Ease of achieving 105–125 DbA		S	–	+	–	+	+	–	–	–	–	S	+	
Ease of achieving 2000–5000 Hz		S	S	N	+	S	S	+	S	–	–	–	S	+
Resistance to corrosion, erosion and water		–	–	O	S	–	–	S	–	+	–	–	–	S
Resistance to vibration, shock, acceleration	D	S	–	T	S	–	S	–	–	S	–	–	–	–
Resistance to temperature	A	S	–		S	–	–	–	S	S	–	–	S	S
Response time	T	S	–		+	–	–	–	–	S	–	–	–	–
Complexity: number of stages	U	–	+	E	S	+	+	–	–	–	+	+	–	–
Power consumption	M	–	–	V	+	–	–	+	–	–	–	–	S	+
Ease of maintenance		S	+	A	+	+	+	–	–	S	+	+	S	–
Weight		–	–	L	+	–	–	–	S	–	–	–	–	+
Size		–	–	U	–	–	–	–	–	–	–	–	–	–
Number of parts		S	S	A	+	S	S	–	–	+	–	–	S	–
Life in service		S	–	T	+	–	S	–	–	–	–	–	–	–
Manufacturing cost		–	S	E	–	+	+	–	–	S	–	–	–	–
Ease of installation		S	S	D	S	S	+–	S	–	–	–	S	–	–
Shelf life		S	S		S	S	–	–	S	S	S	S	S	S
		0 + 6 –	2 + 9 –		8 + 1 –	3 + 9 –	5 + 7 –	3 + 12 –	0 + 11 –	2 + 8 –	2 + 13 –	2 + 13 –	0 + 8 –	4+ 9 –

Source: Institution of Production Engineers, *A Guide to Design for Production* (London, 1984).

5.7.3 Criteria weighting

There is, however, a major shortcoming with either of these methods in that each of the criteria is given an equal weighting (or level of importance). It is therefore quite possible to justify one's subjective desires by selecting criteria that favour one solution more than another. To be more realistic, a system is needed to add some more objectivity to the process by establishing the relative level of importance for each of the criteria.

It can be difficult establishing a ranking order when a large number of criteria are concerned, without attempting some form of simplification first. A possible method is to compare features two at a time. It is normally easy to decide which is the more important of the two alternatives. If a chart is generated recording the decisions, a ranking order eventually becomes obvious. As an example, imagine that we are attempting to rank four criteria: A, B, C and D.

In Table 5.3 the criteria listed horizontally are compared with each of the criteria listed vertically. The relative importance of the features is given by a score of 0, 1 or 2, representing less, equally, or more important. Taking each of the features in turn, column 1 shows A is less vital than B or D, but is equally important to C. Column 2 shows B to be more important than C and so on.

The total row indicates the ranking order of importance. In this example C is the most important, with B and D ranking equal second, and A last.

Table 5.3

	A	B	C	D
A	\	1	2	1
B	0	\	1	0
C	1	2	\	2
D	0	0	2	\
Total	1	3	5	3

This gives us an order of relative importance for each of the criteria, but does not provide any absolute quantification. This can only be added by the designer and as such will be very subjective. The normal method would be to use a scale from 1 to 10, representing respectively at each end of the scale the least and most important criteria. The resulting numbers for each criteria represent the weighting for that criteria. The score awarded to each concept is then multiplied by the weighting factor before being totalled.

Table 5.4 demonstrates how the ratings are applied. In this case the object of the design exercise was a machining cell requiring a number of sophisticated turning machines. The evaluation shown is a small segment of the process used to select the most suitable machine tools. The complexity of the criteria involved has resulted in the table being divided into subsections of each of the criteria. In this instance a rating factor of greater than 10 has been used. As this factor simply relates the relative importance of each of the criteria, this is quite acceptable.

We have covered a number of methods of comparing alternative solutions to a problem. The actual method used and the number of stages will depend on the type of design problem. The point to keep in mind in all cases is the need to progress from a number of potential solutions to the most suitable one in the most objective manner possible.

The result of the idea evaluation will be to discard a number of early ideas that the designer may have known in his heart of hearts were non-starters. These concepts were not a waste of time, for at least two good reasons.

Firstly, they will have helped the free flow of ideas at the generation stage. Unless the idea floodgates are kept wide open for a time the chances of any true innovation will be severely reduced.

Secondly, no matter how good a concept is, the designer will have to 'sell' the idea to someone before it can proceed. This may be to his/her colleagues, to the company directors, or even to the bank manager. If his/her sales pitch can touch on a wide variety of concepts that were con-

Table 5.4

	Rating	*Option 1*	*Option 2*	*Option 3*	*Option 4*
Main drive control, manual	4	5	5	5	4
Axis control, manual	6	5	5	5	3
Leadscrew protection	8	4	5	5	5
Main motor power	10	4	4	5	4
Axis drive thrust	10	5	2	3	4
Main drive torque	10	5	2	3	3
Rapid traverse	4	3	5	4	4
Power subtotal		204	160	186	182
Max turned col free diameter	11	3	1	5	5
Swing over traverse slide	11	2	3	3	4
Max radial drilled OD	11	4	1	3	5
Travel in X axis	8	2	1	4	5
Max chuck diameter	5	5	2	5	5
Capacity subtotal		140	73	178	219

sidered and evaluated objectively, the final concept will have its credibility increased.

5.8 Summary

The starting point for any design is the PDS, which is needed to define the problem and the constraints within which the designer must work. It is an evolutionary document and its contents must be produced with the full agreement of all involved in the product introduction process.

The amount of available relevant background information needs to be maximised. Potential sources are standards, specialist component manufacturers, competitors' products, libraries, and patents.

Solutions should now be considered. The first step is to generate the maximum number of potential solutions, initially ignoring their practicality. Recognising common obstacles to generating ideas, and using techniques such as brainstorming, morphological analysis, analogy, etc. will assist this process.

The ideas then need to be evaluated in as objective a manner as possible. This often requires a number of stages. The first preliminary examination (TEST) should eliminate any impractical ideas. The remaining concepts may need a little further development before sufficient information is available to evaluate them. This is normally done with a rating table, and can take place in stages. At each stage the number of potential concepts is reduced. Between the stages the concepts are developed further (i.e. their viability is investigated in greater depth). The object is to move from an initial set of, say, 20 ideas to 2 or 3 which will be explored in depth. The final evaluation will select the best of these.

5.9 Questions

Now check your understanding of the design process by trying to answer the following questions. Refer to the relevant section in the chapter (shown in brackets) if you have difficulty with any of them.

1. There are eight steps in the design process: what are they? (5.2)

2. What is meant by the term 'brainstorming'? List the fundamental rules of this process.(5.6)

3. There are many sources of information for the designer, including existing standards, component suppliers and through library searches. Name two other important sources of information. (5.5)

4. When evaluating alternative concepts we often need to apply a rating or weighting factor to each of the criteria. Why? (5.7)

5. There are five main areas that provide the boundary conditions for the designer in the Product Design Specification. Name the five areas. (5.3)

6. What is the definition of the design process? (5.2)

7. A product design specification contains the following statement: *The product must be as light as possible.* Explain what is wrong with this statement show how it should be worded. (5.3)

8. Why are patents of benefit to society, and what rights do they grant to the holder of the patent? (5.5)

9. What is the difference between product life, service life, and shelf life? How does the term 'duty cycle' relate to these expressions? (5.4)

10. Name three of the common obstacles that need to be overcome to avoid the creativity process being handicapped. (5.6)

Part II
The Basics of Manufacture

6 The Manufacturing Process

6.1 Introduction

A good starting point in a chapter with this title is to ask what exactly is meant by the term 'manufacturing'. In essence, the manufacturing process is the means by which materials are converted into goods. We could therefore also define the manufacturing process as one in which value is added to materials as they pass through the process. The absolute measure of this added value is, of course, determined by what the customer is willing to pay for the final product. The chosen manufacturing process and its detailed application will determine the actual cost of adding the value. Obviously, this must be such that the manufactured cost (material cost + cost of adding value) is less than that paid by the customer by a sufficient amount for the company to thrive. Does this lead to a definition of the manufacturing process as one in which materials are converted into goods in the cheapest possible way? Well, almost. There is a danger with this view that there is an incentive to deliver shoddy, unreliable goods, albeit at a low manufacturing cost, to the customer, who will then turn elsewhere for his supplies. The process must be capable of converting materials into goods in such a way that the needs of both the producing organisation and the consumer are satisfied on a continuing basis. (See the definition of the Design Process in Chapter 5.) This chapter will start with some material considerations and then introduce the general types of process available.

6.2 Material selection

6.2.1 Importance of materials

If the manufacturing process is essentially a material modification process, we first need to know what material we are dealing with. In fact, selecting the correct material is a task equal in importance to either the design or manufacturing engineering functions. There are many examples from the past that support this. Early airships used hydrogen for buoyancy with disastrous results when the inevitable fire struck the R101 in 1930. The largest single change in the modern airship design is a material change, the use of the much safer helium. Perhaps the best example of material selection is at the start of a formula 1 grand prix where, in dry weather, all cars will be running slicks (treadless tyres). These tyres are manufactured using different rubber compounds, giving alternative compromises between grip, life and rolling resistance. Each team will effectively select the tyre material to suit the driver, car and track conditions for that race. The selection will be based upon trials in the limited time available, and proved during the race.

Often a material change will have a number of related effects. Car design in recent years has concentrated on weight reduction. Lower weight in the essential elements of the vehicle allow improved performance and economy. Many cars now use aluminium alloy castings for the engine block rather than cast iron with, usually, a significant weight reduction. Secondary effects are that aluminium tends to transit noise and vibration whereas cast iron is better at absorbing them. Attention must therefore be paid to greater cabin insulation, redesign of engine mounts etc. Each design finds its own compromise. The 1.5 litre Peugeot/Citroen diesel engine introduced in 1994 uses a cast iron block (rather than the aluminium alloy of its predecessor) with apparently significant benefits in refinement.

In many cases all the consequences of a material change are not apparent at the time of the design. Warship construction during the 1970s made extensive use of aluminium and magnesium alloys in the superstructure. The weight savings over conventional materials allowed much greater design freedom for building features that would have previously resulted in a top-heavy, unstable vessel. It was not until 1982 and the Falklands War, that the result of fire on such materials caused a reappraisal of their benefits. When material changes are made for environmental reasons, cause and effect are frequently very difficult to predict. Patients suffering from Alzheimer's disease tend to show a surplus of aluminium products in their bodies. Does this mean aluminium cooking vessels should be banned? Lead products have been shown to retard the development of young children. Leaded petrol is one source of these products, so car designs have evolved to use unleaded petrol and clean their exhaust gases with a catalytic convertor. Lead emissions from motor vehicles have undoubtedly been reduced, but at the cost of increased emissions in terms of CO_2 (global warming) and benzine products (carcinogen). Is this an overall environmental benefit?

6.2.2 Factors influencing selection

Material selection is likely to be a compromise between many factors – cost, product functionality, manufacturing process, environmental effect, etc. It will be an iterative process involving knowledge of design, materials and manufacture. The following points summarise some of main aspects (see also Appendix B9).

Physical characteristics

- *Strength*: For most metals this is given as the Yield Stress or the UTS (Ultimate Tensile Stress). Brittle materials will tend to have similar values for these characteristics. (See Chapter 14, section 6 for more details).
- *Stiffness*: The slope of the stress–strain curve is called Young's Modulus (E), or the Modulus of elasticity. This indicates the stiffness of the material. The value varies very little between different types of steels.
- *Weight*: This is often an important feature of the final design in determining overall performance. Check the density of the material used but remember there is likely to be a compromise between strength, weight and cost.
- *Corrosion*: The PDS should define the environment under which the material will operate. This influences both the selection of material and its subsequent finishing. Care must also be taken to avoid electrolytic corrosion by poor combinations of materials (see Chapter 14, section 14.7 and Appendix B.8).
- *Thermal properties*: Commonly quoted characteristics are the coefficient of linear expansion for metals, and suitable maximum operating temperatures for non-metals. Thermal conductivity may also be important.
- *Electrical properties*: Often the designer simply needs to know if the material is a conductor or not.

Manufacturing process

All manufacturing processes will not be compatible with all materials – for example, a plastic is unlikely to be suitable for forging! This may not be quite so obvious when selecting steel, where corrosion resistance needs may direct the choice towards 316 stainless steel. This may prove difficult to machine on, say, a multi-spindle automatic lathe where the optimum cutting speed is unlikely to be achieved for all diameters (see Chapter 7, section 7.2.10). Increasing the sulphur content of the material will go some way to relieving the machining problems, but it may result in more care being needed if the component is to be welded. The process and the material must be considered together.

Availability is often another influence. Specifying commonly available forms will normally tend to result in favourable costs and reduced lead times. They will be produced in higher volumes by the supplier, and their ready availability should help minimise any investment in raw-material stocks. Always check with standards and suppliers on the availability of both the form selected (bar, sheet, etc.) and the material specification. Try to avoid specifying anything not readily available.

The number required will affect the viable capital investment, which will in turn influence the process choice, which then colours the selection of suitable materials. We always return to the need for the most suitable compromise. This must meet, but not exceed, all the requirements. (Exceeding the needs implies a cost-inefficient specification.)

Other manufacturing considerations include: safety (fire hazards, hazardous chemicals), ease of quality control, need to involve subcontract operations, finishing processes, hardening, plating, etc.

Environment

Traditionally this has meant considering the conditions that the end product must survive during its service life. Operating temperatures, shock loads, corrosive environment, service conditions, etc. are all environmental factors that the chosen material must be able to withstand.

Today we also need to consider wider environmental aspects, the effect on society and the planet. Energy consumption during manufacture, of both the product and its raw materials, should be a consideration. Are the materials able to be recycled? One of the problems with recycling plastics components used on cars is the variety of materials used. This leads to difficulty in identifying each type of plastic, separating and processing them. A partial solution might be to reduce the variety of plastics used. Can processes requiring the disposal of noxious chemicals be avoided? Volkswagen now use water-based paints for their cars to avoid such chemicals.

Cost

Cost is of vital importance, but as it depends on a great many variables is difficult to quantify in general terms. Always remember that the overall objective is to minimise the total product cost, which will be affected by design, manufacturing and materials choices.

6.3 Manufacturing processes

To help in the quest for repeatable quality at minimum overall cost, the engineer has a wide range of manufacturing processes to select from. In general terms these processes can be used to effect one or more of the following changes:

- Modify the shape or form of the material
- Modify the physical properties of the material, but retain the shape
- Join one or more pieces of (modified) material together.

6.3.1 Shape modifying processes

In everyday terms, materials can be cut, squashed, stretched, heated and stretched, melted and poured into a mould, or hammered, to describe but a few possible processes. In engineering terms the methods of changing the shape will fall into one of two prime categories: those which involve material removal, and those which do not. Figure 6.1 shows some of the commonly used processes.

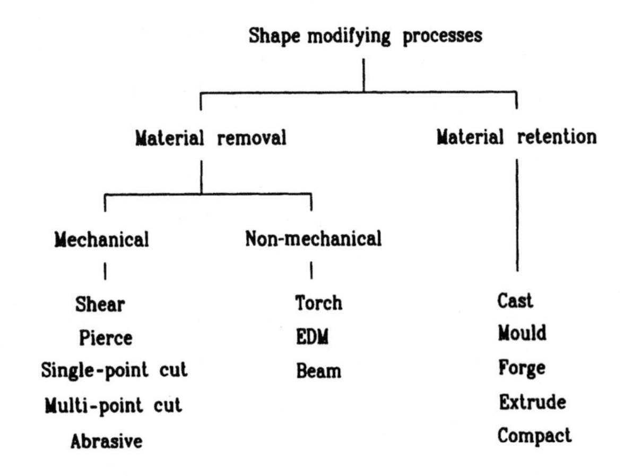

Figure 6.1 *Shape modifying processes*

Extrusion is an example of a typical process that avoids the need for material removal (see Figure 6.2). In essence, a ram is used to force the work through a hole in a die. The material section, while passing through the hole, is modified to the shape of the hole. Aluminium window frames are normally made from material extruded to create a suitable section. These processes will be discussed in Chapter 8.

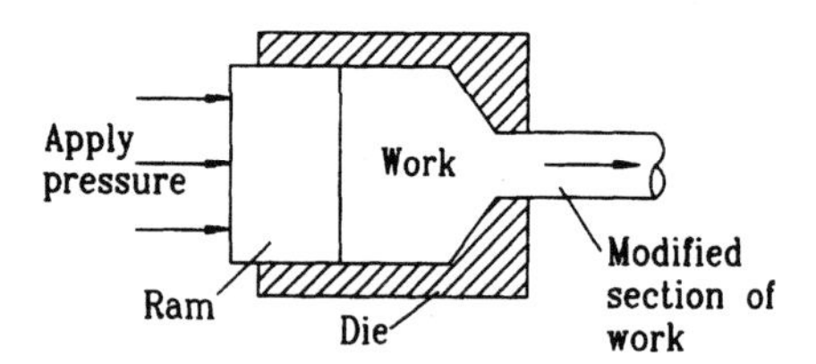

Figure 6.2 *Extrusion*

Material can be removed by mechanical or other means. Plate, for example, can be cut using a high energy beam. This may be in the form of a high temperature flame or a laser beam. The relative movement between the beam and the work is controlled so that localised melting performs the cutting operation (see Figure 6.3a).

An alternative form of cutting is to employ a mechanical means such as shearing (see Figure 6.3b). The mechanism used involves a punch and a die. The die is fixed and the punch is able to move vertically, constrained so that there is minimal clearance between the cutting edge of the punch and the die. The punch is raised, the work placed on the die and the punch swiftly lowered so that the work is sheared. The operation is exactly the same, in principle, as that performed by a pair of scissors. In fact handshears can be used to cut thin sheet metal.

Whilst these methods of material removal would be appropriate for sheet or plate they would not be suitable for the many other desired shapes. One of the most widely used methods is via chip removal. Essentially a mechanical chisel is used to chip fragments of the workpiece away. Put this way it sounds very crude but provided the cutting tool is much harder than the material

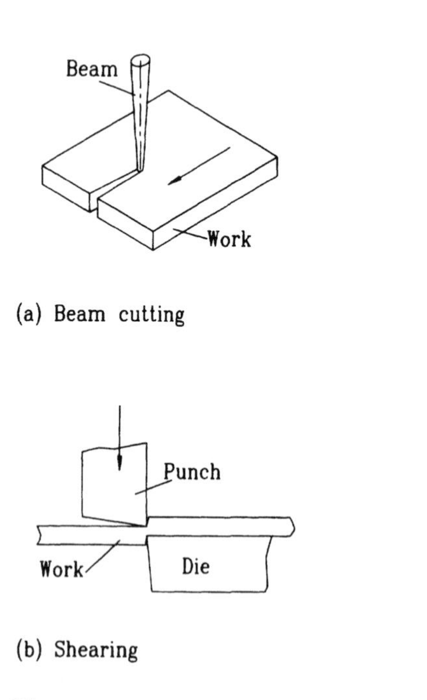

Figure 6.3 *Plate cutting process*

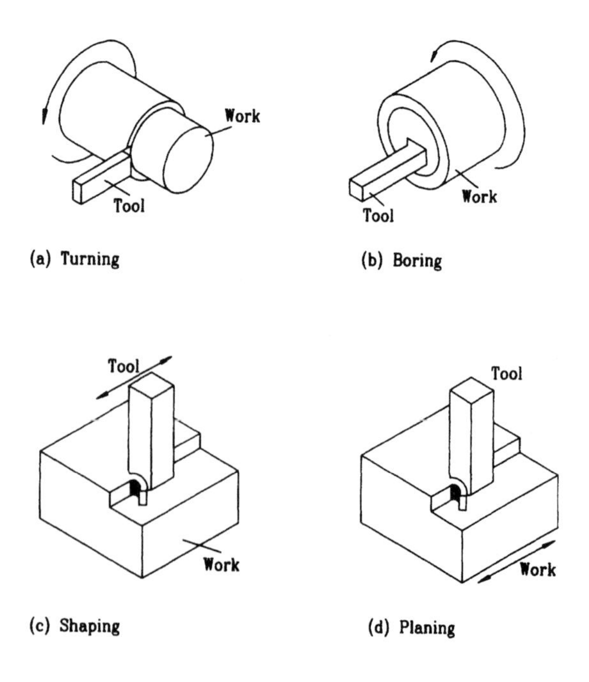

Figure 6.4 *Single point cutting processes*

it is cutting, the depth and speed of the cut is closely controlled and vibration is minimised; accurate dimensions and good surface finishes can be achieved. There are a number of these basic 'machining' processes that you need to become familiar with. The tool used can be a single-point tool (i.e. one with a single cutting edge), a multi-point tool, or an abrasive wheel.

Single-point cutting processes are shown schematically in Figure 6.4. Turning (a) is where the work is held in a chuck and rotated. The cutting tool is slowly moved towards the rotating workpiece so that on contact it scrapes away the surface, forming chips of unwanted material. Boring (b) is a similar process except that the surface produced is an internal diameter. With either of these processes, controlling the tool movements in both the radial and axial directions can generate a variety of different surfaces, such as threads or tapers.

Planing or shaping involves using the single point tool to create a flat surface. If the tool moves and the work is stationary it is called shaping (c). If the reverse is the case it is called planing (d).

Multi-point cutting processes use a tool with more than one cutting edge. A saw blade, for example, has a number of cutting points. Milling is a process where the tool rotates and the work is stationary. In fact, the work is normally fixed to a table that moves slowly across the milling cutter which is rotated, often at high speed. The cutting edges on the tool can be arranged to produce end or slab milling as in (a) and (b) in Figure 6.5. In either case the work can be horizontal or vertical.

Broaching, (c) in Figure 6.5, is the operation used to create non-circular holes. Drilling (e) is a much more commonly used operation that can be performed on a number of different pieces of equipment – lathes and milling machines as well as drilling machines. If the diameter of the drilled hole has to be controlled tightly, the hole may need to be reamed (f) after the drilling operation.

Finally, in this collection of material removal processes, we also have abrasive machining. There are a number of processes in this category but the most common is grinding, two types of which are shown schematically in Figure 6.6 overleaf. In Chapter 7 we will explore the cutting process further.

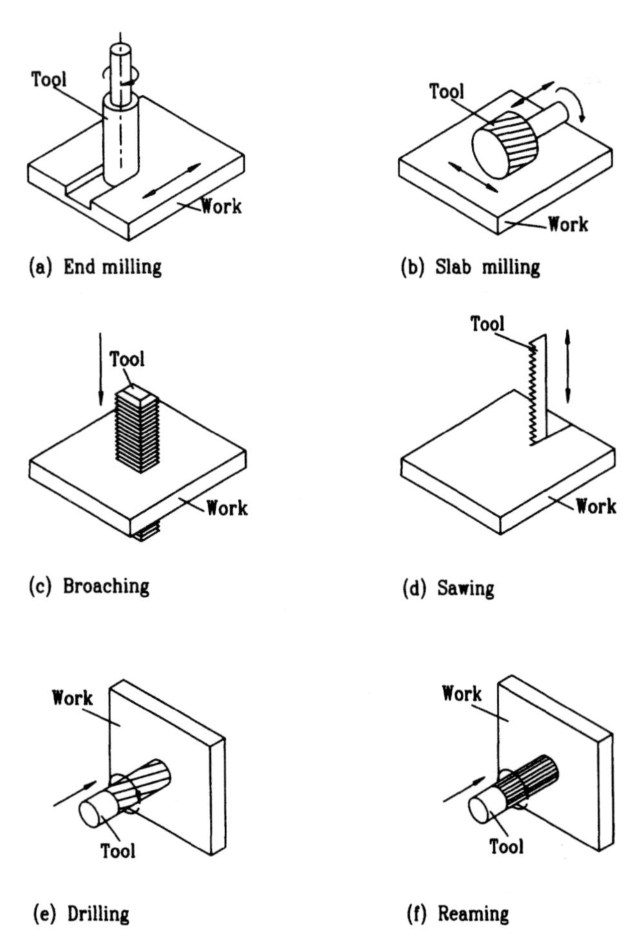

Figure 6.5 *Multi-point machining*

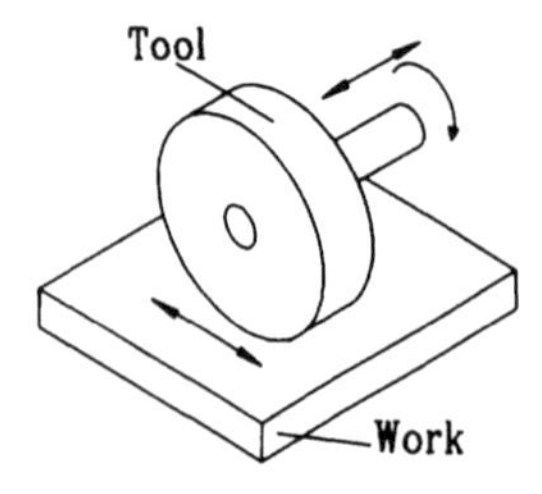

(a) Surface grinding

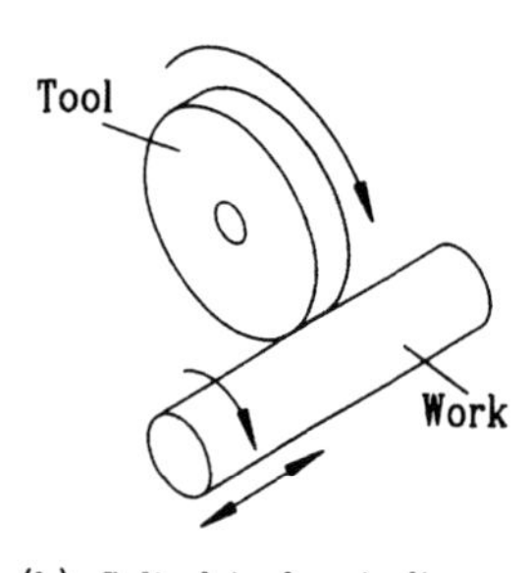

(b) Cylindrical grinding

Figure 6.6 *Abrasive wheel*

6.3.2 Property modification

The selection of a particular material for a component will normally be a compromise between a number of different factors. These can include availability, cost, ductility, hardness, corrosion resistance, and available manufacturing processes, to name but a few. Sometimes there is a need for the properties of the material to vary at different points in the material. For example, a common hand-held chisel will require a hard material at the blade end that is capable of being sharpened to a fine edge. The remainder of the stem does not need to be so hard: in fact, hardness may be a positive disadvantage as this would tend to make the stem more brittle. There are two solutions open to the engineer. One would be to use two materials (hard for the blade and more ductile for the stem) and join the two components. Alternatively, the stem could be made from a suitably ductile material and the blade area locally hardened by using an appropriate hardening process.

Figure 6.7 summarises the processes, by separating them into two categories: heat treatment, and surface finishing. Heat treatment can be used to harden the material, as in the case of the chisel, or to relieve stresses induced by either one of the shape changing or joining processes. There are also a number of other reasons for using heat treatment, such as glazing, or curing.

Surface finishing covers a range of processes where the surface is treated mechanically or chemically so that it is more suitable for the product's final use. These treatments can affect properties such as the ability to bond to other components, wear resistance, surface friction, corrosion resistance, etc. Strictly speaking, coating is the application of another material to the surface of the component, but as the coating layer is so thin that it does not affect the shape of the component significantly we can consider it as a surface treatment. All these processes will be discussed further in Chapter 10.

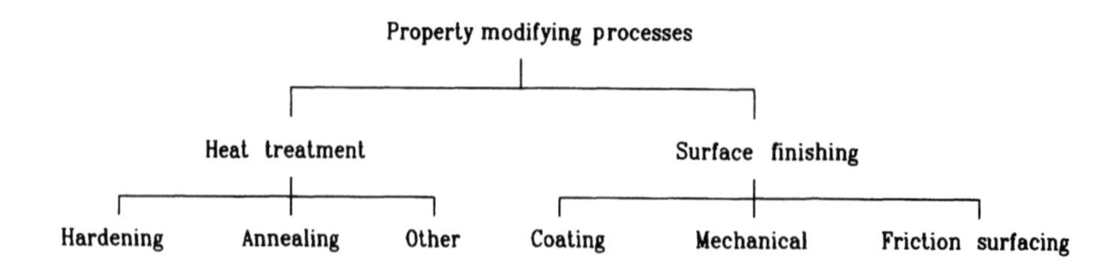

Figure 6.7 *Property modifying process*

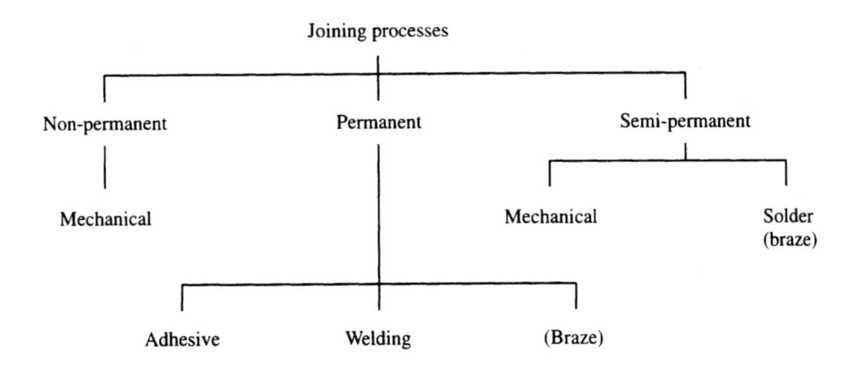

Figure 6.8 *Joining processes*

6.3.3 Joining processes

The third major category involves assembling materials or components to create the product. As Figure 6.8 shows, joints can be categorised by degree of permanency. A bolted joint, for example, would be non-permanent (essential in some cases to allow access for servicing or repair), whereas a welded joint, once made, would normally be expected to remain so for the life of the product. There are also a number of joint types that fall between these two classifications such as pop riveting and soldering. In this type of case, although the joint is permanent, it can be broken relatively easily and remade to effect a repair. This category we will call semi-permanent. Note that brazing falls into two groups, because although it would normally be termed a permanent joint, it could, with heat, be undone, possibly without damaging the component. All these processes will be discussed further in Chapter 9.

6.4 Manufacturing system

Many of these manufacturing processes will involve a number of stages. Individual components may undergo several manufacturing processes, and to further complicate the issue most products will be produced from a number of separate components. Some form of manufacturing system is needed to control the flow of work in an efficient manner. The need for such a system is further enhanced by the fact that most manufacturing organisations produce a range or variety of products, so each process may have to accommodate a variety of components. The manufacturing system must control the flow of material and components from one area to the next in such a way that bottlenecks (traffic jams) are avoided and ensure that all the items needed for an assembly arrive at the assembly point at the correct time. Chapter 16 will discuss in some detail the features of such a system; at this stage you simply need to appreciate that manufacturing involves more than merely selecting an appropriate means of processing the material.

6.5 Summary

On completing this chapter you should appreciate that selecting the correct material is as important as including suitable design features, or choosing an appropriate manufacturing process. The selection process is likely to be iterative and will involve compromise. Today, the effects on the environment must be considered.

Manufacturing is essentially the modification of raw material. This can be achieved by changing the shape (by removing material or deformation), joining one or more pieces together, or by modifying the properties of all or part of the material. We will explore these methods further in the next few chapters.

7 Shape Modifying by Material Removal

7.1 Introduction

In the previous chapter we introduced two primary methods of changing the shape of materials. We will now discuss some of the material removal techniques in more detail. They fall broadly into one of three classifications, as summarised in Figure 7.1. Probably the most common method is via cutting, by creating chips, so we will start by concentrating on this method.

7.2 Metal cutting

From the start of engineering, metals have probably been the most significant single group of materials used. Although this may change with time, their long history of use has created a wealth of information on their properties, their methods of production and ways of adjusting their shape. It is therefore appropriate that we start this chapter with a process, metal cutting (or machining), that nominally relates to metals, but can also be applied to other materials. For example, wood lathes exist, and many plastics can be machined.

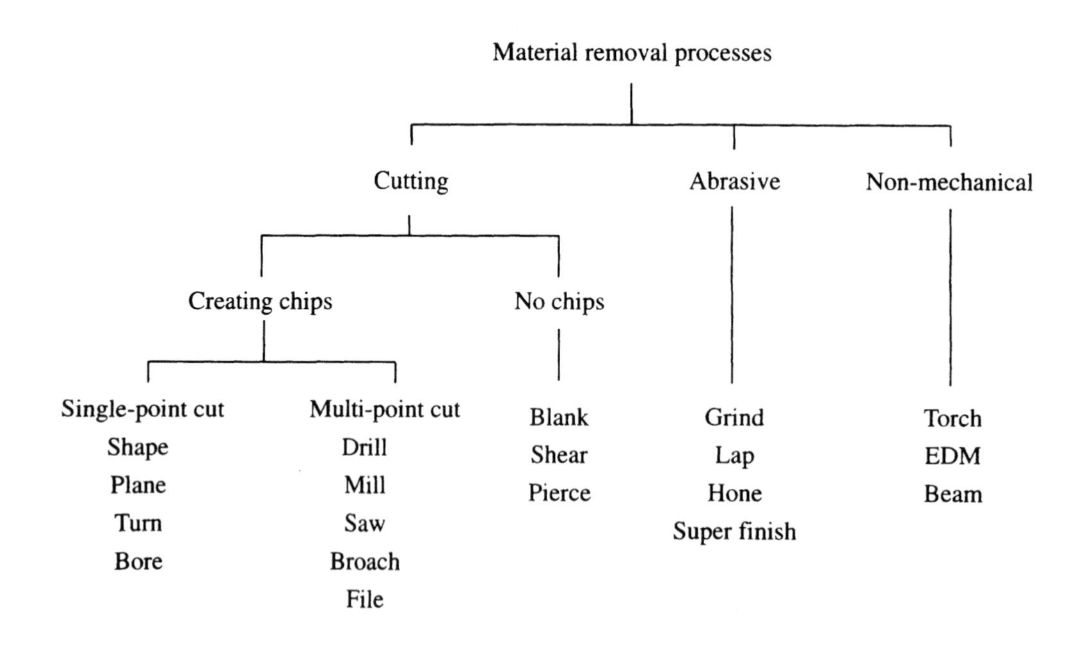

Figure 7.1 *Material removal processes*

7.2.1 Cutting mechanism

The fundamental principle in machining is that a sharp hard tool is drawn across the workpiece, removing chips of material as it progresses. Figure 7.2 illustrates the mechanism schematically. (Although not shown, the workpiece will be clamped rigidly to the base of the machine performing the operation. The tool will also be clamped rigidly, but to part of the machine that will move it relative to the work.)

Figure 7.2 (a) shows the tool creating its first cut on the workpiece, the material in front of the tool curling away and breaking off in chips. When the tool has completed its first cut it is raised slightly and withdrawn to its start position, lowered and indexed in the feed direction and then performs another cut. The situation after several cuts is shown in (b).

The tool used is essentially a rectangular block with one end specially shaped for cutting. Some of the features of this end are shown in (c).

The process as described is called single-point cutting because at any instant of time the cutting action takes place between the single cutting area of the tool and the workpiece. Moving the tool across the work in several stages in this manner is the mechanism used on a shaping machine. A number of other machining processes also use single-point cutting tools, but with different ways of creating the relative movement between tool and workpiece. They include shaping, planing, turning, and boring.

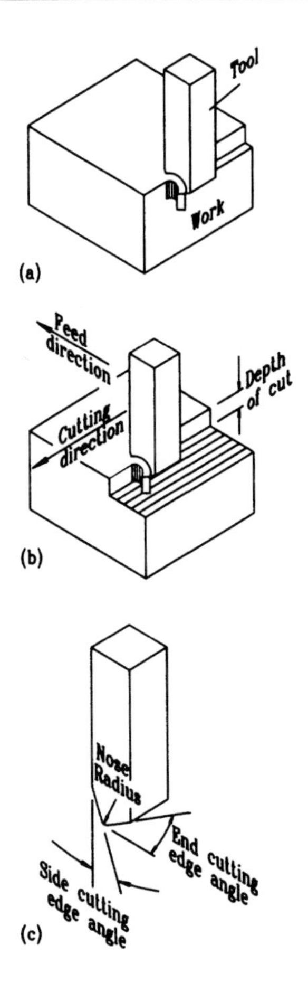

Figure 7.2

7.2.2 Chip formation

As the tool moves across the work (see Figure 7.3), a surface layer of constant thickness (d) is removed. This material slides up the front (face) of the tool and breaks off in chips. The chip thickness (c) is normally greater than that of the material prior to removal and can give an indication of the degree of difficulty of the machining process. A high chip thickness ratio c/d tends to indicate a material proving difficult to machine. With good efficient cutting geometry the chips should be 'C'-shaped.

Ideally during the machining process the material is cut away as described and breaks into small chips which either fall away naturally from the workpiece or can easily be removed. Sometimes the material cut away does not chip, but instead forms long continuous strings of metal.

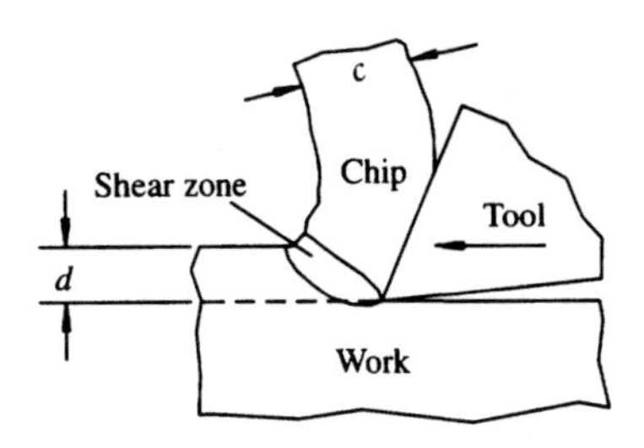

Figure 7.3 *Chip information*

These can cause a number of problems. They can become entangled with the cutting tool, can damage the cut surface of the workpiece, and are a source of danger to the operator (they are very sharp). There are a number of variables that can affect chip creation. These include cutting speed, material composition and tool geometry.

7.2.3 Tool geometry

Figure 7.4 shows both edges of the tool to be angled. The lower surface of the tool slopes slightly upwards away from the cutting point. This is called the cutting edge relief angle. Its purpose is to avoid unnecessary rubbing of the tool on the cut surface. It is usually kept to a minimum to avoid reducing the support for the cutting edges by too much. The face is also angled away from the vertical by the rake angle. Increasing the rake angle will tend to reduce the power required to drive the tool forward, as the cut material is not deflected through such a great angle. Conversely, increasing the rake angle will increase the power requirements as the cut material piles up in front of the tool, but strengthens the tool at the cutting area.

The nose radius influences the final surface finish. If the depth of cut and feed are maintained, reducing the nose radius will tend to worsen the surface finish. However, with machining operations there are many other factors that also affect surface finish, including depth of cut, feed, cutting speed, material, and rigidity of tooling and machine. The nose radius will also influence the angle formed at the end of the cut plane. A small nose radius with appropriate cutting parameters will allow close control of diametral tolerances, but with the penalty of reduced tool life. The depth of cut must always be greater than the tool nose radius.

Sometimes the face of the tool can incorporate a feature such as a groove or a raised portion to encourage the chips to break. The feature forces the cut material into tighter curls and hence to break.

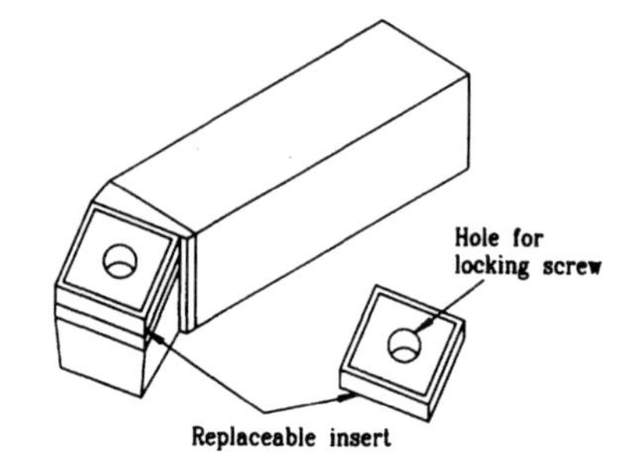

Figure 7.5 *Insert type tool*

As the tool is used it will wear gradually, eventually becoming unable to hold the necessary dimensional tolerances on the component being machined. The tools as described above would then need to be reground and replaced on the machine. An alternative is to use a tool with replaceable inserts (Figure 7.5). The inserts (made from a hard material) normally have three or four sides, so that by rotating the insert, each of the available six or eight cutting edges can be used. The inserts are clamped to the tool holder for quick changing and offer the further advantage that some of the more specialised tooling materials can be used.

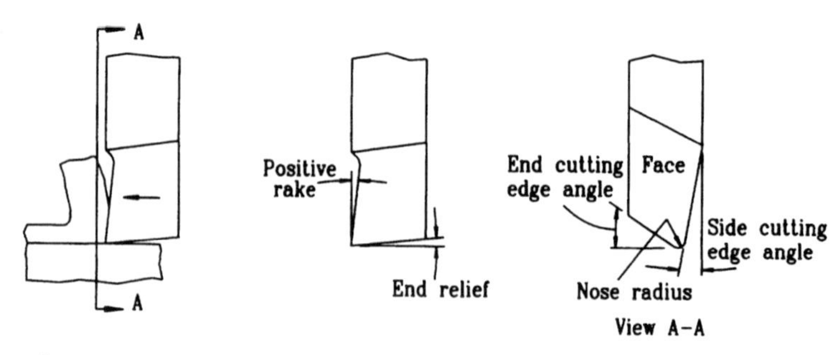

Figure 7.4 *Basic tool geometry*

7.2.4 Tool materials

A cutting tool must be able to survive for an acceptable life in a very hostile environment. Properties that a tool material should have can be summarised as:

- High levels of hardness to resist wear. This hardness must be maintained at high temperatures so that the tool retains its geometry while creating chips.
- Toughness: to be able to withstand the shock loading during an interrupted cut or when localised hard spots appear in the workpiece.
- Ability to withstand thermal shock: the tool may suffer rapid heating and cooling.
- Good chemical stability combined with relative inertness so that the tool does not have any chemical influence on the workpiece.
- Finally, the material should be easy to fabricate and readily available at a low cost.

Initially, cutting tools were almost all made from carbon steel, but over the past 100 years or so, a succession of materials has been developed to withstand higher temperatures and pressures, hence allowing ever higher cutting speeds and material removal rates.

Carbon steel

This is only suitable for very light duties as it softens above about 250°C. It is normally only used for machining soft non-metals although it may be found in hand reamers for metal cutting.

High speed steels (HSS)

HSS are a range of tough, heat treatable materials that can retain their hardness up to about 650°C. They are carbon steels to which alloying elements such as tungsten (W), chromium (Cr), vanadium (Va), molybdenum (Mo) and cobalt (Co) have been added. The tools can be manufactured in the annealed (soft) condition then heat treated prior to final grinding. They can be reground several times. They are used in drills, reamers, taps, dies, milling cutters and single point tools.

Cemented carbides

Cemented carbides are produced by mixing powders, usually carbon and tungsten using cobalt as a binder, then pressing them and pre-sintering at about 800°C. At this stage the blanks can easily be machined. Finally the blanks are sintered at about 1500°C to gain their full strength and hardness. They are available as inserts (Figure 7.5) and are suitable for cutting all materials up to medium strength and hardness. Alternatively they can also be brazed to the tool and ground to shape. Masonry drills are produced in this way. Cemented carbides work best at temperatures over 600°C, so are not suitable for low speed applications.

Coated carbides

The ideal tool material is one that has a very hard, non-reactive surface, and at the same time has a base tough enough to withstand the shock loadings. The coated carbide tool offers this by coating the cemented carbide with a very thin ceramic such as TiC, TiN or Al_2O_3. The coating is only about 5µm thick. The result is a material offering increased tool life. (A TiN coating will be harder than tungsten carbide and about four times as hard as HSS.) The benefits are mainly available at high speeds with high metal removal rates, particularly when machining steels and cast irons. Although TiN is very inert it will tend to gall on titanium so should not be used for machining titanium. Coated carbides are not normally used for non-ferrous materials as the uncoated grades will work just as well in these cases.

Ceramics

It is possible to create a solid ceramic insert with materials such as Al_2O_3 by sintering. The tools are harder than cemented carbides but are more brittle, making them suitable for high speeds, but only at light loads. They are not suitable for interrupted cuts. Their high wear resistance may also make them suitable for machining non-metals, particularly composites containing gritty particles.

Silicon nitride (Si_3Ni_4) is a ceramic with properties that combine the good wear characteristics with much improved toughness and shock resistance. Unfortunately, manufacturing difficulties in pressing the material to achieve its full density means it tends to be expensive. Variants are available with combinations of silicon nitride and aluminium oxide at a lower cost.

A further variation is the ceramic composite where whiskers of silicon carbide are used to reinforce the ceramic matrix. These operate in much the same way that carbon fibres do in an epoxy resin. Claims are made that they can be run at speeds up to ten times faster than for carbides, and speeds some 50% higher than possible with silicon nitride. At present these have yet to become commercially available.

Diamond

Natural diamonds can be shaped to produce an extremely sharp and hard tool suitable for machining non-ferrous metals. Unfortunately, hardness and brittleness tend to go together and this is no exception. Sudden tool failure by edge chipping is a common problem. Polycrystalline diamond (PCD) tools have been developed to overcome this difficulty. These are sintered fine diamond particles normally bonded on to a carbide base.

At high temperatures diamond changes into graphite, which will diffuse into iron. This makes it unsuitable for machining steels. However, PCD tools have high abrasion wear resistance, so are recommended for cutting fibre glass epoxy, graphite reinforced plastics and hard rubber.

Cubic boron nitride

This is effectively an artificial diamond which has a hardness second only to diamond. CBN is not as reactive as diamond and can therefore be used to machine steels.

7.2.5 Cutting fluids

Purpose

Although some cutting operations are performed dry (Figure 7.6 (a)) in the majority of cases a fluid is applied to the cutting area. The fluid performs two primary functions: lubrication and cooling. A correctly applied lubricant (Figure 7.6 (b)) can reduce the friction between the chip and the rake face. This increases the shear angle, creating thinner chips which curl more tightly, thereby reducing the power requirement. (Note the reduction in chip thickness ratio, section 7.2.2.) The cooling abilities of the fluid can reduce temperatures of the workpiece and the tool, often allowing higher speeds to be used. Usually, when

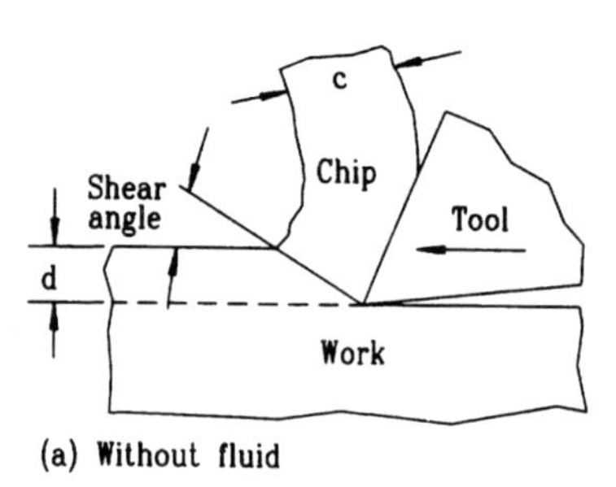

(a) Without fluid

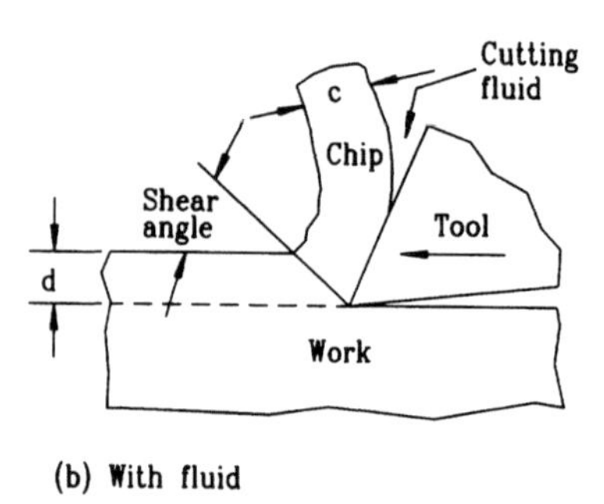

(b) With fluid

Figure 7.6 *Action of cutting fluids*

a cutting fluid is used, there is an improvement in accuracy and surface finish when compared with an equivalent dry cut. Finally, the flow of fluid across the cutting area can perform another function: it can help wash away chips.

Application

With low speed or hand operations the fluid can be applied manually, as a squirt from a can or even in the form of a paste. While this may be satisfactory for a model shop, most production machine tools are fitted with an automatic coolant recirculatory system. The fluid is applied via a nozzle positioned so that the fluid floods the cutting area. Some systems incorporate holes in the tooling through which the coolant can be fed, to ensure that it reaches the cutting edges.

Types

Cutting fluids are available as neat oils, good for lubrication, or water-based fluids containing oils in emulsion, which are good for dispersing heat. Synthetic fluids are also available which do not contain any petroleum oil. A further variant is the semi-chemical coolant which contains small quantities of mineral oil to enhance the lubricating qualities.

7.2.6 Tool wear

Cutting tools can fail in one of two ways. The failure can be sudden and catastrophic – the tool breaks and must be replaced immediately. Alternatively, the process can be gradual, with wear on the rake face and flank of the tool increasing until the tool becomes ineffective.

The sudden failure of a tool can be attributed to a variety of causes: excessive temperature, brittle fracture, excessive stress (e.g. through intermittent cutting), inadequate cooling (leading to thermal stress), fatigue fracture, edge chipping, etc. However, it is difficult to predict the mechanism(s) in a particular situation and hence almost impossible to predict when such a failure will occur.

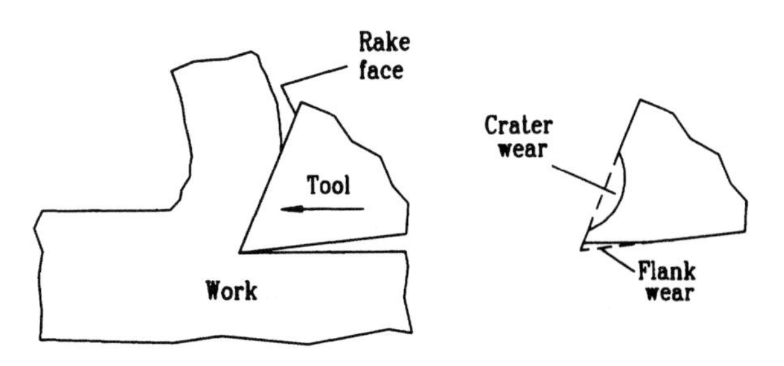

Figure 7.7 *Tool wear*

Provided the tool avoids the premature failure just described it will suffer wear in two main areas (see Figure 7.7). As the deformed chip slides up the rake face, a crater will be worn away on that face. This phenomenon is quite temperature dependent. If the cutting speeds are high, temperatures on the face can reach 1000°C, at which point HSS tools will wear very rapidly. Normally, carbide tips would, of course, be at these cutting speeds but they will also be subjected to crater wear. With cutting time, the size of this crater will gradually increase until, eventually, support for the cutting edge is reduced to such a degree that it breaks.

At the same time as the flank of the tool rubs along the workpiece its surface will gradually wear away. This will reduce the effectiveness of the relief angle, with the result that the rubbing area and friction levels are increased so that even more wear takes place.

Tool wear can never be avoided but its effects need to be controlled if quality in the manufacturing process is to be maintained. Efforts have to be made to optimise the cutting process by careful selection of the appropriate tool, fixturing, toolholding, feed rate, depth of cut, cutting speed, cutting fluid and so on.

7.2.7 Tool life

It is difficult to be specific in defining criteria for establishing tool life, since the point at which a tool becomes worn out depends on the process that the tool is being used for. Finishing operations will have high demands in terms of dimensional accuracy and surface finish, whereas with the initial roughing cuts the ability to remove material rapidly is more important. A roughing tool could therefore tolerate a degree of wear that would create a deterioration in surface finish that would be unacceptable in a finishing tool. The limiting factor for roughing would be when the tool was in danger of breaking under the cutting forces.

The Taylor tool life equation (published in 1907) establishes a relationship between tool life and cutting speed;

$$VT^n = C$$

where T is the tool life in minutes, V is the cutting speed (metres per minute) and C and n are constants depending on the cutting conditions (feed, depth of cut, coolant, material, etc.)

Once the cutting conditions have been established, the formula can be used to predict the effect of changes in the cutting speed on tool life.

Knowing the life of a tool enables the engineer to plan tool changes so that interruptions to the production process are minimised.

7.2.8 Cutting economics

So far, we have concentrated on the mechanics of the machining process, establishing that over the years the technology has improved gradually, allowing ever faster cutting speeds and metal removal rates (MRR). We should now remember that the overall objective is to turn the material into the finished product in a consistent and economic manner. The machining process involves a number of discrete tasks, each with an associated cost, most of which are influenced by the cutting speed.

- *Handling*: The material needs to be loaded onto the machine at the start of the operation and unloaded at the end. This task is unaffected by the cutting speed (although it will be affected by other factors such as the number of machine tools used).

- *Machining cost*: This is normally based on the elapsed time between finishing the loading task and starting the unloading task. An increase in cutting speed will therefore reduce the cost per component. The relationship will not be linear as the cutting speed will not influence the time taken for the tool to index between one cut and the next.

- *Tool cost*: In general, increasing the cutting speed will tend to increase the tooling cost per component, as tools will wear more quickly and more expensive cutting materials will be needed.

- *Tool changing*: The more rapidly tools wear, the more frequently they need to be changed, so costs will rise with an increase in cutting speed.

The relationship between these factors is shown in Figure 7.8. Note that the minimum cost occurs at below the maximum material

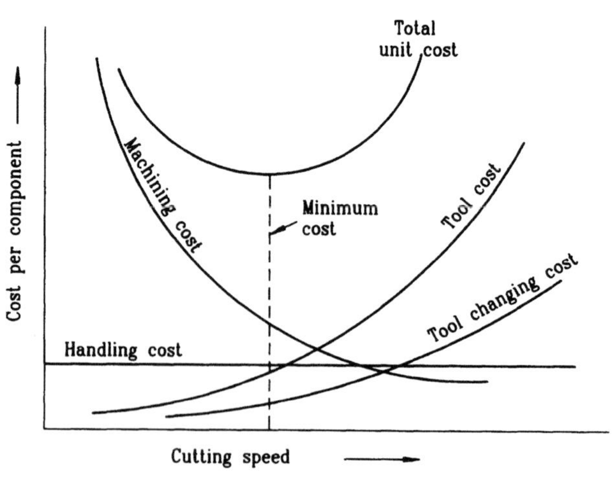

Figure 7.8 *Influence of cutting speeds on cost*

removal rate (i.e. highest cutting speed). This is because as tools wear out faster, the extra time spent changing them loses more time than is gained by using the higher cutting speeds. The aim is to find the optimum MRR. This is not easy as it depends on finding the ideal mix of feed, speed and depth of cut.

7.2.9 Cutting forces

Final adjustments to find the ideal, or at least an acceptable, combination of feed, speed and depth of cut really need to be on the actual machine concerned. In many cases, experience with previous similar components will minimise the need for final adjustment. In the absence of this experience, a knowledge of the forces the tool is subjected to, combined with tables of standard cutting information, are used to guide the engineer.

In section 7.2.1 we used the shaping process to describe cutting with a single point tool. During the cutting process the tool will be subjected to a cutting force (in the cutting direction), and a force normal to this (in the direction of the depth of cut, see Figure 7.2), holding the tool down, maintaining the depth of cut. There will also be a force in the feed direction, but remembering that the actual feed movement only takes place at the beginning of each stroke, before any cutting starts, this force simply maintains the position of the tool in the feed direction.

A more commonly used machining operation is turning, where the workpiece rotates and the tool is moved only to adjust the feed and depth of cut. The depth of cut often remains unchanged during each cut, as with shaping, but the feed is created by a slow, continual movement of the tool in the axial direction. Imagine a bar being turned to reduce its outside diameter. The forces acting on the cutting tool will be as shown in Figure 7.9.

1. Cutting force, F_c: This is normally the largest force on the tool and accounts for almost all the power requirements.

2. Thrust force, F_t: This acts radially and maintains the depth of cut. Although this may be some 25% in magnitude of the cutting force, the lack of movement in this direction means it has virtually no effect on the power requirements.

3. Feed force, F_f: This acts in the direction of feed, in this case axially. In magnitude it may be about 50% of the cutting force, but as the feed rate will be low compared with the cutting speed, it only accounts for a very small portion of the power requirements.

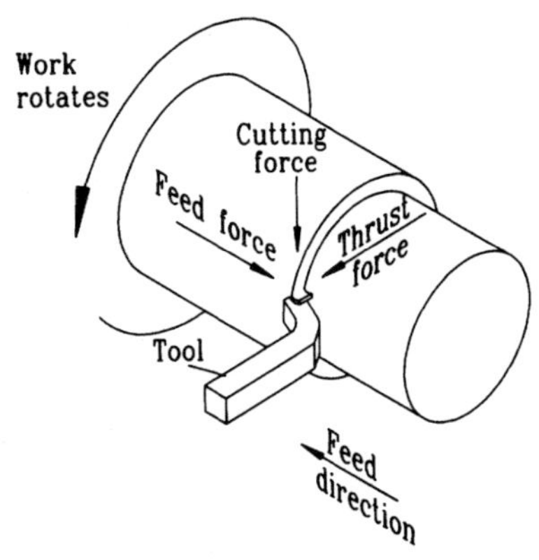

Figure 7.9 *Cutting forces*

From these forces we can conclude that the power needed is a function of the main cutting force (F_c) and the cutting speed (V); i.e.

$$\text{Cutting power} = \text{function}\ (F_c, V)$$

In section 7.2.8 we introduced the concept of material removal rate (MRR), which depends on the feed rate (f_r), depth of cut (d) and the cutting speed (V); i.e.

$$\text{MRR} = \text{function}\ (f_r, d, V)$$

A useful parameter is specific power (P_s), which is the power required to remove a unit of volume of material per unit of time

$$\text{Specific power} = \frac{\text{actual power}}{\text{MRR}}$$

One would expect the specific power to be dependent on material properties such as shear strength and hardness. It is also apparent from the above that it is influenced by cutting parameters, feed rate, depth of cut, and speed. Figures giving approximate specific powers for some common materials are shown Table 7.1.

Combining this information with some data on suitable cutting speeds and depth of cut, we are able to make some predictions about the machining process. This is best illustrated by an example.

A steel bar (200 BHN) of 60mm diameter is to be turned to 50mm diameter in one cut on a lathe using a cemented carbide tool. If the axial feed rate is to be 0.2mm per rev, estimate:

(a) the power required
(b) the cutting force at the mean diameter
(c) the time to machine an 80mm length of the bar.

1. First we need to decide on a suitable cutting speed. We will assume from the depth of cut (5mm) that this is a roughing operation. From the table of typical cutting speeds, select 150m/min.
   This enables us to set the rotational speed of the component, which will have a mean diameter of 55mm.

   $$\text{Rotational speed} = \frac{150 \times 10^3}{55 \times \pi}\,\text{rpm}$$

   i.e. 868 rpm

   (Note that, in practice, because many machines only have a limited range of spindle speeds available, we may have to choose the nearest speed available on the lathe concerned.)

2. Now calculate the volume of the material cut per revolution. As the feed rate is 0.2mm per revolution, the volume removed per revolution will, effectively, be a hollow cylinder 0.2mm long with diameters of 60 and 50mm.

   Volume cut per rev =

   $$\left(\frac{\pi\, 60^2}{4} - \frac{\pi\, 50^2}{4}\right) \times 0.2\text{mm}^3$$

   i.e. 172.79mm^3

3. Combining this with the speed will give us the material removal rate (MRR)

   $$\text{MRR} = 172.79 \times 868\text{mm}^3/\text{min}$$
   $$= 150 \times 10^3\text{mm}^3/\text{min}$$

4. From Table (7.1(a)), typical specific powers, we see that 55×10^{-3} W/ mm^3/min should be an appropriate figure, so knowing

   $$\text{Actual power} = \text{specific power} \times \text{MRR}$$
   $$\text{Required power} = 55 \times 10^{-3} \times 150 \times 10^3\text{W}$$
   $$= 8250\text{W}$$

   i.e. *power requirement is 8.25kW*

5. Now we will use this to estimate the cutting force, using

   $$\text{Cutting power} = \text{function}(F_c, V)$$

   Power is the rate of doing work, force × distance per unit time, so

   $$\text{Force} = \frac{\text{Power}}{\text{speed}}$$

Power required	=	8250 W
	=	8250 Nm/sec
Cutting speed	=	150 m/min (as used at the start of this example)

   Hence

   $$\text{Force} = \left(8250\frac{\text{Nm}}{\text{sec}}\right) \times \left(\frac{1}{150}\frac{\text{min}}{\text{m}}\right) \times \left(\frac{60\text{sec}}{\text{min}}\right)$$

   This gives a cutting force of 3300N.
   (Note the care taken with units and use of unity brackets. Use Appendix A as a reference if you need a reminder on units.)

6. Finally, we need to calculate the time to machine an 80mm length of bar.

   In one revolution the tool moves axially 0.2mm.

   Hence 400 (80/0.2) revolutions are needed for the tool to cut an 80mm length.

   At a speed of 868 rpm this will take 27.6 seconds.

Table 7.1

(a) Typical specific powers

Material	*Hardness (BHN)*	*Specific power W/mm³/min (×10⁻³)*		
		Turning HSS or carbide tools Feed 0.1 – 0.5 mm/rev	*Drilling HSS Feed 0.05 – 0.25 mm/rev*	*Milling HSS or carbide tools Feed 0.10 – 0.20 mm/tooth*
Mild steel	85–200	55	45	55
Plain carbon steels	330–370	70	65	70
Alloy steels	370–475	91	55	91
Tool steels	475–560	100	100	100
Stainless steels	135–175	65	55	65
Cast iron	100–190	45	45	45
Cast iron	190–132	80	75	91
Copper alloys	60–130	36	27	36
Copper alloys	130–200	55	46	55
Aluminium	30–150	15	9	18

(b) Typical cutting speeds

Material	*Turning (m/min)*		*Drilling (m/min)*	*Milling (m/min)*
	Roughing	*Finishing*		
Steels	130–200	180–250	35–40	115–180
Stainless steels	80–100	100–140	7–17	70–115
Cast iron	80–120	120–150	25–35	70–100
Copper alloys	160–200	200–250	65–85	150–300
Aluminium	100–150	150–250	210–430	230–430

(c) Typical feed rates. (These are applicable to the above cutting speeds.)

Material	*Turning (mm/rev)*		*Drilling (mm/rev)*	*Milling (mm/tooth)*
	Roughing	*Finishing*		
Steels	0.4–0.8	0.1–0.4	0.2–0.5	0.1–0.2
Stainless steels			0.1–0.3	0.1–0.18
Cast iron			0.1–0.3	0.08–0.25
Copper alloys			0.2–0.5	0.1–0.25
Aluminium			0.3–0.8	0.15–0.25
Depth of cut (mm)	2.5–5.0	0.4–2.5	Depends on drill diameter above feeds based on 6.0–25.0 dia	0.5–1.5

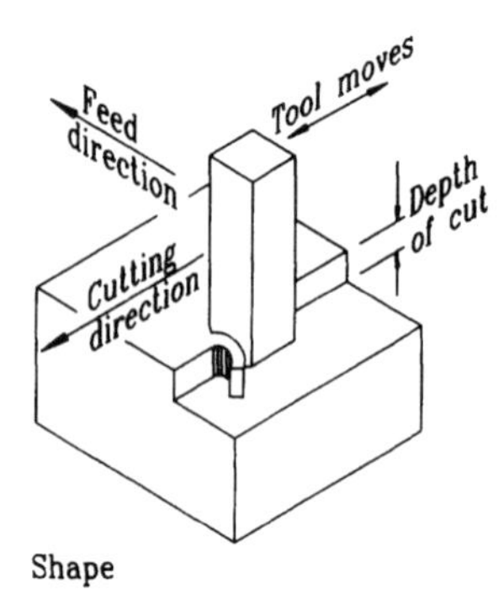

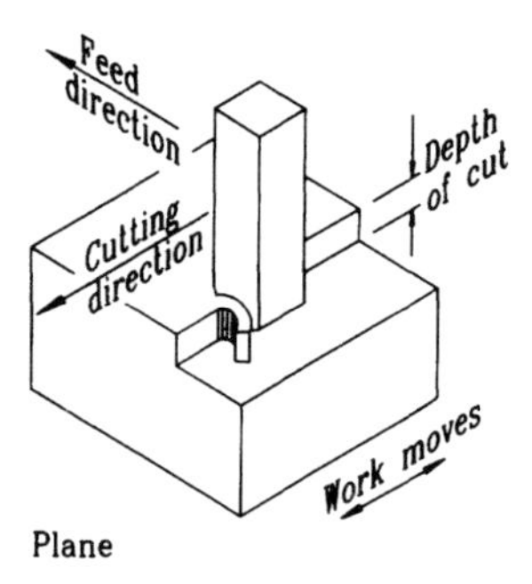

Figure 7.10 *Shaping and planing*

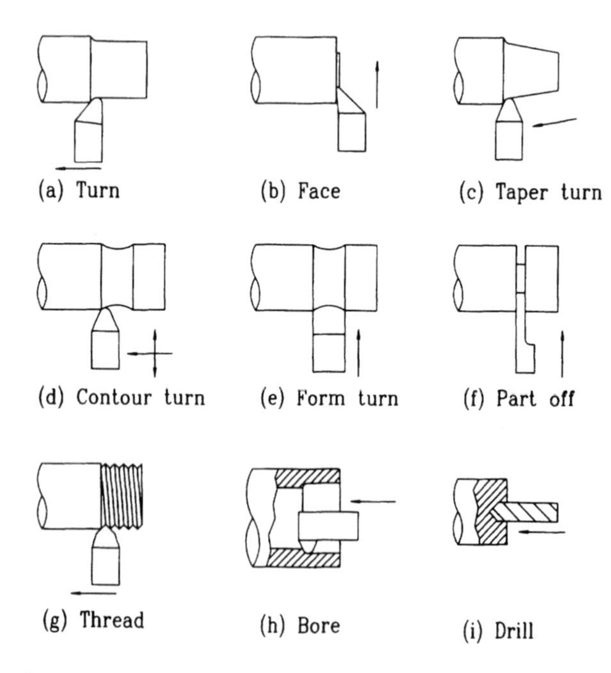

Figure 7.11 *Typical lathe operations*

7.2.10 Machine tools

The machine tool is the device that holds the workpiece and the cutting tool and generates the relative movement (cutting process) between the two. Most (though not all) machine tools are only able to perform one type of cutting process. We now discuss those processes using single-point tools.

Shaping or planing

These processes are used to create flat surfaces. Shaping has already been introduced in section 7.1.1. Here the work is held stationary and the tool is drawn across it, creating the cut. Planing is essentially the same process, performed on a different machine, where the work is moved past one or more stationary tools. These machines are shown in Figure 7.10. Planing tends to be used for workpieces that are too large for shaping machines. Both processes are slow, and are now largely being superseded by other, faster processes, using multi-edge tools, on more versatile machines.

Lathe

Probably the most widely used machine tool is the lathe. Here the work is rotated, usually about a horizontal axis, and the tool moved to create the cutting action. (We saw this in Figure 7.9.)

The primary machining process on the lathe is turning but there are a number of related processes that can also be performed. Some common operations are shown schematically in Figure 7.11. In each of these diagrams the work is rotating and the tool moves in the direction of the arrow.

An external diameter will be created if the work is rotated and the tool fed along the axis of rotation, maintaining a constant depth of cut. This is basic turning (a). If the tool is close to the end of the workpiece and fed radially, a flat surface will be created on the end of the work. This is called facing (b). By using a combination of the axial and radial feeds for the tool, a movement approximating a straight line at an angle to the work axis will result. This will generate a conical surface (c) and is called taper turning.

Varying the axial and radial components of the tool feed can control the finished surface to a more complex shape (d). This facility is often used to demonstrate the flexibility of a computer-controlled lathe by turning symmetrical chess pieces such as pawns. This is called contour turning. A similar effect can be created on less sophisticated machines by using a tool specially shaped to produce a particular surface. This is called form turning (e).

The surface created depends on a combination of the movement and the shape of the tool. A very narrow tool, fed at 90° to the work axis, would create a groove. This could, depending on the tool shape, be suitable for a circlip or an 'O' ring seal. If the tool was fed through to the centre of the work, a cut through the work would result. This is called parting off (f) and would be used if more than one separate component is to be produced from a single piece of material.

External threads (g) can also be cut on most lathes. All the operations discussed so far have cut material on the external surface of the work. If the tool is arranged as in (h), virtually any of these surfaces can be created on an inside diameter. Internal operations are called boring. Note that the boring does not create the hole, it bores the hole to a particular size or adds a particular feature to the inside surface.

All the tools considered so far are single-point tools: they have a single cutting edge (see Chapter 6). A drill would be a multi-point tool as it is cutting on more than one edge at any one time. A drill can be used on a lathe, but normally only along the axis of the component. Unlike a drilling machine, where the work is stationary and the tool rotates, on a lathe the work rotates and the tool is stationary: however, the relative movement between the work and tool is identical (i). Other operations, such as reaming and tapping, can also be performed on a lathe, but usually only along the axis of the component.

The work is normally held on a diameter in a three-jaw self-centring chuck. Any subsequent diameters that are created by the machining operations will therefore be concentric to the holding diameter. If this concentricity is not appropriate, other types of chuck with independently adjustable jaws can be used. Note that if a self-centring chuck is not used, the time to load the work to the machine is considerably increased as the operator has to adjust the axis of rotation of the work carefully. Some lathes have a hollow section through the chuck area, allowing bar material to be used. The bar is held in collets and is fed through to the work area in steps, enabling a number of components to be produced from the single bar.

At one end of the machine is the headstock, incorporating the chuck and its drive mechanism. At the other end is the tail stock, which can be used to support long workpieces (see Figure 7.12).

The tool is held in a tool post, which allows the tool to be set at the appropriate angle to the workpiece. This can then be moved vertically, across the axis of the work or along the axis of the work, and so create the required cutting path.

Lathes have been in existence for many years and, although the basic principles remain the same, there are a wide range of variants. The capstan lathe, for example, has provision for holding a number of tools already set up in a turret that can be rotated about a vertical axis. When a new tool is required the turret is simply indexed to align the tool in its operating position and the machining can proceed. Once the tools have been set there is little skill required by the operator, so a number of automatic lathes have been developed that only require loading and

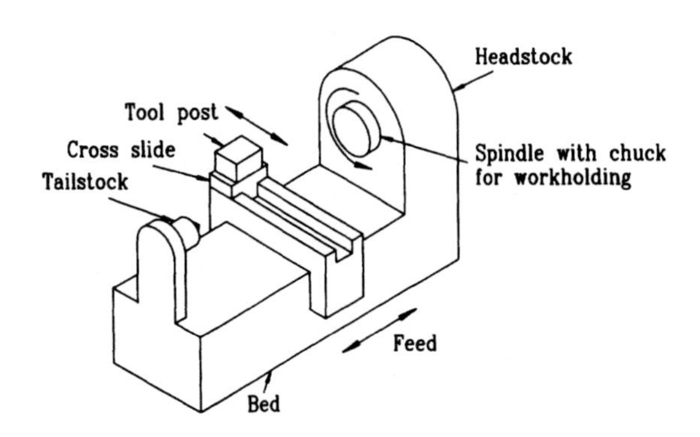

Figure 7.12 *Schematic for a basic lathe*

unloading to be performed manually. With a bar-fed machine even this is partly automated.

Automation of lathes has progressed a long way over recent years in an effort to improve productivity. One variant that offers high levels of output is the multi-spindle automatic lathe. This machine features up to eight spindles, each fitted with a chuck (or collets in the case of a bar-fed machine). These are in turn mounted on a drum which indexes to bring each spindle to one of the (eight) working positions. Each of these positions has its own cutting tool complete with feed mechanism. The cutting operations on the work are split into separate components, so that the work passes from one cutting position to the next as the drum rotates. At any one time the machine will hold eight pieces of work, all of which can be machined simultaneously. The indexing of the drum and the various tool movements are controlled by a series of cams so that the operation of the machine is automatic. Effectively, the multi-spindle auto offers the output of eight single lathes.

There are, however, a number of disadvantages. Firstly, the complexity of the machine is such that it can take a very long time to set the machine – up to 20 hours is not uncommon. This means that it is only suitable for large batch production work, so that the setting time can be spread over a large number of components. Secondly, all of the spindles are linked mechanically, so must run at the same rotational speed. If there is a significant difference between the diameters being machined at each of the cutting positions there will need to be compromise on the ideal cutting speed. The author encountered this problem with a material that failed to produce chips unless close to the ideal feed was used. Machining on a single-spindle machine caused no problems because the correct cutting speed could be used for each cut. On the multi-spindle auto long strings of swarf were generated that clogged the machine. The problem was eventually solved by changing slightly the specification of the material being machined.

Many machines today use servo motors to control the tool movements. These are in turn computer controlled, so that once the machine is set and loaded, the operation is completely automatic. Tools are normally held in a turret, the indexing of which is also controlled by the computer. This enables a sequence of operations to be performed. For example, a number of cutting operations could be performed by an individual tool, and the turret would then rotate to bring a specially shaped tool into play and perform the next operation. It also enables the use of sister tooling where, provided there are enough stations in the turret, duplicate tools can be held for those tools with the shortest life. The computer is programmed to switch to a sister tool once the defined cutting life of a tool has been reached.

Recognising that the machine is only productive while it is cutting metal, quick-change tool holders have been developed. If the tool holder is removed from the turret and then replaced, the accuracy and design of the location is such that the cutting tip will return to its original position and cutting can continue. This means that tools can be set up away from the machine on a tool pre-setter. A pre-set group of tools can be delivered to the machine, and the tool-changing operation will be very straightforward. The main task for the operator is to ensure that the correct tool is fitted to the correct location in the turret, and typing into the control any offset corrections.

Milling machine

Milling, like turning, involves a rotary motion, except that this time the work is held stationary and the tool rotated.

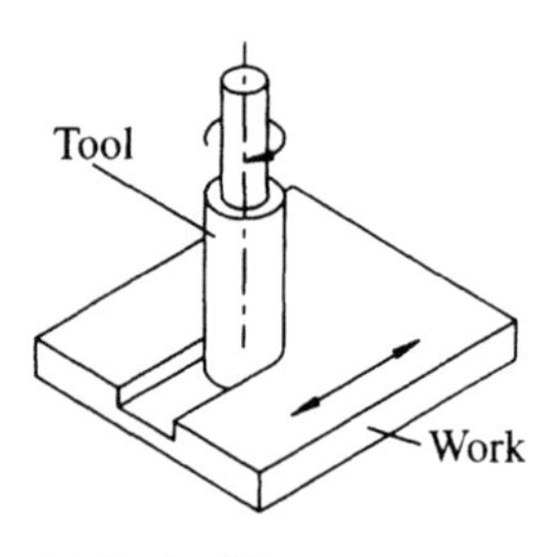

(a) End milling

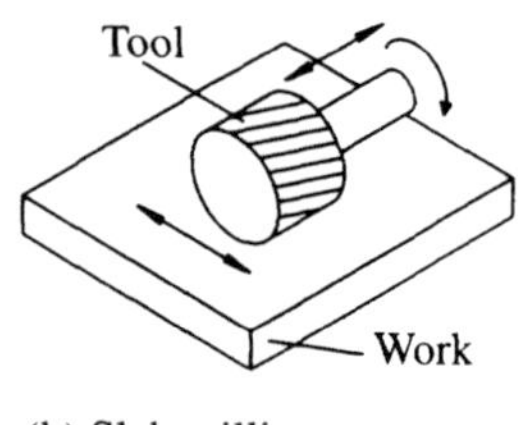

(b) Slab milling

Figure 7.13 *Types of milling*

The tool is also different in being a multi-point cutter, i.e. it has a number of cutting edges. The cut can be made normal (end milling) or parallel (slab milling) to the tool axis (see Figure 7.13). As the tool rotates, the cutting edges will each only be employed in cutting material for part of each revolution. The cutting action therefore involves an interrupted cut action. The cutting edges are often arranged helically so that they engage the work smoothly, usually with more than one tooth in engagement at any one time.

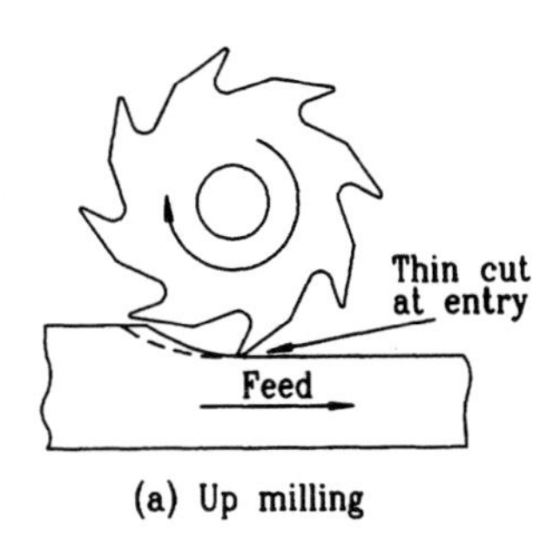

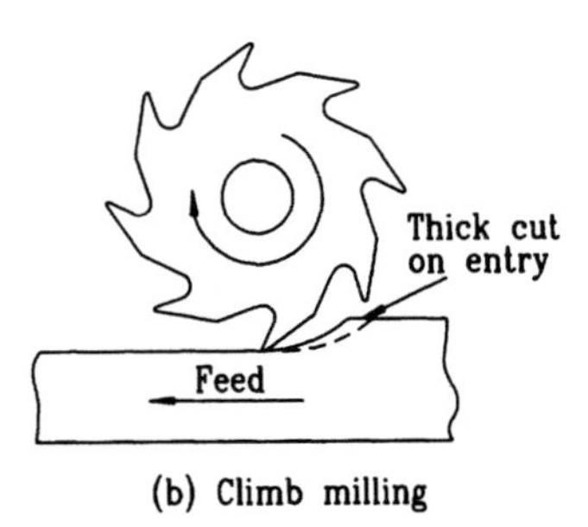

Figure 7.14 *Milling*

Surfaces can be created by one of two different methods. *Up milling* (Figure 7.14(a)) is the traditional method where the cutter rotates in the opposite direction to the feed of the workpiece. The chip is very thin where the tooth makes its initial contact, increasing in thickness to a maximum when the tooth leaves the work. As a result there is a tendency for the cutter to be deflected away from the work, causing rubbing, work hardening in the workpiece and increasing cutter wear. The cutter tends to push the work along and lift it up from the table. This can have a loosening effect on the workholding system, but should overcome any problems with backlash in the feed mechanism. *Climb or down milling* (Figure 7.14(b)) involves the tool rotating in the same direction as the feed. This time the chip thickness is at its greatest as the tooth enters the work, the forces tending to pull the cutter into the work. The cutting process tends to be smoother and have less chatter than up milling, but as the initial force is high there is a force reversal in the direction of the feed, so a rigid machine is required without any backlash in the feed drive. Virtually all modern milling machines are capable of climb milling.

Horizontal milling machines have the axis of the cutting tool parallel to the work, usually supported at each end. *Vertical mills* have the cutter axis perpendicular to the work with the cutter only supported at one end (Figure 7.15). In each case the work is clamped to a table that provides the feed movement in two directions, the third normally generated by movement along the spindle axis.

The modern *machining centre* (see Figure 7.16) is replacing both types of dedicated mills. These are CNC machines with increasing levels of sophistication. They can have up to five axes of freedom, with the computer control combining these movements, creating the ability to machine quite complex surfaces. Many are fitted with

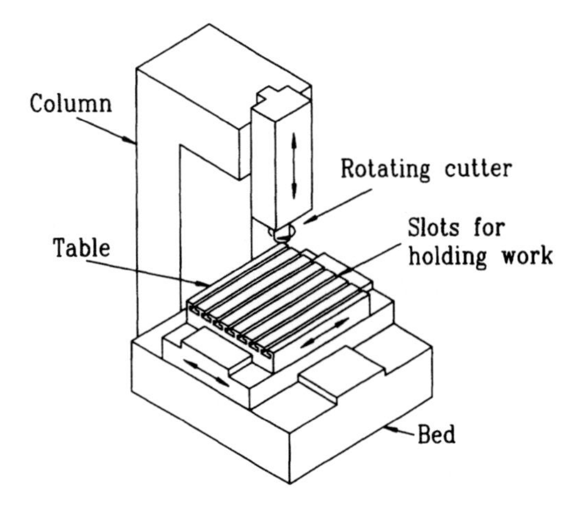

Figure 7.15 *Vertical mill schematic*

automatic tool changing equipment and probes to check alignment component identification and tool wear. They are often also equipped with cutter torque controls and tool life monitoring systems. As the work has to be clamped to some form of fixture, provided sufficient fixtures are available, a bank of work can be prepared and combined with an automatic loading system to enable the machine to run with minimum attention for several hours.

Turning centres

These are the lathe's equivalent of the machining centre. These are in effect automated lathes that utilise driven tooling. The turret will contain a number of conventional lathe single-point cutting tools, and will also have some small milling cutters, drills or taps. These enable flats to be cut, holes drilled, or holes drilled and tapped, perpendicular to the spindle axis. Some machines offer the facility to orientate the chuck, enabling off-centre holes to be drilled on the end of a part, or profiles machined. To date, very few machines have automatic tool changing, or component loading, as although it is technically possible it adds considerably to the cost of the machine.

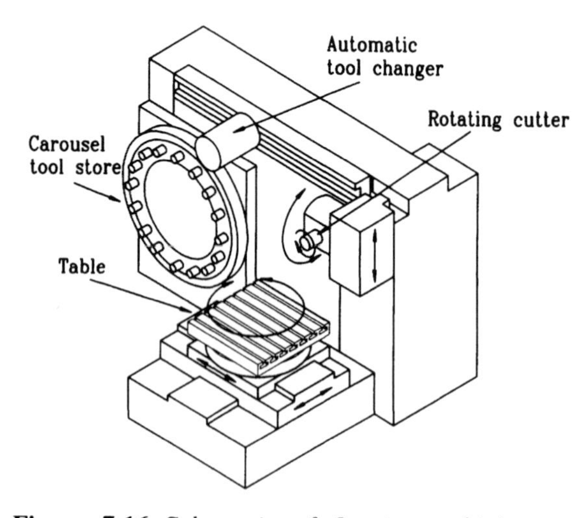

Figure 7.16 *Schematic of 5-axis machining centre*

7.3 Abrasive machining

Single-point and multi-point metal cutting as discussed so far could more accurately be termed single-edge and multi-edge cutting. Abrasive machining is true multi-point cutting, where the tool uses the many cutting points on abrasive particles or grit.

7.3.1 Grinding

The abrasive particles are bonded into a grinding wheel which removes small amounts of material from the work, creating a smooth and accurate surface. It is primarily a finishing process. *Surface grinding* (Figure 7.17) is used to create flat surfaces. The wheel can be dressed (shaped) to create other features such as grooves. In all types of grinding the wheel rotates at high speed and coolant is used to keep the workpiece cool and remove the chips. *Cylindrical grinding* is used to finish cylindrical surfaces. There are various machine configurations. In some the work is held

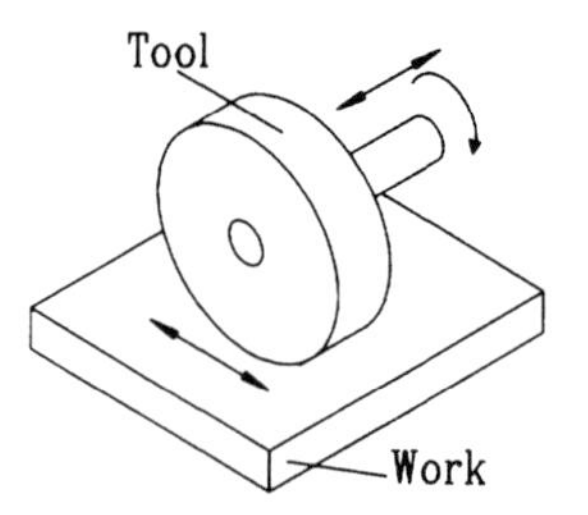

(a) Surface grinding

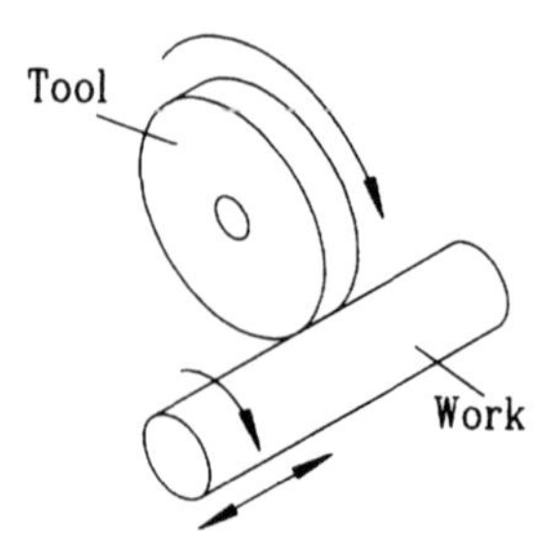

(b) Cylindrical grinding

Figure 7.17 *Types of grinding*

between centres, rotated and traversed axially against the rotating grinding wheel. In others the work is held in a chuck and a small wheel rotating at high speed is arranged so that it can grind internal surfaces. *Centreless grinding* avoids the need to mount the work in a chuck or between centres. Two wheels are employed. The larger one does the grinding, while the smaller (regulating) wheel controls the rotation and longitudinal motion of the work. It is mounted at an angle to the plane of the grinding wheel. This process is rapid, requires little skill from the operator, can be automated, and allows accurate size control. However, dedicated machines are needed, features such as flats, keyways or more than one diameter cannot be accommodated, and concentricity with other diameters cannot be guaranteed.

7.3.2 Honing

Honing uses a bonded abrasive stone for removing small amounts of material. The process involves a combination of rotary and reciprocating motions, with a lower cutting speed than grinding. It is often used to size (reduce out of roundness or waviness) and finish the bores of internal combustion engines. Material removed is normally less than 0.1mm, and diameter tolerances of 0.02mm are typical.

7.3.3 Superfinishing

Superfinishing is a variation of honing. It uses an abrasive stone moved across the work in short (less than 6mm), rapid (over 400 per minute) movements under light pressure with plenty of coolant/lubricant present. This creates a high quality finish by removing surface fragmentation from previous finishing processes. Material removed is normally less than 0.01mm.

7.3.4 Lapping

This differs from the abrasive processes described so far in that the abrasive particles are carried in a liquid or paste and applied to a shaped lap, usually made from a soft metal. The relative motion between the work and the lap removes tiny particles of material, slowly modifying the size and improving the surface finish. In some cases lapping may take place between two components, e.g. a valve and its seat. The process is slow and expensive. It is vital that the work is cleaned and free from all traces of the abrasive after lapping. Optical flats can be used to check flatness of the lapped surface.

7.4 Other methods

Although most machining involves chip creation you should at least be aware that other processes exist.

7.4.1 Electrochemical machining (ECM)

Electroplating (Chapter 10) involves the deposition of material onto the work which is the cathode in an electric circuit. ECM is effectively the reverse. The work is the anode, and a shaped tool, the cathode, is moved towards it (both being in a bath of electrolyte). As the tool advances, the workpiece erodes. The electrolyte is usually a dilute acid which dissolves the eroded metal particles. A high-pressure flow of electrolyte between the tool and the work is needed to dissipate hydrogen created at the cathode. The shape of the tool dictates the shape of the surface created.

A variant of the process (*electrochemical turning*) involves rotating the work, as in a lathe, but with the cutting tool being a cathode, separated from the work (anode) by an electrolytic film. This is normally used as an alternative to grinding where very hard materials are involved.

7.4.2 Electrical discharge machining (EDM)

In this process (sometimes called spark erosion) a high potential is applied between the tool and the work, which are immersed in a dielectric fluid. Provided the potential difference is maintained, an arc will be created and sustained between the two conductors. The temperature of the arc (over 5000°C) is sufficient to create local melting of the metals. The erosion rate is greater for the cathode, which is the work.

The process is normally used for producing holes or cavities in hard metal components. Press dies and mould tool cavities are often produced this way as this can obviate the need for subsequent hardening and the associated potential for distortion.

7.4.3 Laser profiling

Plate or sheet material can be cut by shearing, blanking, piercing (Chapter 8), or by using one of the welding heat sources: gas, plasma arc (Chapter 9). A high energy beam, usually a laser, can also be used. Complex profiles can be created avoiding problems of sharp edges requiring deburring, or high heat inputs causing distortion.

7.5 Summary

On completing this chapter you should

- understand the mechanism of metal cutting, the reasons for the shape of the tool, types of material used for cutting tools, and the factors affecting tool wear
- appreciate the difference between single- and multi-point tools
- be able to make simple calculations re: the cutting time and power required by a single-point tool
- understand the concepts of common machine tools
- appreciate that although most metal cutting is performed by creating chips, other more specialised methods do exist.

7.6 Questions

Now check your understanding by trying these questions.

1. What is the difference between single-point and multi-point cutting tools? (7.2)
2. You should always aim for the maximum MRR. True or false? Explain. (7.2)
3. List the basic properties that an ideal cutting tool should have. (7.2)
4. Describe some of the operations that can be performed on a lathe. (7.2)
5. A steel bar (200 BHN) is to have its diameter reduced in one finishing cut from 45mm to 43mm diameter over a length of 50mm using a carbide tool. Suggest a suitable feed rate, calculate the cutting time and estimate the power required. (7.2.9)
6. Describe the difference between climb milling and up milling. (7.2.10)
7. Lathes and turning centres can both produce holes by drilling. What are the limitations on the lathe? (7.2.10)
8. A die for a press tool has to be produced in a hard metal. Suggest, with reasons, a suitable material removal process. (7.4)
9. What is the main difference between lapping and the other abrasive machining processes? (7.3)
10. MRR is dependent on three main cutting parameters. Name them. (7.2)

8 Shape Modifying: Retaining Material

8.1 Introduction

Modifying the shape of a component by removing material cannot, in many cases, be the best way of achieving the final format. Machining, for example, normally requires the use of expensive equipment to turn the unwanted material into chips of swarf. Although the swarf may be recovered and recycled, its value, weight for weight, will be considerably less than that paid for the original raw material. There are a number of alternative processes that can be employed, which we will discuss in this chapter. Figure 8.1 lists some of the more common alternatives. Later in Chapter 12, we will expand on the reasons for selecting a particular process.

8.2 Casting

Almost all metals at some stage in their production are cast – i.e. they are in a molten state, poured into a mould, cooled, removed and then often subjected to further treatments such as rolling to create sheet material. Here we are concerned with subsequent processing, starting with an ingot of material, and ending with a component close to the required final component form. There are several alternative processes. They vary in complexity, cost, and capability of achieving the final component shape. All casting processes require a mould. In some cases this is a permanent part of the tooling: in others, the mould has to be recreated for each cast component.

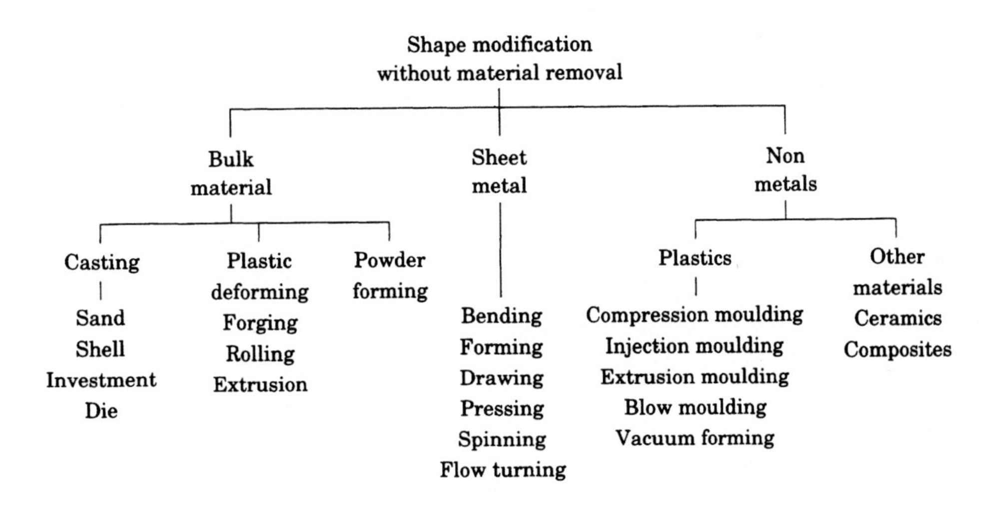

Figure 8.1 *Shape modifying processes*

8.2.1 Terminology

Before examining some of the alternative processes, an introduction to some of the commonly used terminology would be useful:

Pattern

This is needed to create the mould. It is a replica of the final product, but with some important differences. Dimensionally it must include allowances to take account of shrinkage, as the hot metal, after solidifying, cools. The required dimensions of the final casting must also include an appropriate machining allowance if the cast item is to be machined. The pattern may be in one or more pieces, depending on the process and shape of the component. It may also incorporate separate core pieces so that hollow features can be produced. In some cases it is reusable; in others it is expendable and has to be remade for each component cast.

Draft

To enable the mould and pattern to be separated easily those surfaces normal to the joint (or parting line) are usually slightly tapered, i.e. they feature a draft angle. This will obviously also appear on the final casting.

Mould

The vessel into which the molten metal is poured, and from which the shape of the casting is derived. The mould cavity in Figure 8.2 represents the shape of the final casting.

Pouring basin

This receives the melt before it flows down the sprue. The basin sometimes incorporates a weir to trap any heavy inclusions, or even a filter to trap lighter impurities.

Sprue

The passage through which the melt passes on its way down to the mould. The constant mass flow of the fluid means that as its velocity increases it will tend to pull away from the sprue walls, potentially drawing air into the mould. To avoid this, the sprue narrows with depth. At its base a well of a larger cross-section is provided in order to dissipate some of the melt's kinetic energy. This can also act as a further trap for unwanted inclusions.

Runners

The melt is distributed to the mould via passages called runners. These are designed to fill the mould cavity as evenly as possible.

Gate

Once the metal has solidified, any material in the runners will be connected to the main casting. To simplify removal of these unwanted features, the runner narrows as it connects to the main mould cavity. This narrowed section is called a gate.

Riser

Finally, an additional void, a riser, is included. It provides a reservoir of molten metal that can flow into the mould cavity during the solidification process to compensate for any shrinkage. The idea is that any shrinkage hollows will be in the riser rather than the casting.

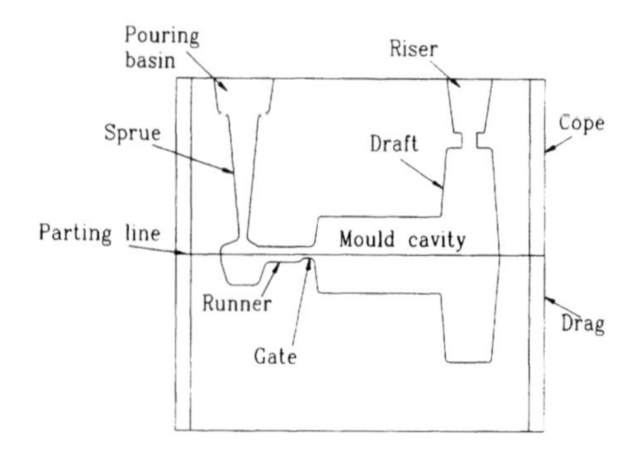

Figure 8.2 *Typical sand mould*

8.2.2 Sand casting

Sand casting employs a reusable pattern to create an expendable sand mould. The pattern is usually made from wood, resin, or sometimes metal, the choice being influenced by cost and the number of times it is expected to be used.

Figure 8.3 schematically illustrates the basic steps of the sand casting process. The pattern (a) in this case is made in two pieces, with dowels ensuring correct alignment. Step 1 is to place one half of the pattern on a mould board (b) and then add the bottom half of the moulding box (c), which is called the drag. The box is then filled with sand mixed with a bonding agent (d) which is compacted tightly around the pattern. Excess sand is removed and the box inverted (e), revealing the joining face of the pattern. The other half of the pattern is carefully put in place, the dowels ensuring alignment (f), and parting compound applied. The parting material is needed to prevent the second half of the mould (which we are about to produce) bonding to the half just created. The top part of the moulding box, the cope, is added (g) and more sand and bonding agent compacted in place (h). The two halves of the sand mould can be separated and the pattern removed (i).

Although not shown on these illustrations in the interests of simplicity, additional pattern pieces would have been included in the above process to provide an entry path for the molten metal (i.e. the pouring basin, sprue, runners, gates and risers). The mould halves are reassembled (j) and the melt introduced. After allowing time to cool, the sand is removed, any surplus feed metal fettled away and we have the cast component (k).

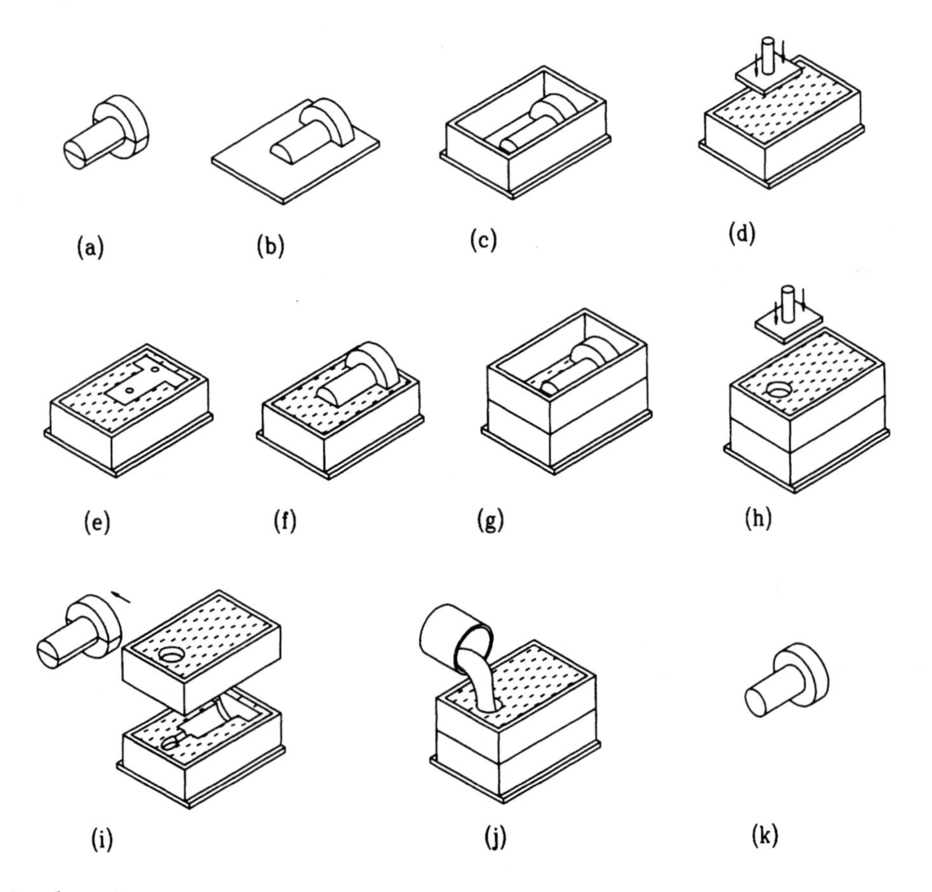

Figure 8.3 *Sand casting process*

The process can vary slightly according to the choice of sand and binding agent. The cheapest, called *green sand moulds*, use clay as the binder. Problems can arise through lack of strength and the presence of moisture. As the hot metal arrives in the mould any moisture present may turn into steam and impair the surface of the casting. To overcome these problems the mould can be baked prior to pouring the melt, leading to a *dry sand moulding*. This tends to be expensive and time consuming, so an alternative is to dry the sand locally at the mould cavity with a torch. This is called *dry skin moulding*.

Another common variant is *CO_2 moulding*, where the sand is mixed with about 4 per cent sodium silicate (water glass). The sand remains soft and pliable until exposed to CO_2 gas, when it hardens rapidly. The result is a firmer sand mould, enabling greater accuracy to be achieved.

Ceramic moulding uses a ceramic slurry in place of sand, offering benefits of improved surface finish, and a lower risk of inclusions. The mould is fired after removing the pattern to strengthen it, and is often heated prior to pouring the melt. This ensures good flowing characteristics, enabling the use of long, thin sections. *Plaster moulds* are also sometimes used, giving good accuracy and surface finish, but are only suitable for lower temperature metals: aluminium, magnesium and copper alloys.

8.2.3 Shell moulding

This process can be thought of as a development of the sand casting process, offering improvements in dimensional accuracy (up to 0.5mm per 100mm) and easier adaptation for mechanisation. The permanent pattern is made from metal in the form of a plate (Figure 8.4 (a)). Note that the runners are included as part of the pattern. The pattern is heated to about 450°C before being clamped to the open end of a dump box (b) containing a mixture of sand and uncured synthetic resin. The box is then inverted (c), dumping this mixture on to the hot pattern. The heat from the pattern cures the resin to a depth dependent on the pattern temperature and the curing time. As the box is turned upright (d) the uncured sand mix falls away. The pattern with the cured mould attached is removed and ejector pins are used to separate the mould from the pattern (e). The process is repeated to produce the other side of the mould. The halves of the mould are clamped together, often being further

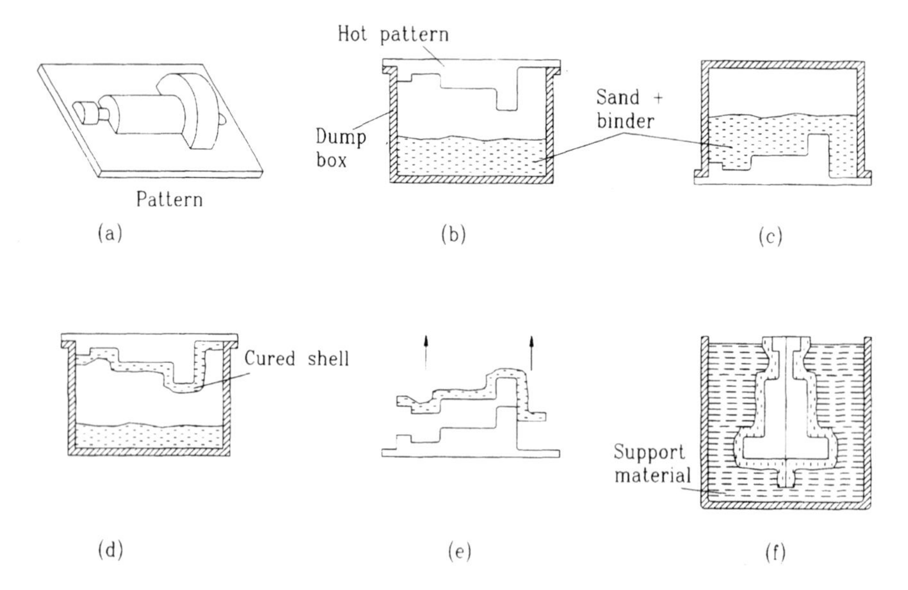

Figure 8.4 *Shell moulding process*

supported by placing in a flask of sand or metal shot (f) ready for the melt to be poured. Once cool, as with the sand casting, the mould is broken away to reveal the cast component.

8.2.4 Investment casting

The lost wax, or investment casting process, despite being one of the oldest casting processes, is able to produce components to much tighter tolerances than either sand or shell moulding techniques.

In this method, both the pattern and the mould are expendable. Figure 8.5 outlines the process, where the first step (a) is to produce a wax pattern. This is cast under pressure in a permanent metal die. This is usually a fairly expensive piece of tooling, as its tolerances and surface finish will be reflected directly in the final casting. The wax, therefore, has to be handled with care. A number of wax patterns are attached to a wax sprue to create a 'Christmas tree' of patterns. This operation is done by hand, using a soldering iron to melt the wax locally at the joint. One of the factors affecting the cost of the process is the number of items that can be attached to one tree; the fewer the items, the higher the cost. The size of the tree, in turn, is normally limited by the handling system used, i.e. size and weight.

The next stage of the operation can be highly automated. The tree passes through a series of dips of ceramic slurry. The first dips contain very fine slurry (c) to provide a coating that closely follows the contours of the wax pattern. The later dips (e) contain a more coarse mix, their purpose being to build up the strength of the mould. Between dipping, the coating partially dries in the air (d, f).

Next the whole assembly is baked (g) to both harden the ceramic mould and melt the wax, which drains out. Finally, the molten metal can be poured into the mould (h). When cold, the ceramic is broken away and the castings separated from the sprue and runners to leave the final castings (i).

As investment castings offer good dimensional accuracy and surface finishes, what may be dismissed as minor surface imperfections on a sand casting are not acceptable on the investment product. Such defects can arise through damage to the wax, dirt on the wax, or impurities in the melt. The first two are tackled by good housekeeping. Automating the dipping sequence can also help. Impurities in the melt can arise from a variety of sources: the more sophisticated systems minimise air contact with the melt. Pouring is sometimes done with vacuum assistance, both to avoid air contact and help ensure that the mould cavities are full.

Investment castings are normally used on small, intricate components where the high investment in tooling is justified by the saving in subsequent machining.

A variation on the lost wax process is the *full mould* or *lost foam* casting. The principle is similar, with the pattern being made of expanded polystyrene. It may consist of a number of components glued together. This is then dipped in a water-based ceramic to add a thin coat (0.08mm thick) but little rigidity to the pattern. Green sand is then compacted around it to form the mould. The polystyrene pattern remains in place as the melt is poured. The heat instantly vaporises the pattern, which as a gas relies on the permeability of the mould to escape. Obviating the need to withdraw the pattern means that there is no need to allow any draft angles and parting lines are eliminated.

8.2.5 Die casting

Die casting is suitable for those intricate components where strength needs suggest that a low melting point alloy would be a suitable material. The process involves metal moulds that are reusable.

Gravity die casting

This is sometimes called permanent mould casting and uses a mould normally made of cast iron, in two or more pieces. The mould components are clamped together and the molten metal poured in, as in sand casting. After solidification the casting is removed and the mould prepared for the next casting. Components have good

Figure 8.5 *Investment casting process*

dimensional accuracy (typically 0.8mm per 25mm for small items) and surface finish (R_a values 3.2 to 6.3μm). A wide potential range of castings is possible, from 50g to 100 kg. Capital costs are lower than for pressure die casting.

Low pressure die casting

Rather than using gravity to deliver the melt this process has the mould positioned above the melting or holding furnace and uses air pressure (typically 0.5 to 1 bar) to force the fluid through a vertical tube into the mould. Advantages are less oxidation of the melt, and good yields, since the pressurised tube acts as a riser and its contents can be used for the next casting. Often used for aluminium castings such as beer barrel halves.

High pressure die casting

This differs from the previous die casting processes in that the pressures used are very high, up to some 150 MPa. In the *hot chamber process*, as with the low pressure machines the mould is positioned close to the melting equipment. When the dies are closed, hydraulic pressure forces the melt into the mould. This is generally used for the lower melting temperature zinc-based alloys (about 420°C). The *cold chamber process* uses separate melting facilities. A measured quantity of melt is ladled into a pouring hole, from which a ram forces it into the mould cavity. This is normally used for alloys with higher melting points, such as aluminium (about 660°C). Die life for the higher melting point alloys can be a problem, so the melt temperature should be kept as low as possible. Some brass die castings, which melt at around 900°C, are cast in a semi-molten state. As a result, the final casting will not be as dense as a pressing and may be porous. The high pressures involved with this process mean the tools are made from hardened tool steels and are thus expensive. It is suitable for high volume production of those components requiring an excellent surface finish, and for which subsequent machining operations can be eliminated by virtue of the good dimensional accuracy of the casting.

8.3 Powder forming

Some materials are difficult to shape by conventional means. They may be too hard to cut easily, too brittle for possible material flow in the solid state, or have such a high melting point that casting is difficult. Powder metallurgy (PM) can offer a solution for such materials, along with some potential advantages. These include good surface finish and dimensional accuracy, an ability to combine otherwise immiscible materials (e.g. tungsten carbide and cobalt), and an inherent porosity in the product.

The process involves three stages:

1. Powder manufacture
2. Compaction
3. Sintering.

8.3.1 Powder manufacture

A number of processes are used, the most common being melt atomisation, which is the mechanical disintegration of molten metal by a high pressure jet of liquid or gas. Reduction of the metal oxide is often used to produce iron powders. Other methods include electrolytic deposition from fused salts, and precipitation from vapours.

The output from these processes is then subjected to further treatments. These may include crushing and grinding to create the powder, or annealing to remove work hardening. If the end product is a filter, spheroidal particles are required. These can be created by allowing the particles to fall through a heated space.

8.3.2 Compaction

This stage involves forming the powder into the shape of the component. Mechanical or hydraulic pressure is used to force the powder into a die. The pressures applied depend on the required properties in the end product: up to 140 MPa may be needed if low levels of porosity are a requirement. The main difficulty in pressing is that the powder does not behave like a fluid, with pressure equally distributed about its volume. Friction between the particles themselves and between the particles and the walls of the die means the pressure tapers off away from the punch, with a subsequent density variation in the product. To minimise this effect, the walls of the die are polished and waxes or soaps used to lubricate the powder. The press design also has an effect; applying pressure at opposite ends of the product improves pressure distribution.

Isostatic pressing is where the powder is placed in a flexible mould which is then subjected to high hydrostatic pressures (70 MPa). This method can achieve fairly uniform densities, but is expensive and cannot hold close tolerances.

8.3.3 Sintering

This is the process that converts the fairly fragile green compact from the pressing stage into an engineering component by heating. After compression, the particles are held together by interatomic bonds at the contact points. Heating produces diffusion between the particles, producing a stronger bond. As the particles diffuse, any spare spaces between them disappear, the overall density increases and the volume decreases (i.e. the component shrinks).

Sintering takes place below the melting point of the powder (typically half the melting temperature), normally either in a reducing atmosphere or in a vacuum. Mechanical properties tend to improve with sintering time, although for most metals the rate of improvement after about 40 minutes is small.

8.3.4 Hot isostatic pressing (HIP)

The HIP process is a special case of PM where the powder is sealed in a ceramic mould which is then placed in a special vessel where it is heated and subjected to hydrostatic argon gas pressure. The high pressures and temperatures compact and sinter the component in one step. This process has been used to produce high strength alloy components used in aircraft engines.

8.4 Plastic deformation

This group of shape-changing processes involves the flow of material in the solid state: that is, stresses are induced above the yield point, but below the point of fracture, causing permanent plastic deformation of the material. The metal can be worked either 'hot' or 'cold'

8.4.1 Hot and cold working

In cold working processes, plastic deformation results when sliding occurs along the slip planes. There is a limit to this sliding. When a dislocation is reached, the planes become locked together and the slipping continues along another slip plane, until that also locks. Eventually, when all the planes are locked, the structure is said to be fully work-hardened. Any further force applied will fracture rather than deform the material. If the material needs to be deformed further, it will first have to be annealed. This involves heating to above the recrystallisation temperature, thereby recreating the pre-deformation crystal structure.

Hot working involves the deformation taking place with the material heated to above the recrystallisation temperature. This means that the deformation and recrystallisation occur simultaneously without any work-hardening effects. A larger amount of deformation can therefore be performed without risk of fracture.

Cold working offers good dimensional accuracy, smooth clean surfaces, and increased strength and hardness. However, it makes the material too brittle for further work unless heat-treated.

Hot working offers higher levels of deformation without risk of fracture, but the heating results in poor dimensional accuracy, a scaly surface, and no improvement in material mechanical properties.

8.4.2 Forging

Forging normally involves hot working. The work is heated then beaten by a hammer to the required shape.

Open die forging

In its simplest form the traditional blacksmith hammering the hot metal against the anvil to produce the horseshoe is called open die forging. Today a large mechanised hammer would be used. The end of the hammer and the anvil are often flat, the shape of the component being created by the operator, repositioning the workpiece between blows. Sometimes a slightly shaped die may be included between the hammer and anvil to assist the shaping process. As it is slow and relies on the skill of the operator it is not suitable for large-scale production. Normally this is used as a preliminary operation to preshape metal for subsequent processes.

Drop forging (closed die)

This process involves hammering a block of metal into shape between two dies (see Figure 8.6). A piece of heated metal is placed on the lower die block which is attached to the anvil. A heavy weight (called the tup) holding the upper die is raised and drops under gravity on to it. A number of blows may be needed to force the metal into the die. If the required shape change is large a series of dies may be called for. This will almost certainly mean that the work has to be reheated between dies.

The amount of metal used is usually slightly greater than that required by the end product; the excess will appear as flashing around the part.

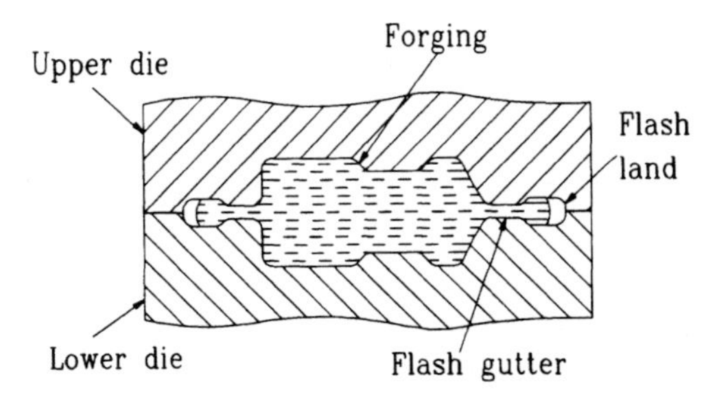

Figure 8.6 *Closed die forging*

Press forging

As with drop forging, this is a closed die process, but instead of a sudden impact by a hammer, a squeezing action between the dies takes place. The hammer action tends to concentrate the forces on the surface of the work, the energy being dissipated through the surrounding structure. The squeezing process tends to penetrate through the metal and can produce a more uniform metal flow. This process is needed if a large section is to be forged.

Upset forging

This involves increasing the diameter of bar material by compressing it along its length (see Figure 8.7). This can be performed either hot or cold. *Cold heading* is such a process involving cold working, used to create heads on fasteners such as rivets and bolts. There may be a number of stages to achieve the final form. Special-purpose high-speed machines are used for volume production. The main benefits are through material savings, there being very little wastage.

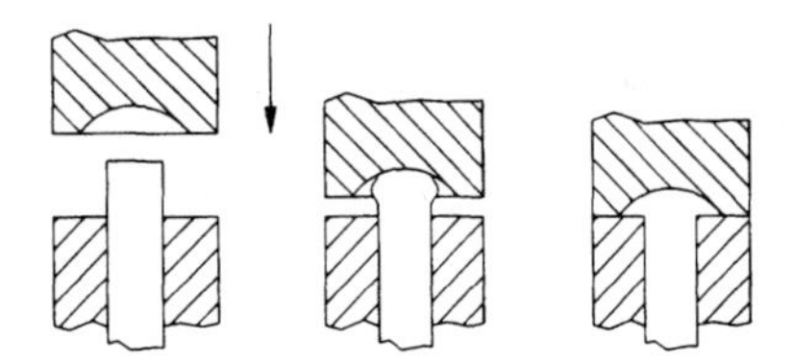

Figure 8.7 *Upset forging*

8.4.3 Rolling

Much of the metal raw material used by industry has been rolled as part of its production process. At the foundry, after casting, the ingots are rolled either into slabs (rectangular sections) or blooms (square sections). The hot material passes between a pair of rollers on a machine called a mill. Their direction of rotation is reversible, so that after the material has passed through the mill the rollers are reversed and the metal passed through again. The gap between the rollers is gradually reduced until the required section has been achieved. To produce a bloom the work needs to be turned to maintain a square section. The blooms are used to form billets and the slabs plate or strip.

Plate and strip are produced by further hot rolling, in a hot strip mill where after heating to some 1300°C they pass through a series of rollers, each set reducing the material's thickness and increasing its width. At the same time the length (and thus the speed) of the material increases each time it passes through a set of rollers, giving the fascinating effect of a single piece of material at one end travelling at walking pace and at the other at some 30 mph! Accurate speed control of each set of rollers is critical. At the end of such a strip mill the material is coiled.

A large proportion of this steel is then cold rolled to improve the surface finish and dimensional accuracy, along with mechanical properties such as strength and hardness, albeit with some loss of ductility. After hot rolling it is pickled to remove scale from the surface by immersing in a dilute acid bath. It is then washed and dried before cold rolling.

Rolling can be used to produce other products such as structural steel sections (I beams, etc.). With suitable dies it can also be used to roll external threads onto bars.

8.4.4 Extrusion

A ram is used to force the material through a hole in a die. The material section, while passing through the hole, is modified to the shape of the hole (see Figure 8.8). If large section changes are involved the process takes place hot to reduce the extruding force and minimise any work-hardening effects. Typical products include hexagonal bar, curtain rails, aluminium window frame sections, etc.

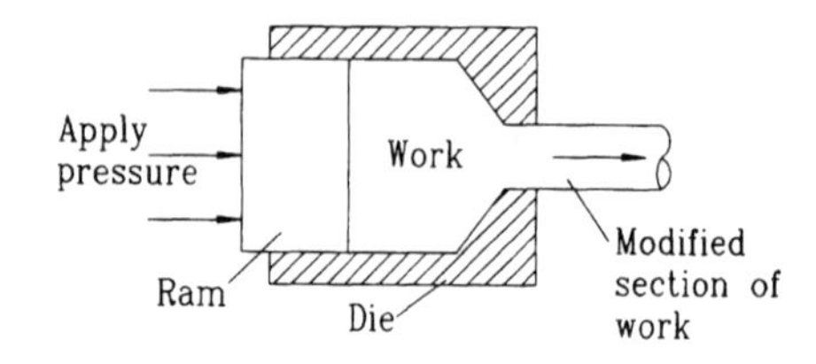

Figure 8.8 *Extrusion*

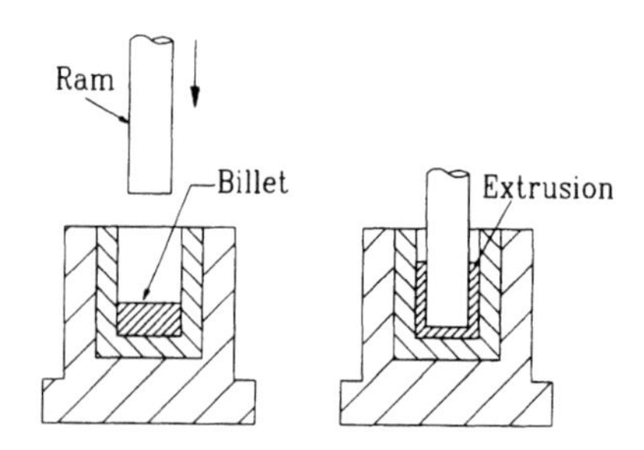

Figure 8.9 *Back extrusion*

A variation on the process is *backward* or *impact extrusion*, as in Figure 8.9. This is always performed cold and usually only on the more ductile non-ferrous metals. A punch enters the die at high speed, the metal being extruded through the gap between the punch and the die. Typical components produced in this way are toothpaste tubes and battery cases.

A related process is *wire drawing*, where the diameter of wire or bar is reduced by pulling it through a die. The process can involve a number of stages to ease the rate of reduction. Drawing offers good dimensional accuracy and mechanical property improvements. It is a cold working process (hot wire would have insufficient tensile strength). Note that extrusion is a pushing operation while drawing is a pulling operation.

8.4.5 Sheet metal processes

So far we have introduce a number of bulk deforming processes. There is also another group of operations that can be performed on metals in the sheet form.

Bending

This is a simple operation where a punch forces the material into a die (Figure 8.10 (a)). The material is raised to its yield point on both sides of the neutral axis, one side being in tension, the other in compression. Once the punch is removed the residual forces in the material tend to cause

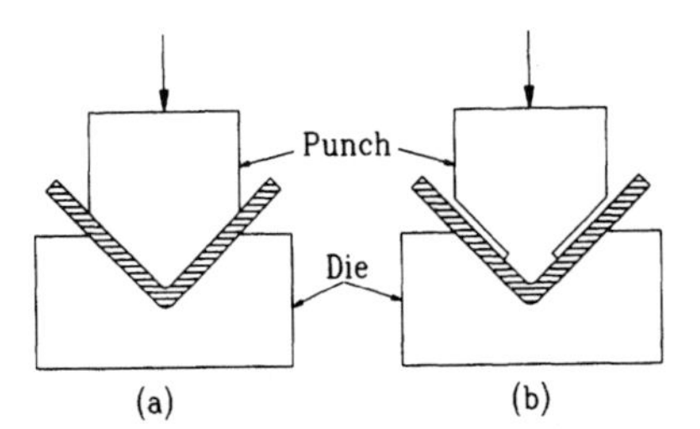

Figure 8.10 *Bending*

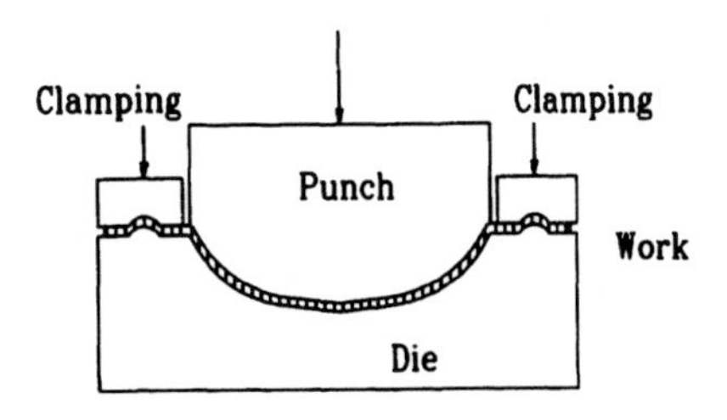

Figure 8.12 *Stretch drawing*

the bend to open, i.e. spring back. This can be overcome in a number of ways such as over-bending, i.e. bending to a greater angle than required, or shaping the nose of the punch (b) so that it indents the work, creating plastic compression throughout its thickness. Alternatively, prior to bending, the sheet could be clamped so that the whole operation stretches the material, causing a tensile yield throughout its thickness.

Stretch forming

This process uses only one mould, over which the sheet is stretched with sufficient force to exceed its elastic limit (Figure 8.11). Only simple shapes can be generated in this way. The wall thickness reduces with stretching but tends to remain constant as the stress is evenly distributed across the tool. The process is suitable for large parts and is used in the aircraft and coach building industries.

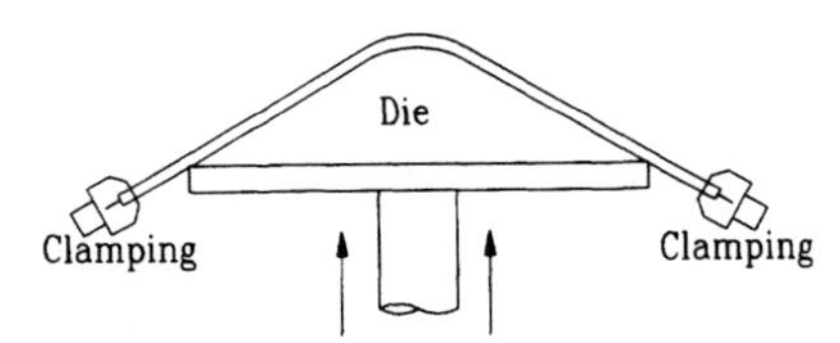

Figure 8.11 *Stretch forming*

Stretch drawing

This is similar to stretch forming in that the material is clamped around its perimeter, but a female die is used to control the final shape (Figure 8.12). Again, the shape is attained by a reduction in the material thickness.

Deep drawing

Unlike stretching, deep drawing allows the blank to be drawn into the die. This allows a much deeper pressing to be achieved. The process often takes place in a number of stages (see Figure 8.13), with annealing between them to minimise the risk of tearing. Sometimes a blank holder is used to prevent wrinkling as the material is drawn into the die. The pressure on the blank holder is sufficient to control the rate of material movement into the die rather than restrict it. Deep drawing can normally only be performed on very ductile materials – brass, copper, aluminium and some steels. Typical products include oil filter cases, fire extinguisher bodies, military shells, etc.

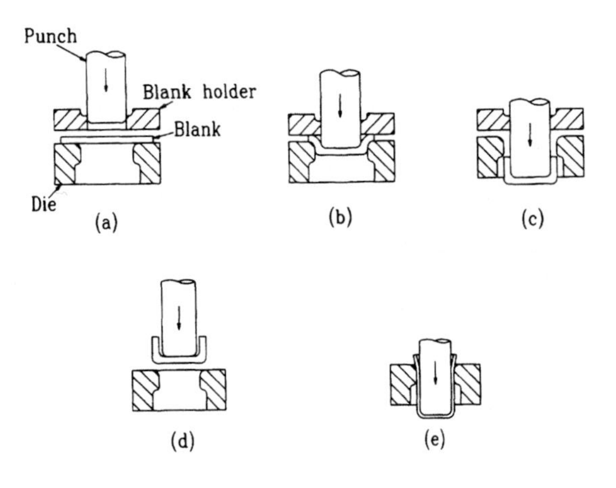

Figure 8.13 *Two-stage deep drawing*

Pressing operations

A number of basic operations can be performed on a press tool in addition to those involved with shape forming. *Blanking* or *shearing* (Figure 8.14) will cut the sheet to size and shape prior to deformation, and *piercing* will add holes as appropriate. Mass-production products such as sheet-metal car-body parts are known under the generic term, pressings. The shape deformation process is usually a mix of stretching and drawing. To help achieve the high production rates needed, some press tools are designed to perform two or more of these operations simultaneously: the holes may be pierced in the same stroke of the press as the final deformation operation but with the tool designed to complete the deformation before piercing.

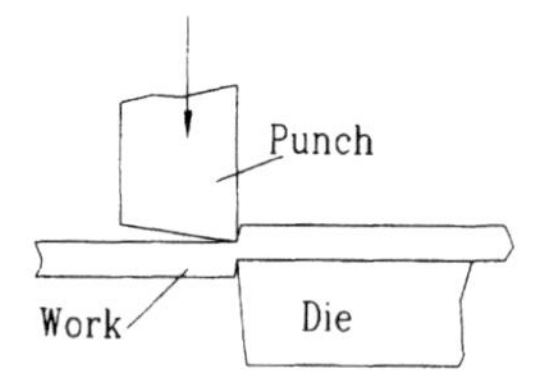

Figure 8.14 *Shearing*

Spinning

This is a versatile process used for forming sheet metal parts with an axial symmetry. A circular blank is held against a male die which is rotated using a mechanism similar to a lathe spindle (Figure 8.15). A specially shaped tool forces the material gradually against the die. The blank thickness is largely unchanged through spinning. As the die is not subjected to material movement across it, it is often made of wood. The low tooling cost can mean that spinning can be used for low volume production runs. It can be used for a wide range of component sizes, from the domed end of a petrol tanker to a copper lamp shade.

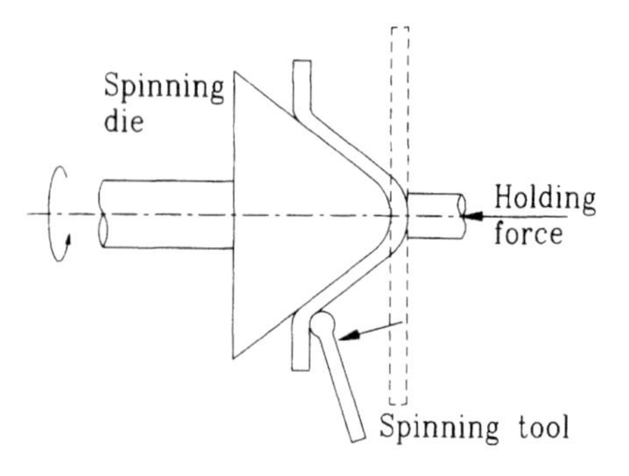

Figure 8.15 *Spinning*

Flow turning

Related to spinning, but in this case the diameter of the blank remains constant, the shape being generated by a thinning of the material (see Figure 8.16).

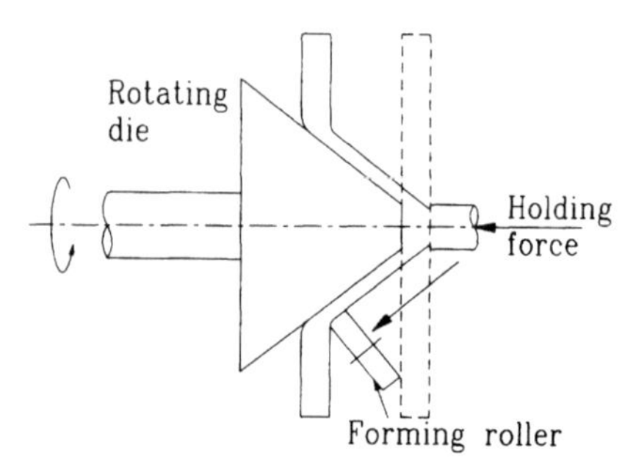

Figure 8.16 *Flow turning*

8.5 Processes for non-metals

8.5.1 Plastics

The processes discussed so far apply to metals, although many can also be used with plastics – machining and casting, for example. Here we will introduce some of the common methods of creating shapes in plastics that are specific to the material.

Compression moulding

Solid granules of the raw, unpolymerised plastic are placed in a heated die. A plunger (also heated) is lowered into the die, creating sufficient pressure on the plastic as it melts, to fill the cavity (Figure 8.17). Some designs of mould (a) allow for a small amount of flash (similar to the closed die forging of metals). This will necessitate an additional operation to remove the flash, but avoids the need for close control of the quantity of material entering the die. The landed plunger type (b) requires precise measurement of the raw material, but avoids flash, and provides good dimensional density control. The process is normally used for thermosetting plastics, as the alternate heating and cooling of the mould would make it uneconomical for thermoplastics. (See also section 12.6.)

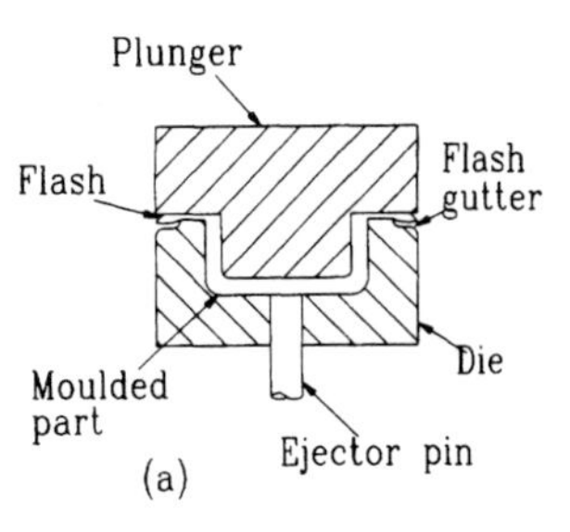

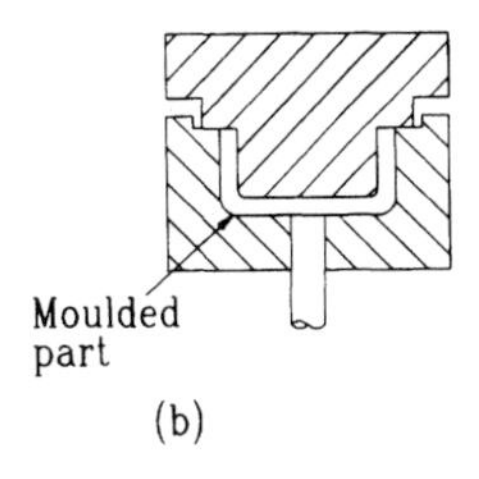

Figure 8.17 *Compression moulding*

Injection moulding

This is the most common method of shaping thermoplastic products. The plastic as a powder is placed into a hopper which gravity feeds into a pressure chamber ahead of a plunger (Figure 8.18). The plunger forces the material through a preheating chamber. From here it is forced through a section fitted with a central torpedo shape which deflects the powder to the edges of the chamber, close to the heating elements, to ensure even heating. In this area it is melted and heated to 200–300°C. The melt then feeds through a nozzle into the mould. The plastic solidifies almost as soon as it enters the mould, resulting in a very quick process. As the mould remains cool it is suitable for thermoplastics.

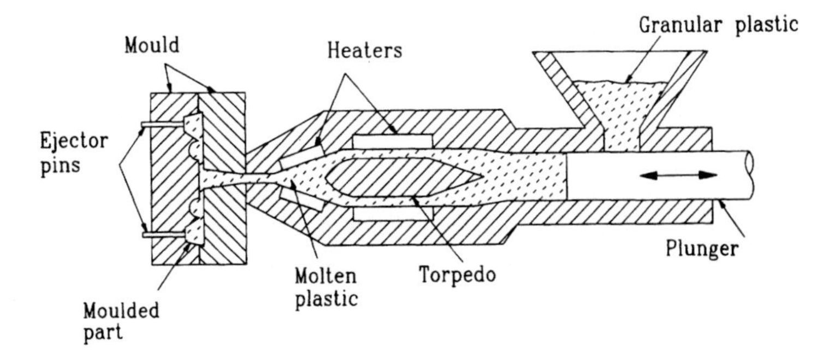

Figure 8.18 *Injection moulding*

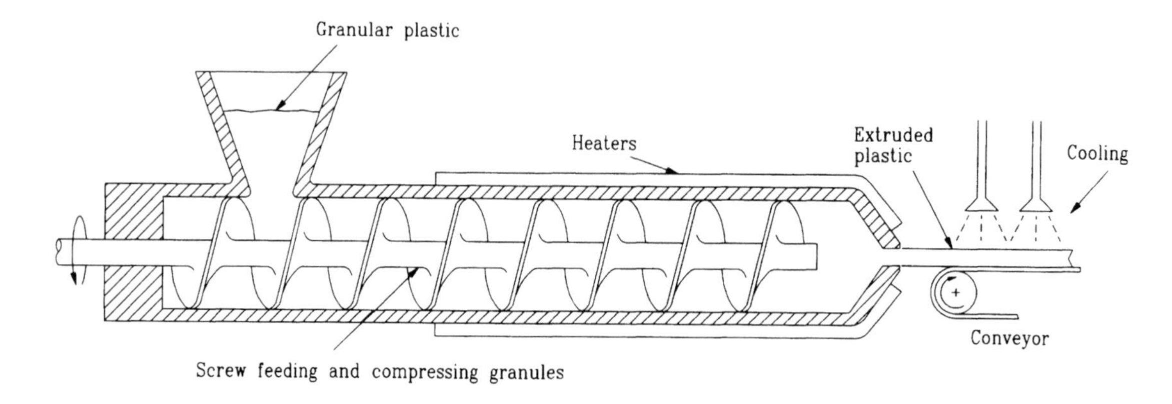

Figure 8.19 *Extrusion moulding*

Extrusion moulding

The raw material, in the form of powder or granules, is fed via a hopper into a rotating screw, which carries them forward to a heated zone where they melt. The action of the screw forces the molten plastic through a hole in the die. The hole is shaped to give the required section. As the plastic emerges from the die it passes onto a conveyor belt where it is cooled by air or water sprays (Figure 8.19). It continues to cool as it passes along the belt, eventually being cut into lengths or coiled as appropriate. It can be used for thermosets, but the risk of thermosets setting in the screw pump mean that thermoplastics are more commonly used. Typical products are pipes and angle sections. If a multi-hole die is used, plastic fibres can be produced.

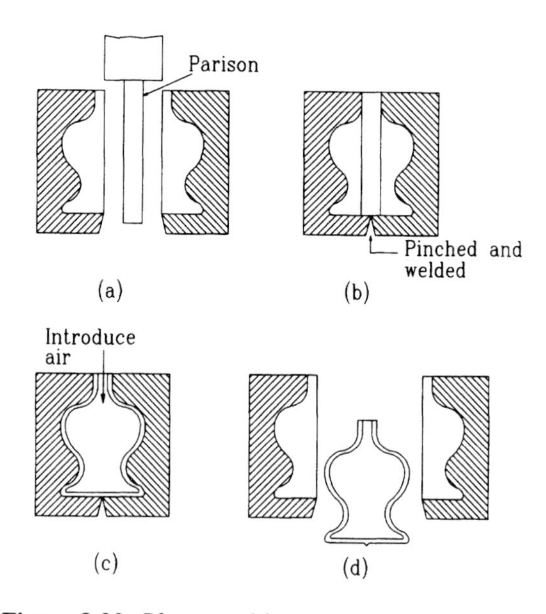

Figure 8.20 *Blow-moulding*

Blow moulding

A hollow plastic tube (the parison) produced by an earlier extruding operation is passed between two halves of a mould: see Figure 8.20 (a). As the mould closes (b) it pinches and welds the bottom end of the parison. Air (or a non-reactive gas such as argon) is blown in through the top of the tube and heat applied. The tube (c) is softened by the heat and expanded by the air pressure to rest against the mould wall.

This an inexpensive process with potential for very high productivity. It is commonly used for producing small plastic bottles and containers, but can also be used to produce large parts such as drums of up to 2000 litre capacity. It is suitable for thermoplastics.

Stretch blow-moulding is similar, except that the expansion is in both the axial and radial directions.

Vacuum forming

Sometimes called thermoforming, vacuum-forming is used to create shapes from sheets of thermoplastic materials. As shown in Figure 8.21, a sheet of plastic is placed over a die, and held around its edges. Heat is applied to soften the material, and then air is extracted from the space between the sheet and the surface of the die, thereby pulling the sheet into position. Cold air is introduced, the vacuum dropped and the completed component removed.

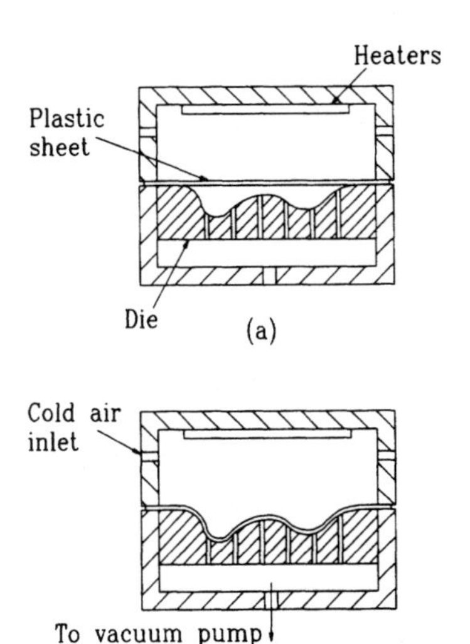

Figure 8.21 *Vacuum-forming*

Foam plastics

A number of plastics are produced with internal voids. These can be the foam used to insulate buildings, packaging materials or even as structural elements such as car dashboards. The 'air bubbles' can be introduced in a number of ways: by mechanical whipping, by adding hollow spheres, or by introducing a gas which diffuses through the plastic.

8.5.2 Ceramics

In terms of materials processing, ceramics fall into two distinct categories: glasses and crystalline ceramics.

Glass

Glass is generally shaped by raising its temperature to the point at which it becomes a viscous liquid and can be made to flow to the desired shape. The processing techniques are then similar to those for thermoplastic polymers. Sheet or plate glass is normally produced by casting onto the surface of a bath of molten tin in a controlled atmosphere. This is called float glass. The lower surface is kept smooth by the tin while surface tension effects keep the top smooth. Components such as bottles or light-bulbs are produced by blow-moulding similar to that used for plastics. The main difference is that the parison is produced by pressing a gob of hot glass in a blank mould to an approximate shape before passing to the blow-moulding stage. Light-bulbs can be blow-moulded on high-speed rotary machines with outputs of 2000 pieces a minute. Fibres for insulation or reinforcement can be produced, as with plastics, by hot extruding through multi-holed dies. The fibres are flowed onto a rotating mandrel and mechanically stretched as they are cooled. They can be further spun to create glass wool.

Crystalline ceramics

These are generally hard and brittle, making them difficult to form or machined, and have high melting points, making casting difficult. Therefore they are normally processed in the solid state using the powder metallurgy techniques described in section 8.3. The dry pressing uses high pressures, but permits the mass production of parts to high tolerances. Spark-plug insulators, for example, are produced in this way.

If the ceramic powder is mixed into a liquid slurry it can then be cast into a mould (normally plaster of paris) with fine pores that allow the liquid to be removed by capillary action. The remaining green casting can then be sintered. This type of casting is called *slip casting*. In a similar way it is possible to blend the ceramic with various additives to create a mixture that can be formed with pressure and heat. Under these conditions ceramics can be extruded or even injection moulded. The additives are then removed, usually by controlled heating, before the ceramic is sintered.

Finally, some of the engineering ceramics can be machined, but they require unconventional techniques, such as diamond abrasives, high-energy beams, or chemical machining.

8.5.3 Composites

The term 'composite' refers to structures made from two or more distinct materials, the identities of which are maintained, but with the physical properties of the final structure offering advantages over those of its components. The manufacturing technique depends on the types of composite. Powder metallurgy (section 8.3) can be used to create composite items such as cutting tools or self-lubricating bearings, or electroplating would be used to create a galvanised sheet (Chapter 10, section 3).

The most common perception of composites is the use of fibre reinforced material. Here strands of fibre in materials such as glass, carbon, kevlar etc. are combined with resins to create a structure with a high strength to weight ratio. This can be done with a low level of capital equipment, but with a high labour content.

A mould would first be coated with a gel coat, which may be coloured to avoid later painting. Next a layer of fibre, in the form of woven matting or chopped strands, would be added, and coated with resin, making sure that none of the fibres are left dry. Further layers of fibre and resin are added until the desired thickness is reached. The resin will contain a hardener so that after a suitable time delay the structure will stiffen and can be removed from the mould. The quality of the final product is dependent on the skill employed during the lay-up process. Any air pockets or air inclusions will reduce the strength of the final product. This is a low-cost method of manufacture only suitable for components that are not highly stressed, such as boat hulls, or replacement wings for cars.

If the full potential of the strength to weight ratio is to be achieved, the above process needs to be refined. Prepregs are fabric woven from the reinforcing material, which has been infiltrated with resin under conditions in which the resin is partially cured. These sheets are normally stored in refrigerated containers to increase their shelf life. The prepregs are placed in layers around the mould, with care being taken to meet the design requirements of fibre alignment. They are pressed by hand into shape, then put in a plastic bag from which the air is evacuated. This pulls the layers together tightly onto the mould, removing any air pockets. The whole assembly is then placed in an autoclave where it is subjected to increased pressure and temperature to a closely defined cycle. The increased temperature causes the resin to cure fully. At the end of the cycle the final product can be removed from the mould. This is a very slow and expensive method of manufacture, but the close controls allow much greater predictability of the properties of the final product. Typical products would be structural components of aircraft or racing cars.

8.6 Summary

On completing this chapter you should have an understanding of the ways in which the shape of material can be changed by making the material flow. This flow may be in the liquid state (casting), in the solid state (plastic deformation), in the particulate state (powder metallurgy), or in the viscoelastic state (moulding of plastics).

Casting can involve permanent or expendable moulds, and permanent or expendable patterns, with various combinations of each. You should understand the main differences between the various processes and be capable of selecting an appropriate process, taking account of factors such as accuracy, size, and number required.

Plastic deformation can either involve cold or hot working. You should appreciate the benefits of each and understand the mechanism of common bulk material or sheet metal forming processes.

Some materials are too hard or brittle to be deformed without fracturing, and have too high melting points for easy casting. These are candidates for powder metallurgy. You should understand the process and some of its benefits and disadvantages.

Finally, you should also be aware of a number of common manufacturing methods for non-metallic parts.

8.7 Questions

Now try answering these questions to check your understanding. Refer to the relevant section in this chapter if you have difficulty with any of them.

1. Describe three variants of a casting process using an expendable mould and a permanent pattern. (8.2)
2. What happens to the pattern in the full mould casting process? (8.2)
3. Name some areas for potential quality problems with investment castings. What factors would you consider when deciding to use the investment, die or sand casting methods? (8.2)
4. What is the material shaping process that results in shrinkage of the part as it is heated? Explain why this happens. (8.3)
5. Name and describe the three stages in powder metallurgy. (8.3)
6. Describe the difference between hot and cold working, giving the relative advantages and disadvantages for each. (8.4)
7. Compare the open and closed die forging processes. (8.4)
8. Describe the deep drawing process, explaining the purpose of the blank holder. Why is annealing sometimes needed as part of the process? (8.4)
9. What is the difference between spinning and flow turning? (8.4)
10. It is normally better to use thermoplastic materials with extrusion moulding, but thermoplastics with compression moulding. Explain why. (8.5)

9 Joining Techniques

9.1 Introduction

The previous two chapters have introduced a number of manufacturing techniques for producing single components. Wherever possible, it makes sense to reduce the number of components needed in a product as this simplifies both the manufacturing control and assembly processes. However, the vast majority of engineering products require a number of components and therefore need a manufacturing technique to create the assembly. The technique chosen will depend on a variety of factors, such as the permanency of the joint, the materials used, or the ability of the assembly to withstand a heat input. Figure 9.1 summarises the main techniques available. In this chapter we will discuss the techniques and later, in Chapter 13, we will expand upon the reasons for selecting a particular method.

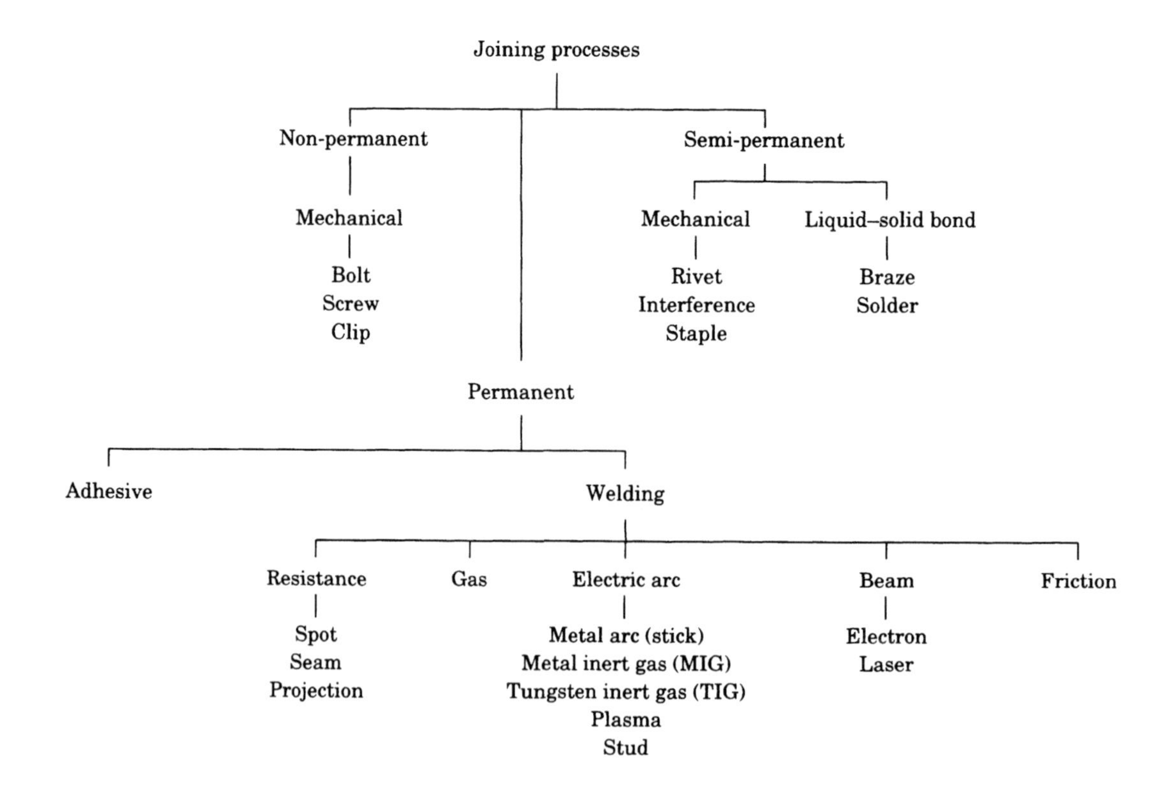

Figure 9.1 *Joining processes*

9.2 Non-permanent joints

Joints that need to be broken and remade a number of times during the product's life are likely to utilise a mechanical joint such as a bolt, screw or clip. An integral thread would normally be produced by turning, rolling, or tapping (see Chapters 7, 8). In some cases it may even be possible to mould the thread, for example as on the lid of a coffee jar. Where third-party fasteners such as bolts, nuts or screws are used, the engineer would normally select those available from a specialist supplier, so from the manufacturing point of view the engineer needs to concentrate mainly on the assembly implications (see Chapter 13).

Interference

In this type of joint the relative dimensions of the mating surfaces are tightly controlled to ensure that on assembly each surface suffers a slight deformation, which in turn creates a pressure between the surfaces. The resulting friction holds the components together. Normally the mating surfaces are cylindrical; examples can be found in cylinder liners or bushed plain bearings. The tolerances on the diameters are close and the surface finish should be good (see Chapter 13, section 13.5.2). When assembling, care must be taken to ensure correct alignment. Often the outer component may be heated or the inner cooled to facilitate assembly.

9.3 Semi-permanent joints

In this category we include those joints that would not normally be broken by the user, but may need to be undone when the product is serviced or repaired.

9.3.1 Mechanical

Rivet

This is probably the most common type of joint in this category. The items to be joined are clamped between the two heads of the rivet. A solid, single-headed rivet is passed through pre-drilled holes in the items to be joined and the second head generated by upsetting the non-headed end. Where large rivets are used for joining plates (e.g. in shipbuilding) the rivets are normally heated prior to application, so that on cooling the joint tightens. Lighter gauge materials can use pop rivets (Chapters 4 and 13), which are hollow, the second head being formed by flaring the non-headed end of the rivet. The operation is performed using a rivet gun, and has the advantage of only needing access from one side of the assembly. Rivets can be removed by drilling out one of the heads. This is fairly easy with a pop rivet.

Staple

The common wire staple is frequently used in the furniture industry as a quick and cost-effective fastening method.

9.3.2 Liquid–solid bonding (brazing and soldering)

In this case the joint relies on adhesion of the components to be joined to a filler material. Sufficient heat is applied to the joint to melt the filler but not the base metal of the parts being joined. Compared with welding, where the applied heat locally melts the items being joined, this method offers a number of advantages: the lower heat input involves less distortion, differing wall thicknesses can be joined, and, most important of all, full access to the joint area is not required, as surface tension and capillary action will draw the filler material into the joint area. The main difference between soldering and brazing is the melting temperature of the filler material. Soldering takes place below about 430°C and brazing above this temperature.

If this type of joint is subsequently taken to a temperature above the melting point of the solder or braze metal, the joint will part.

Brazing

The brazing filler is placed in position prior to heating the joint. It can be either inside or outside the joint. Obtaining the correct clearance between the mating surfaces is crucial. A typical clearance is 20–120μm. If the gap is too large the joint will lose strength; if it is too small the filler penetration of the joint by capillary action will be poor.

The success of a brazed joint depends on good adhesion between the filler and the base metal. The mating surfaces must therefore be clean and free of any scale or contaminants such as oil or grease. Oxides generated during the heating process can also adversely affect adhesion. To minimise this problem the process can take place in a neutral atmosphere, or under vacuum. Additionally, the mating surfaces are normally coated with a flux which reacts with the surfaces to facilitate wetting by the braze material. The flux melts at a low enough temperature to prevent oxidation of either the surfaces or the braze, and has a low viscosity so that it is easily displaced by the molten braze. Fluxes are normally in powder or paste form and are of a composition tailored to a particular application. As most fluxes are corrosive the residue must be removed from the work as soon as the brazed joint has been made. The majority of fluxes are soluble in hot water so can be removed by immersing in a tank of hot water for a few minutes.

There are a number of ways by which heat can be applied to the joint. *Torch brazing*, using a gas flame torch to heat the area selectively, is a common method for manual repair work, as well as in production applications. The process relies largely on the skill of the operator, particularly in terms of temperature control. Greater control can be achieved with *furnace brazing*, where the items to be joined are pre-loaded with braze metal, and placed in a furnace. This not only enables the temperature to be closely controlled but also the content of the ambient atmosphere can be adjusted to reduce any oxide films. Vacuum furnaces are often used for particularly reactive materials. Ideally, components should be designed so that their joints are self-locating. Where this is not possible, jigs and fixtures will be required to maintain correct alignment while in the furnace. The process is suitable for mass production.

Another technique, *dip brazing*, involves immersing the assembled components in a bath of molten salt, whose temperature is maintained at slightly above the melting point of the braze metal. Direct contact with the heating medium provides protection against the formation of oxide films and ensures a rapid and even temperature rise for the component. It is particularly suited to brazing aluminium components, where precise temperature control is needed. As with furnace brazing, the braze must be preloaded and the assembly loaded into a jig or fixture.

Resistance brazing involves pressing the components to be joined between two electrodes through which a current is passed. Most of the electrical resistance is in the electrodes, usually carbon or graphite, which become hot and conduct their heat to the workpieces. This process is mainly used to join electrical components such as cable connectors.

Induction brazing employs high-frequency induction currents. The process is rapid, and the operation can be semi-automatic, obviating the need for skilled labour. It is not suitable for metals whose melting point is close to that of the braze metal, such as aluminium or magnesium alloys.

Braze metals are normally copper or copper alloys.

Soldering

Soldering is effectively a brazing operation that takes place at below about 430°C. It is frequently used to make electrical connections, and as a filler to disguise joints on motor-vehicle body parts. The joint strength is less than that of a brazed connection, but the parts can be joined without exposing them to excessive heat. The surfaces to be joined must be free of any dirt, oil or grease.

Fluxes are needed to facilitate wetting and avoid problems with oxide films. They can be corrosive or non-corrosive. When making electrical connections a non-corrosive flux must be used to avoid any local corrosion causing a high electrical resistance.

Heat can be applied using any of the brazing methods. The low temperatures involved enable soldering irons to be used. Wave soldering is a common technique for joining a number of electronic components to printed circuit boards, where the boards are passed over a bath of molten solder. The solder is pumped through a nozzle so that fresh solder moves in a slow wave under the board. As the solder is still molten when the work is removed it is largely only retained in the joints.

Most solder materials are alloys of lead and tin.

9.4 Permanent joints

In this category we include joints that are not intended to be broken during the service life of the product. There are two main methods: welding and adhesives.

9.4.1 Welding

Welding is a process in which two materials are joined by the formation of an interatomic bond. In most cases this is achieved by melting the joining surfaces, sometimes adding some filler material, and allowing to solidify. The resulting joint will be much stronger than its soldered or brazed equivalent; indeed, the joint should be as strong as the parent material. However, the greater heat input may lead to distortion of the components as the weld shrinks during cooling. Care needs to be exercised in selecting a suitable process. We will now consider some of the more common options.

Gas welding

Oxygen and acetylene are burnt in a welding torch, producing a flame temperature of around 3500°C. This flame is played on the joint so that the metals locally melt and fuse together. As there is normally a slight gap at the joint, filler metal is often added in the form of a wire or rod which is melted in the flame or pool of weld metal. Fluxes can be used to improve the bond by minimising any contaminating oxides. The welding rod may be pre-coated or can be dipped in a flux paste. Gas welding tends to be slow, and can involve distortion, as the rather unfocused heat source tends to input more heat to the workpiece than some other welding techniques. Whilst it may be suitable for heavy components on, say, a farm tractor, it is not now considered suitable for use on the light gauge steels used in car-body manufacture. It tends to be used more for repair work than as a main production process.

Arc welding

When an electric arc is struck between two electrodes it creates a concentrated heat source of some 3900°C. Arc welding uses such an arc to melt the joint surfaces. The simplest form of *stick welding* uses a rod of filler metal as one electrode, and the workpiece as the other (Figure 9.2). As the arc is struck, the end of the filler rod melts and passes drops of molten metal to the joint. To maintain the gap between the consumable electrode and the workpiece (and thus a stable arc) the electrode has to be moved gradually towards the joint. The consumable electrode is pre-coated with a material that vaporises as the

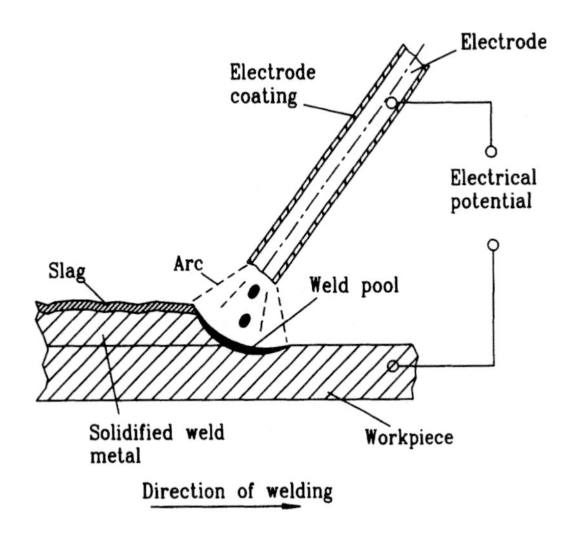

Figure 9.2 *Stick welding*

electrode melts, so creating a protective atmosphere to protect the molten metal from contamination. The electrodes are supplied in 'sticks', each about 450mm long. These melt during the welding process at a rate of about 250mm per minute, so need to be replaced frequently. The completed weld is covered with slag, which has to be removed (usually mechanically) to give a clean component. If the welding process has to stop to allow a new stick to be fitted, or the weld requires more than one pass, the slag must be removed before the welding can continue. This means stick welding tends to be slow and only suitable for manual operation.

The flux coating used in stick welding is brittle and would crack and break away from the rod if the stick was bent. *Flux core welding* uses a hollow wire, fed from a coil, as the electrode. The wire is filled with a granular form of flux which is used to protect the molten weld metal, either by creating a protective gas shield, or by covering the weld with slag. The automatic feeding of the wire means that the welding head can be kept a constant distance from the workpiece, and removes the need to continually replace the electrodes. This process is suitable for automating.

Submerged arc welding (Figure 9.3) offers yet another variant on the method of applying the flux. The consumable electrode is a plain wire of the filler material, with the granular flux being applied separately through a hopper, so that a thick layer covers the weld as it is created. This is usually an automated process where high welding speeds and deep penetration are required. Welding speeds of 300mm per minute in plate up to 40mm deep is quite possible. Typical applications are in the construction of large structures such as in shipbuilding.

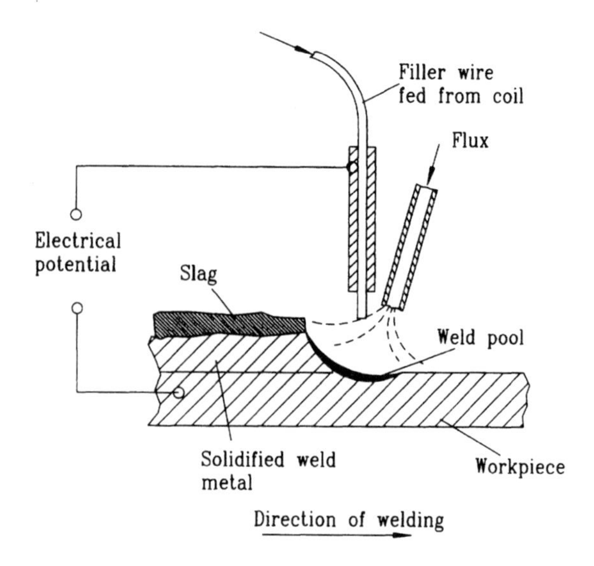

Figure 9.3 *Submerged are welding*

As an alternative to the use of a flux an inert gas can be employed as a shield. *MIG (metal inert gas) welding* (Figure 9.4) uses as the consumable electrode a wire fed continuously through the centre of the welding gun. An inert gas (normally argon, or an argon and CO_2 mix) is fed into the welding head and exhausts with the wire providing an invisible shroud shielding the weld from the effects of the atmosphere. The result is a clean, neat weld requiring no slag removal. The process requires less skill than gas or stick welding and is therefore eminently suitable for automation. When operated manually, the variables that need to be set are the wire feed rate, electrode voltage, and gas pressure. An automated system would also need to control the parameters covered by the skill of the manual

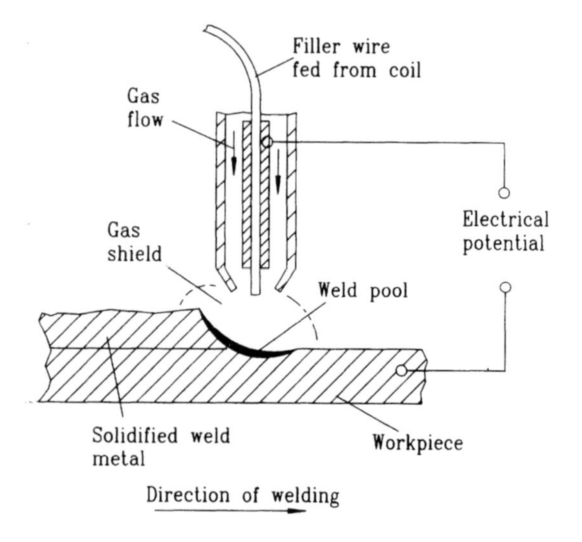

Figure 9.4 *MIG welding*

operator, i.e. welding speed, and arc length. This process is well suited to the repair of relatively thin car-body panels because the speed of welding minimises the heat input and hence any subsequent distortion.

TIG (tungsten inert gas) welding (Figure 9.5) is similar to the MIG process in that an inert gas (normally argon or helium) is used as a shield. However, the tungsten electrode in the welding head is virtually non-consumable. The arc is initiated with MIG by striking the electrode against the workpiece, but if this were done with the TIG welding the tungsten electrode would become contaminated with the metal being welded. A high-frequency, high-voltage current is superimposed on the normal welding current to create a spark which initiates the arc. This arc is of constant length and fairly easy to maintain. If there is a close fit between the items being joined it may not be necessary to use any filler material. Alternatively, filler can be provided by a separate wire feed. For instances where a large amount of filler is required, the filler wire is preheated. This allows higher deposition rates. A skilled operator can use the TIG process to produce small, neat welds that do not require any subsequent slag removal. Typical applications are sheet metal or tubular materials up to about 4mm thick. Its use is fairly common in the fabrication of equipment for the food and chemical processing industries where a good standard of finish without any inclusions is required.

Welding very thin sheet with TIG involves low currents, which tend to reduce the stability of the arc, and hence the controllability of the welding. This can be overcome by using *pulsed TIG welding*. Here a low background current is used to maintain the arc with minimum heat input. On this is superimposed a higher pulsed current (typically one to ten pulses per second) which tends to stabilise the arc while creating a series of overlapping spots in the melt. The heat input is dependent on the mean current flowing, and this can be closely controlled by varying the peak current, and the frequency and duration of the pulses. The greater control of the energy inputs results in an energy- and cost-efficient weld with minimum distortion or discolouration that is suitable for thin materials. An application of this process is in equipment for welding fittings automatically on to stainless steel valves, pipework, etc. for the food process industry.

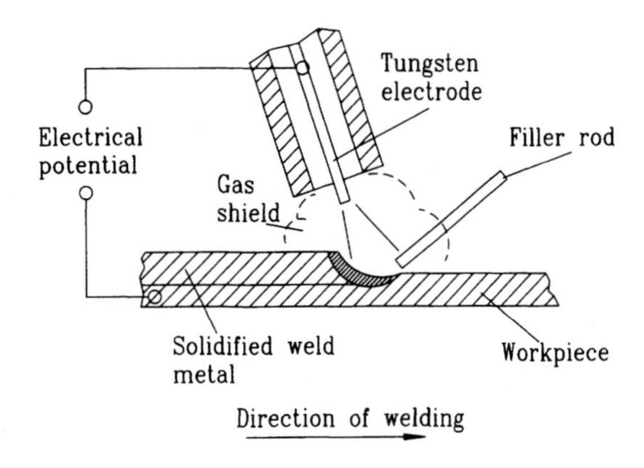

Figure 9.5 *TIG welding*

Plasma arc welding is another process that uses a non-consumable electrode. The arc is created between the electrode and either the welding gun or the workpiece. An inert gas is directed through a small opening in the welding gun (through which the arc passes), where it is heated to a high temperature, forming a plasma. The flow of hot gases transfers heat to the workpiece to create the weld. The result is a process capable of high welding speeds, and deep penetration due to the good energy concentration.

Finally, there is a specialised form of arc welding which can be used as an alternative to riveting to hold fasteners in place. This is called *stud welding*. The end of the stud to be welded has a pip in its centre. It is placed in the welding head and brought towards the workpiece where an arc is struck from the slight projection, causing the joint surfaces to melt. The components are held together under light pressure as the metal solidifies. The process is fast, requires little skill on the part of the operator, and avoids the need for creating the holes that rivets would need.

Resistance welding

Resistance welding involves passing an electric current through the joint, heat being generated by the electrical resistance. At the same time, the joint is held under pressure by the electrodes, the combination of heat and pressure creating the weld. The process is normally very rapid, so is well suited to automated manufacture of sheet metal components. It does, however, normally require access to both sides of the joint.

Spot welding (Figure 9.6) is widely used in industries such as white-goods and motor-vehicle manufacture. A pair of, normally water cooled, electrodes press the two sheets together, the current is applied, and the joint heats, creating a molten pool or weld nugget. The current is turned off, and the pressure released once the nugget has solidified. The whole process normally only takes about a second. Spot welding machines come in all shapes and sizes, ranging from a fixed machine (about the size of a pillar drill) with a pair of electrodes, to portable welding guns used on a car assembly line. These are held on counterbalanced wires with the electrodes fitted into wide opening jaws that allow access to both sides of the joint. Today many of the portable guns are built into robots which perform the welding tasks. The characteristic of a spot weld is a slight dimple in the surface at the point of the weld, with slight discolouration.

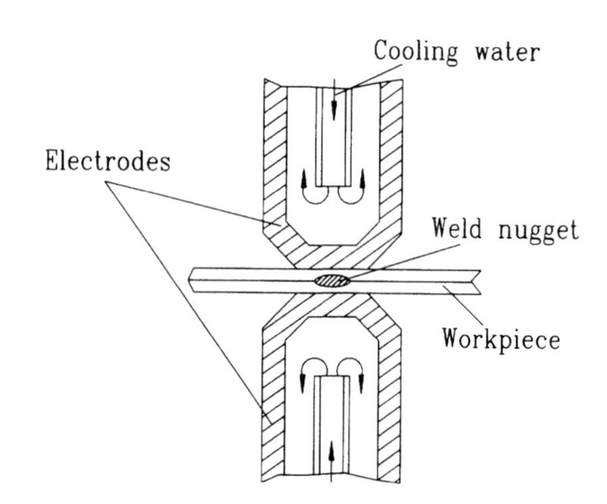

Figure 9.6 *Spot welding*

Seam welding allows a series of spots to be welded along a seam much more quickly by incorporating the electrodes in rollers. The current is continually switched on and off during the welding process, thus creating a series of evenly spaced spot welds. If the welds are placed closely enough together a leak-proof joint will be created. A typical application might be the joint between the roof section of a car and its windscreen header rail.

Projection welding (Figure 9.7) involves embossing a dimple on one of the items being joined at the place where the weld is to be located. The components are then placed between large electrodes in a projection welding machine, and pressure and current applied as with spot welding. As the current flows from one electrode to the other through the workpiece, the point of greatest resistance will be at the dimples as this is the least area of contact. Consequently this will concentrate the heating at the required place for the weld. The dimples, of course, disappear as the weld is made. Several projection welds can be made simultaneously, by increasing the number of dimples, the only limiting factor being the ability of the projection welding machine to deliver sufficient current and pressure.

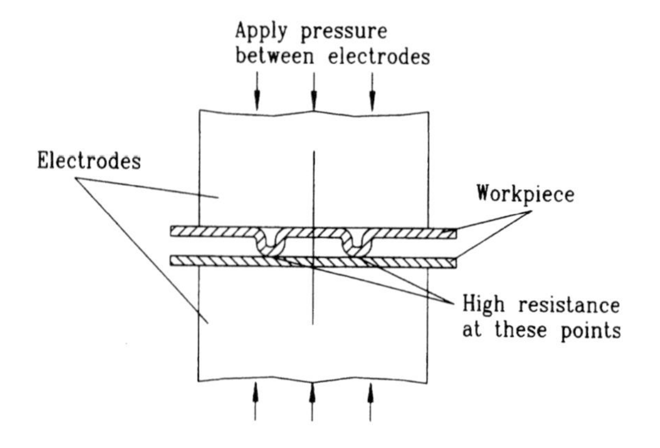

Figure 9.7 *Projection welding*

Beam welding

Beam welding involves passing the joint under a high-energy beam to obtain the welding conditions. There are two basic types.

Electron beam welding (EBW) equipment has two elements: the gun and the chamber (Figure 9.8). The gun creates a beam of electrons by using a high voltage current to heat a tungsten filament to around 2200°C so that it emits electrons. These are accelerated through an anode, and focused by coils into a concentrated beam. This beam, although of high energy concentration (up to 30 kW over a circle of 1mm diameter) will be dissipated in air. In the gun it is created in a high vacuum (10^{-4} torr). The workpiece also needs to be in a vacuum, although this is usually only about 10^{-2} torr. Normally the gun is attached rigidly to the top of the work chamber and the joint passed under the static beam. This can be done either by rotating the work in a chuck, or by the use of an XY table. The weld penetration depends on the beam parameters (energy, focus, etc.), the material, and the time spent under the beam. No filler is used, and as the beam is very narrow, gaps at the joint must be kept very small (0.2mm max). The high voltages employed (up to 120kV) mean the equipment needs shielding to protect the operators from X-rays.

Heating is very localised and the welds very narrow (depth to width ratios of 25:1 are common), so distortion is minimised. Disadvantages are that the process is slow (pump down time to create the vacuum) and expensive. It is used for a wide range of products, from large machines able to repair aircraft engine parts to small equipment used for instrument manufacture.

Laser beam welding also uses a high-energy beam, this time a focused laser, and also creates a narrow weld. It does not require a vacuum and no X-rays are created. The energy concentration is somewhat lower than EBW so this equipment tends to be used for thin gauge materials (0.2mm to 20mm).

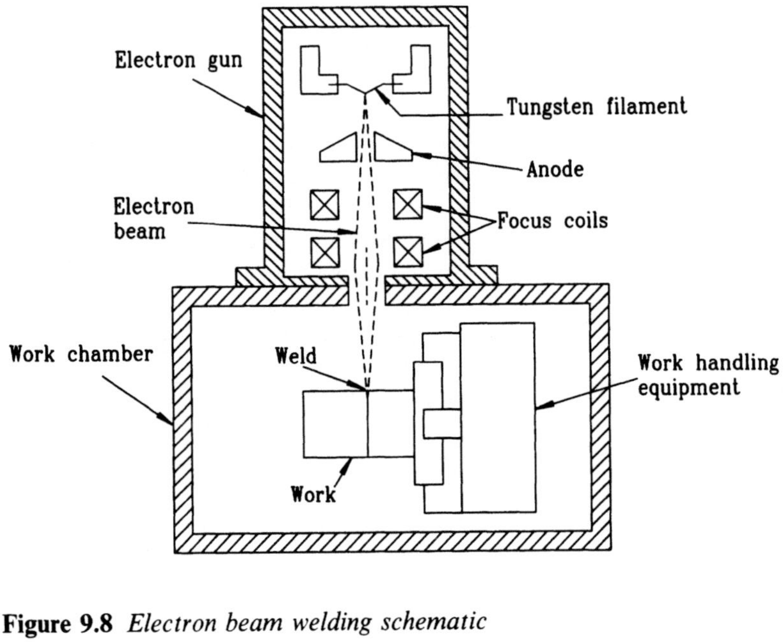

Figure 9.8 *Electron beam welding schematic*

Friction welding

Friction welding is used to create joints on components with a circular section. One part is rotated, and the other held stationary. The parts are pressed together and the resulting friction generates sufficient heat for the welding process. Once the joint temperature is high enough the rotation is stopped, and the pressure across the joint maintained to complete the weld.

9.4.2 Adhesive bonding

Adhesives have been in use, in one form or another, for many thousands of years. Prior to the twentieth century, these were mainly either animal product or vegetable glues (resins, gums, waxes, etc.), and were used for bonding porous materials such as paper and wood. The somewhat narrow limitations of these glues provided the stimulus necessary for the significant advances since the 1920s of the development of new adhesives based on synthetic resins and other materials. Modern adhesives can be tailored for particular end uses and can be resistant to such environmental problems as moisture, mould growth, heat, vibration, shock, and so on.

Today, the technology of adhesives has advanced to the point where they are used in many and varied industrial applications such as the paper, footwear, furniture, building and construction, road-vehicle and aircraft manufacturing and electrical/electronics industries.

Adhesives are now available to bond most metals, plastics, composites, ceramics and woods.

The components of adhesives

The majority of adhesives today comprise mixtures of several complex materials which may be organic, inorganic or hybrid.

The components of adhesives are determined by their end use – the basic component being the binding substance which provides the adhesive and cohesive strength of the bond. This is usually an organic resin, but may be an inorganic compound, a rubber or a natural product.

To perform other functions, more constituents are included, such as:

- the *catalyst/hardener* – the curing agent for the adhesive system;
- the *modifiers* – chemically inert ingredients that are added to alter the adhesive properties;
- the *diluent* – the solvent that provides the adhesive's viscosity control; and
- the *accelerator/inhibitor/retarder* – the controllers of the curing rate.

Adhesives available

Broadly speaking, an adhesive will fall into one of the following types:

- two-part, room temperature cured, paste adhesives;
- one-part, moisture-cured adhesives;
- one-part, heat-cured paste adhesives;
- one-part, micro-encapsulated adhesive systems (toughened);
- anaerobic adhesives (i.e. which cure in the absence of oxygen); or
- heat-cured film adhesives.

The selection of the most suitable type will depend on the substrates to be bonded, the design of the joint, the forces present, the environmental conditions and the physical constraints.

Joint design

It is important to remember that the design of a bonded joint will not necessarily follow that of other fixing methods. See Chapter 13, section 13.4 for more details.

Surface preparation

In order to ensure a satisfactory result, it is imperative that not only is the correct design of joint and type of adhesive used, but also the surfaces of the adherents are suitably prepared – different surfaces require different preparation techniques. For maximum strength, all surface contaminants such as paint, grease, dusts, oxide films, moulds, releasing agents and others must be removed.

Note: the use of some adhesives and surface preparation compounds can, if not handled correctly, be hazardous to health. Always follow the instructions on the relevant containers and COSHH (Control Of Substances Hazardous to Health) statements.

There are three basic methods of removing contaminants:

- *abrasion*: includes the use of abrasive paper, sand blasting and vapour honing (in which powered abrasive is propelled by high-velocity water or steam against the surface)
- *degreasing*: surfaces are cleaned with either a hot alkaline solution or a solvent (e.g. acetone)
- *chemical cleaning*: etches the surfaces to form highly adhering oxides or deposit complex inorganic coatings.

Testing methods

The destructive testing of bonded joints is frequently carried out for evaluation, performance prediction and quality control reasons.

Parameters of interest to the designer/manufacturer include the ability to withstand:

- tensile forces
- shear forces
- peel forces
- cleavage
- shock.

In the past, these tests were carried out in laboratory conditions and did not satisfactorily replicate the environment in which the bonded joints would work. Recent years have seen moves towards carrying out these tests in more lifelike conditions, thus bringing more credibility to the results.

Advantages

Adhesive bonding is not necessarily the answer to all our joining problems. Nevertheless, recent advances in this field have provided many opportunities for their use. Advantages include:

- can be used to bond dissimilar materials
- can be used to bond thin sheet materials that would distort using other joining methods
- bonded joints can damp vibration
- stress within the joint is more uniformly distributed than in some other joining methods
- the bonding process is easily automated
- bonding can offer higher strength joints at lower cost than other joining methods
- adhesives can offer good insulating properties against electricity, sound and heat, and good sealing properties against moisture and chemicals
- refinishing costs are lower than for alternative joining methods
- bonding offers weight reduction – important in the aerospace and defence industries
- bonding allows heat-sensitive materials to be joined
- bonding can prevent electrolytic corrosion between two dissimilar conducting materials
- modern adhesives are fatigue resistant
- less skilled labour is required
- the technology is available now.

Disadvantages

- the surface preparation may be complex and hazardous
- the curing mechanism may be impractical
- very careful joint design is required;
- the bond may be degraded by heat, cold, chemicals, etc.
- unlike some other joining techniques, the full bond strength is not reached immediately (curing time).

9.5 Summary

On completing this chapter you should have an overall appreciation of the variety of ways in which components can be joined, i.e. mechanically, by brazing, welding or by adhesive bonding, and an understanding of the manufacturing process in each case. Table 9.1 summarises the main characteristics of these categories of processes.

Table 9.1

Joint characteristics	*Mechanical joints*	*Brazing and soldering*	*Welding*	*Adhesive bonding*
Permanent	No	Semi	Yes	Yes
Appearance	Normally visible bolts, but some clips can be hidden	Good	Normally acceptable can be polished to improve appearance	Almost invisible
Temperature tolerance	Normally high	Limited by filler material	Very high	Poor
Stress distribution	Local points of high stress	Good distribution	May need stress relieving	Good distribution, but poor peel resistance
Preparation	Varies, often drilling and tapping	Normally need to apply flux	Edge preparation for thick edges	Cleaning
Post processing	None	Remove corrosive fluxes	Varies, none, heat treat, polish.	Normally none
Capital cost	Low	Manual cheap, furnace brazing expensive	Varies, manual gas cheap, EBW expensive	Normally low
Consumable cost	Relatively expensive	Varies	Low	Quite expensive for structural adhesives
Production time	Including preparation, quite slow	Automated systems quite fast	Varies, can be fast	Very variable according to adhesive
Inspection	Simple for bolted joints	Difficult	Various NDT methods available	Limited NDT methods

9.6 Questions

Now test your understanding of this chapter by trying the following questions:

1. Describe the brazing process. What is the difference between brazing and soldering? (9.3)
2. Suggest suitable joining processes for
   (a) joints that need to be regularly broken for servicing,
   (b) joints that need to be undone for repair,
   (c) joints that should remain for the full service life
   (sections 9.2, 9.3, 9.4)
3. What advantages does MIG welding offer over basic stick arc welding? (9.4)
4. Why is EBW expensive, and when would it be used? (9.4)
5. Compare MIG and TIG welding. (9.4)
6. Why is the fit between parts that are to be brazed more critical than for a welding joint? (9.3)
7. Name four advantages and four disadvantages of using adhesives. (9.4.2)
8. List the main constituents of an adhesive. (9.4.2)
9. What welding process might you select for the manufacture of a diesel engine piston where the material at the top of the piston is different from that used for the skirt? (9.4)
10. Fluxes are used with brazing, soldering and some types of welding. Explain why. (9.3, 9.4)

10 Property Modification

10.1 Introduction

The techniques discussed so far have all been involved with changing the shape of the workpiece, by removing part of it, by deforming some of it or by joining two or more pieces. In this chapter we will introduce the third fundamental method of changing the raw material: modifying its properties.

You might well ask why we need to modify the properties: surely a material with the correct properties could have been chosen at the outset! There are in fact a number of good reasons. In some cases the properties needed to enable the shape to be changed may well not be the same as, or could even conflict with, those required by the end product. For example, a chisel blade that is to be forged must be sufficiently ductile for the shape-forming process to take place. The material properties required for this will conflict with those needed by the blade which needs to be hard to remain sharp. Possibly the property of the product needs to vary across the material. Another good reason is cost. For example, a sophisticated, and therefore expensive, material may be needed to resist wear when two components make sliding contact. As the wear occurs only on the surface of the two materials, these surfaces rather than the whole component need the wear-resisting properties. The cost-effective solution is to coat a low-cost backing material with the more expensive wear-resistant material.

By way of a summary, Figure 10.1 separates some of the more common processes into two groups, surface finishing and heat treatment.

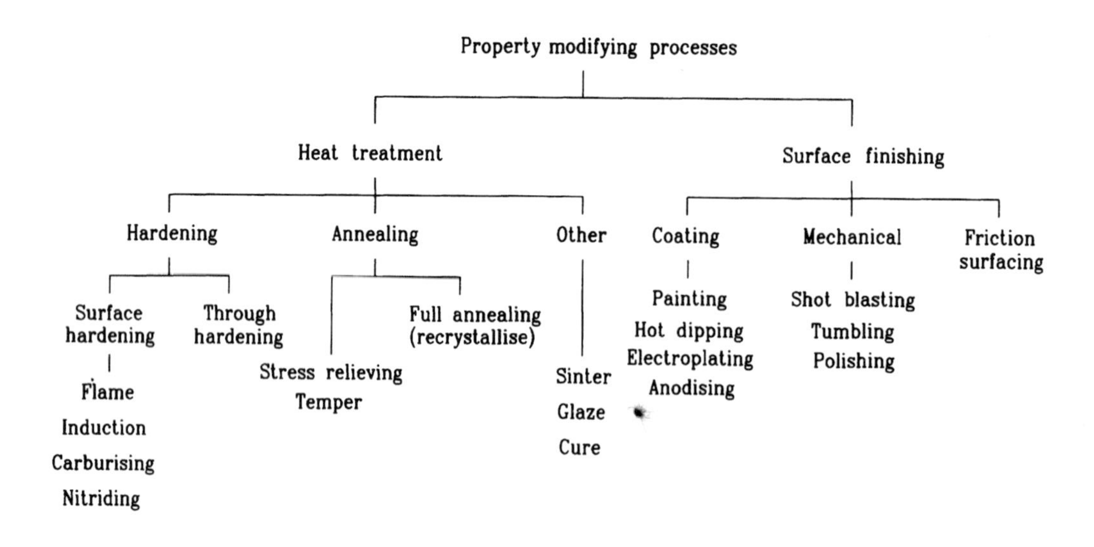

Figure 10.1 *Property modification processes*

Surface finishing, as the name implies, involves modifying the properties of the surface of the component. This may be by adding a thin layer of another material, by painting, by electrical treatment, or even by friction welding. Alternatively, the properties of the main material of the component may be modified only on the surface of the component. Examples might be shot peening to increase fatigue resistance, or anodising to improve both appearance and corrosion resistance.

Probably the most widely used technique for material property modification is heat treatment. It can be used to adjust the hardness and ductility of materials, or to relieve residual stresses built up in the material during the forming process. It can be applied to the surface, or through the body of the material. We will start by introducing the basics of heat treatment.

10.2 Heat treatment

Heat treatment is a process whereby the material is subjected to one or more cycles of temperature changes in order to obtain physical properties. Although the treatment can be applied to different materials we will concentrate on the treatment of metals: in particular, steel, as this is the most commonly heat treated metal. To understand the process we need first to examine the structure of steel.

10.2.1 Structure of steel

We will start by reminding ourselves how steel is manufactured. The first stage is to reduce iron oxide, by heating it and allowing carbon monoxide to remove the oxygen from the iron ore:

$$FeO + CO = Fe + CO_2$$

This reaction is somewhat simplified as the iron ore is far from pure FeO and there will be a range of impurities present. The pig iron so created is very brittle. To convert it into a more useful material, steel, oxygen is passed over the pig iron. Oxides of the impurities are created and absorbed into slag floating on top of the molten metal

Table 10.1

Element	*Pig iron (%)*	*Steel (%)*
Carbon	2–4	up to 0.2
Silicon	1.5–4	up to 0.4
Sulphur	0.05–0.2	0.05 max
Manganese	0.5–2.5	up to 0.6
Phosphorous	0.1–2.0	0.05 max
Iron	87–95	99

which is steel. The steel is a tough, ductile material suitable for a range of engineering uses. A comparison of typical compositions of pig iron and mild steel are shown in Table 10.1.

Of all the elements listed in Table 10.1 carbon is the one that has the most influence on the properties of steel. The best way to understand heat treatment is to imagine steel as being an alloy of iron and carbon. Iron is weak and soft, and the addition of carbon makes it stronger.

Steel is an allotropic material – it changes its crystal structure with temperature. At high temperatures its structure has a face-centred cubic pattern, enabling it to dissolve anything up to 1.7% of carbon. When it has this structure it is called austenite, or γ iron. At lower temperatures the structure has a body-centred cubic pattern and is called ferrite or iron. Ferrite can only hold very tiny quantities of carbon in solid solution, so it follows that if austenite containing a lot of carbon is cooled and becomes ferrite, then some of the carbon must be rejected. This happens by the formation of Fe_3C, cementite. The cementite is very hard and brittle. It precipitates in the form of plates or layers, with the iron forming a laminated structure called pearlite. At a carbon concentration of 0.83%, there will be enough cementite to create 100% pearlite; below this level of carbon the material will be a mix of ferrite and pearlite. Most steels contain no more than about 0.5% carbon.

The whole process is reversible, so that if a mixture of ferrite and pearlite is heated above a critical temperature the structure will change and austenite will form. When heating, the temperature at which the changes start is 723°C. The temperature at which the conversion to austenite is complete varies with the amount of carbon present. The relationship between temperature, carbon content and structure is shown in Figure 10.2. Note that all of these changes are occurring with the metal in the solid state; steel needs to be at about 1500°C to melt. We are now in a position to consider some of the heat treatments.

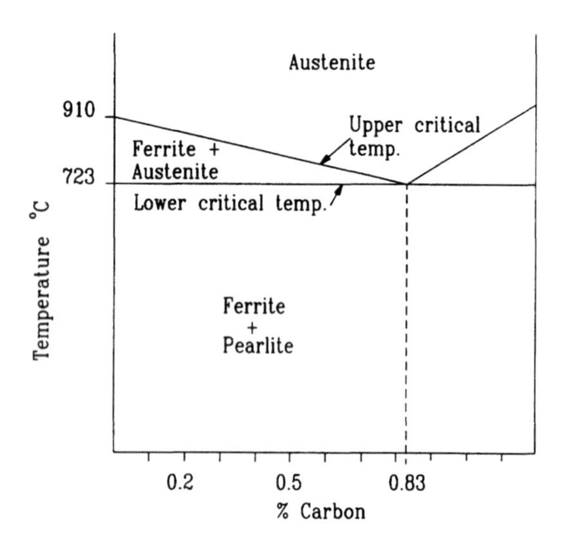

Figure 10.2 *Iron carbon equilibrium*

10.2.2 Stress relieving

When steel is heated to a temperature just below the recrystallisation temperature (723°C for steel), and allowed to cool slowly, any small residual stresses are freed. These unwanted stresses may be present as a result of cold working, casting, welding, etc. The crystal shape will be unchanged, so there will be no significant change in the hardness or strength.

10.2.3 Annealing

If, however, the steel is slowly heated, at 723°C the pearlite starts to change into austenite. Remember, the pearlite is a ferrite–cementite layer, so as this ferrite turns into austenite it proceeds to dissolve the cementite. The change begins at a number of spots within each pearlite grain, each spot producing a new austenite grain. The result is that each pearlite grain produces a number of small grains of austenite: this is said to be grain refinement. If the heating continues, by the time the upper critical temperature is reached all the remaining ferrite grains will become small grains of austenite. If the heating is slow enough, the carbon from the cementite will be evenly distributed about the austenite.

If the steel is kept at this temperature for any length of time, or it is heated much above the upper critical temperature, the small austenite grains will tend to absorb each other. This grain growth is normally to be avoided as it will reduce the toughness and strength of the steel.

If the steel is allowed to cool slowly back to room temperature, the structure will revert to ferrite and pearlite, but the grain size will be strongly influenced by that at the upper critical temperature. The treated material will have a finer grain size, and as a result will be stronger and tougher. The changes as described will be triggered by the critical temperatures on the graph, but as they require a significant rearrangement of the atoms, the time for this may mean some overshooting of the theoretical temperatures. Annealing normally involves allowing the metal to cool within the furnace.

10.2.4 Normalising

This involves heating to the upper critical temperature, and soaking to allow the carbon to disperse evenly, then cooling in air. Less time is spent in the austenite condition than with annealing, so there is less grain growth. The result is that normalised steel will have a smaller grain size, making it harder and stronger. Typical values for the yield stress for mild steel would be

216 MN/m^2 for annealed and 325 MN/m^2 for normalised.

10.2.5 Hardening

If austenite is cooled too rapidly to allow time for the carbon to precipitate out, it becomes trapped, forming a supersaturated solution of carbon in iron called martensite. This is very strong and hard, but the distortion of the crystal structure results in brittleness. Tempering is usually required to reduce the brittle characteristic.

10.2.6 Tempering

The hardened steel is reheated to just below 723°C to allow some of the trapped carbon to come out of the solid solution and reside as a fine dispersion in the ferrite mix. This is not laminated as in pearlite, so most of the hardness is retained, but the distorting effects on the structure are reduced, thereby increasing the toughness and ductility.

10.2.7 Surface hardening

Sometimes it is desirable to harden only the surface (as with gears to improve the wear characteristics). *Flame hardening* involves heating the outside with a torch, raising the temperature high enough to form austenite, then quenching and tempering to achieve the required properties. *Induction hardening* uses an induction coil to generate current and hence heat in the part. This method allows good control of the heating effect and minimum distortion.

10.2.8 Case hardening

If the steel has too low a carbon content for surface hardening to be successful, case hardening enables some chemical modification of the surface. *Carburising* involves packing carbon around the part and heating to a high temperature (900°C) for a number of hours to enable some of the carbon to diffuse into the outer surface of the part. This is normally followed by quenching and tempering. *Gas carburising* is a variation where, instead of packing with solid carbon, methane gas is used to provide the carbon. *Nitriding*, used for alloy steels containing Cr and Vn, involves heating the part at around 550°C in a nitrogen atmosphere (often ammonia gas). A nitrided part does not need to be quenched rapidly, so minimising any distortion problems.

10.3 Surface finishing

10.3.1 Mechanical

Shot blasting

An abrasive material – sand, steel, or glass beads – is blown in a high-velocity stream of air at the component. Possible applications include the cleaning of the remnants of ceramic from an investment casting, the removal of rust, or as a cost-effective alternative to polishing for the exterior of some food processing products. The process is sometimes called *shot peening*, when steel balls are used to improve the fatigue resistance of a surface. The shot peening tends to smooth the surface, reducing the number of potential stress concentration points that might form the start of a fatigue crack.

Tumbling

Sometimes called *barrel finishing*, this involves placing the parts to be finished in a large drum containing finishing media (specially shaped pellets). The drum then rotates, creating a sliding motion between the parts and the media inside it. The finish achieved depends on the length of tumbling and the type of media and lubricant employed. Apart from loading and unloading, the operation is labour free, and hence fairly low cost.

Polishing

Polishing is normally performed manually with hand-held power tools that can be fitted with a variety of abrasive or buffing bits. It is often used as a cosmetic treatment for welds. It is difficult to detect the presence of a weld in a well-polished part. The polishing action can be far better targeted than with other mechanical operations, but the manual content is high, making it expensive. Some automated systems are used but they are less common.

10.3.2 Coating

Painting

This is the most widely used process applied to finish manufactured products, offering a good combination of aesthetic appeal and corrosion protection. Prior to painting, the parts need to be degreased, and, if steel, are often passed through a *phosphate dip* for added corrosion resistance. Normally at least two coats are used. The first, a primer, serves to aid corrosion resistance, help adhesion and fill in minor blemishes in the surface. The final coat serves mainly to achieve the required cosmetic effect. The paint is normally applied by *dipping* or *spraying*. Dipping is simple and economical if the whole component is to be covered. Spraying can be manual or automatic, the paint being atomised either by air or mechanical means. *Electrostatic spraying* involves giving the paint particles an electrostatic charge, so that they are attracted to the work which is included as an electrode. This method saves on paint wastage, but can result in a build-up of paint around holes and on sharp edges. After coating, most painted components are then baked to dry the surface.

Hot dip coatings

Corrosion resistance can be improved by dipping into certain molten metals. *Hot dip galvanising* is commonly used with steel. After cleaning, the steel is fluxed by dipping in a solution of zinc chloride and hydrochloric acid, and finally dipped in a bath of molten zinc.

Tin plating can be done by dipping clean steel in a bath of molten tin which has a layer of zinc chloride floating on its surface. The work is fluxed by passing through the zinc chloride. Electroplating is more common as it results in a more uniform coating.

Electroplating

This can be used with a wide variety of materials, plastics as well as metals. If plastics are to be plated they must first be coated with an electrically conductive material. Care needs to be taken with surface preparation, as any defects such as scratches or pin holes will cause blemishes on the plated finish. The surface must also be chemically clean to ensure good adherence of the coating. The process is very simple; the parts to be plated become the cathodes and are suspended in a solution that contains dissolved salts of the material to be deposited. The metal to be deposited is the anode. When a DC voltage is applied, the metallic ions move to the cathode where they lose their charge and are deposited on it.

Common plating materials are tin, cadmium, chromium, copper, gold, platinum, silver, and zinc.

Anodising

This is a process used to provide decorative and corrosion-resistant finishes to aluminium. Aluminium normally has a very thin layer of aluminium oxide on its surface. Anodising increases the thickness of this layer, by an electrolytic process whereby the work is the anode. Colours can be produced in the surface by using suitable dyes in the electrolyte. Note that this process differs from most of the coating techniques where another material is deposited on the base material. In this case, the naturally formed aluminum oxide layer is enhanced.

10.3.3 Friction surfacing

Friction surfacing is a technique based on friction welding technology. The coating material, in rod form, is rotated under pressure whereby a hot plasticised layer is generated in the rod, at the interface with the substrate. By moving the substrate across the face of the rotating rod, a plasticised layer some 0.2–2.5mm thick is deposited (see Figure 10.3).

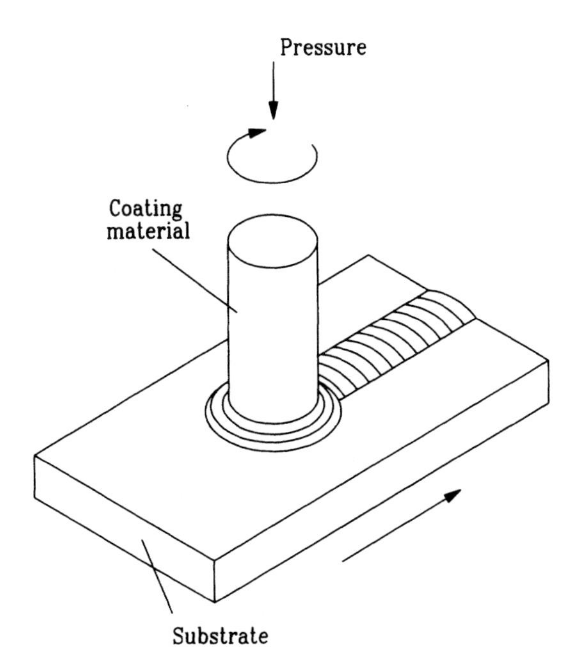

Figure 10.3 *Friction surfacing*

During the coating process the applied layer of metal reaches a temperature near the melting point whilst simultaneously undergoing plastic deformation. The coating, then, is a product of a vigorous hot forging action, as opposed to the casting mechanism of welding processes. This important difference means that many of the defects (porosity, slag inclusion, etc.) commonly associated with these techniques are avoided.

A wide range of component sizes and geometries can be coated with a broad range of alloys.

10.4 Summary

By the end of this chapter you should have an understanding of some of the ways in which the physical properties of metals (particularly steel) can be modified by various heat treatments. You should be aware of the meaning of terms such as annealing and tempering, and understand how they can affect the structure of steel.

Normally one of the least desirable properties of a metal is its ability to corrode. You should be aware of some of the common basic processes that inhibit corrosion as well as enhancing the appearance.

10.5 Questions

Now check your understanding of this chapter by trying to answer the following questions. Refer to the relevant section (shown in brackets) if you have difficulty with any of them.

1. What is the essential difference between anodising and electroplating? (10.3)
2. Describe the differences between martensite, austenite, pearlite and ferrite. (10.2)
3. Name two methods of applying tin plate to steel. Which is the more commonly used and why? (10.3)
4. Describe the essential mechanism by which steel can be hardened. How would you harden the surface of a steel that contains only a small amount of carbon? (10.2)
5. State three reasons why you might wish to modify the properties of a material. (10.1)
6. Describe the basic process of friction surfacing. (10.3)
7. Why are aluminium components sometimes anodised? (10.3)
8. Shot blasting is sometimes used to remove remnants of ceramic from an investment casting. What else can it be used for? (10.3)
9. When would case hardening be used? (10.2)
10. Describe the process of annealing. (10.2)

11 Quality Control

11.1 Introduction

The manufacturing processes discussed so far have all been techniques for making changes to raw materials. Although these form the basis of manufacturing, which normally involves the production of goods in quantity, in practice they will only produce satisfactory products if they are performed in a controlled and consistent manner. There needs to be a means of ensuring that this consistency is maintained. The means is called quality control.

Quality is an important feature that needs to be applied to the whole product development process including design, marketing and production – in fact, the whole business of the company. However, here we will examine three aspects that are directly applicable to the manufacturing process.

11.2 Inspecting geometric features

In Chapter 3 we introduced the need for tolerances and methods of applying them to dimensions, in a way that took account of both the functional and manufacturing requirements of the product. Now we need to look at some of the ways in which checks can be made during the manufacturing stages, to ensure that these tolerance limits are maintained.

11.2.1 Inspection equipment

Measuring devices

The simplest device is the machinist's steel rule. The feature being measured is directly aligned with the markings on the rule. It can be used with square or protractor attachments to help with measuring angles. Accuracy is limited to about 0.5mm.

Vernier callipers offer improved accuracy, up to about 0.05mm. These are indirect reading devices in that the calliper jaws are closed on to the feature to be measured, with the actual size being read from a vernier scale. The jaws are moved manually, with a fine adjusting screw being employed for the final part of the movement. The jaws can measure both internal and external features. Variants are also available as height or depth measuring devices. To ease the problems of reading the vernier scale, versions have been developed incorporating a dial gauge, or a digital readout on a liquid crystal display.

The micrometer provides the next step in improving accuracy, normally up to about 0.01mm. As with the vernier, jaws are closed on the feature being measured. However, jaw movement is via a fine screw thread, and the readings are made via a rotating vernier scale. Modern micrometers often incorporate a digital readout.

The dial indicator gauge has a plunger, emanating from a circular graduated dial, with a centrally mounted indicator needle. Movement of the plunger is translated into movement of the needle, which can be read from the graduated scale. This device is not designed for absolute measurements: it is used to measure movement in a single direction. A typical use might be to check the deflection under load of an area of a component.

Angles are normally more difficult to measure than linear dimensions. Protractors can be used, or a sine bar could be employed. This is essentially a special bar resting on a pair of rollers. The rollers must each be of the same known diameter, and be perfectly cylindrical. Their axes must be parallel and a known distance apart. Finally, the bar must have a flat upper surface which is parallel to and equidistant from the roller axes. Figure 11.1 shows how it may be used to check the angle between two sides of a component. Note that this diagram introduces another piece of equipment, the gauge block. The

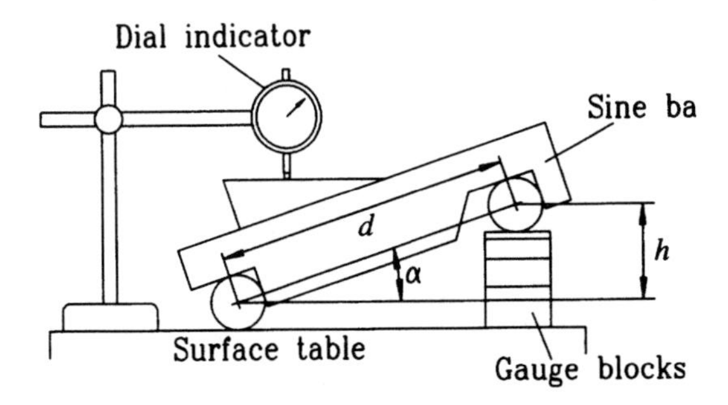

Figure 11.1 *Using a sine bar to measure an angle*

gauge blocks or slip gauges are standard-width blocks of material used as in this illustration, or for calibration work. They are manufactured to very tight tolerances, in terms of flatness, parallelism, length and surface finish, and are available in five grades of accuracy (BS4311). The dial indicator as shown would be used to determine when the surface of the item being measured is parallel to the surface table. The object is to measure the height h, so that knowing d the angle α may be found.

Probably the most sophisticated of the mechanical measuring devices is the coordinate measuring machine (CMM) (Figure 11.2). This normally fairly large piece of equipment consists of a stable bed (usually granite) fitted with a bridge type gantry mounted on air bearings to enable it to move to and fro along the length of the bed. The bridge is fitted with a vertical column that is able to move in two directions, vertically and horizontally, across the bridge. At the base of the column is a touch trigger probe. (This is a very sensitive switch that generates an electrical signal as it contacts the item being measured.) It is normally able to be rotated. The machine records any movement in each of the three planes and is therefore able to identify the location in space of the end of the probe when it is triggered. Accuracies of the order of 0.004mm are possible with a CMM. In its simplest form the probe position is adjusted manually, with read-outs of the coordinate positions in space being recorded. However, many of these machines feed their output data into a PC (personal computer) which will interpret the data, enabling checks to be made fairly quickly on characteristics such as roundness or concentricity. More and more CMMs are now equipped with servo motors and encoders to move the probe, the motors being controlled by the computer. Machines can be taught a measuring sequence which, when loaded with the component in question, enables the measuring sequence to be performed automatically – the operator simply waits for a printout identifying any defects! It is even possible (although very rare at present) to link the output of such a machine to a CNC machine tool that may be producing a batch of components, so that it can make corrections to avoid potential defects.

The astute reader may have noticed a certain commonality of the available movements on a machining centre and those on the CMM. Essentially they are the same, and one manufacturer at least offers the facility to use the machining centre as a measuring device. The accuracy of such a machine would, of course, be somewhat less than a dedicated CMM, which would normally be kept in a temperature-controlled environment.

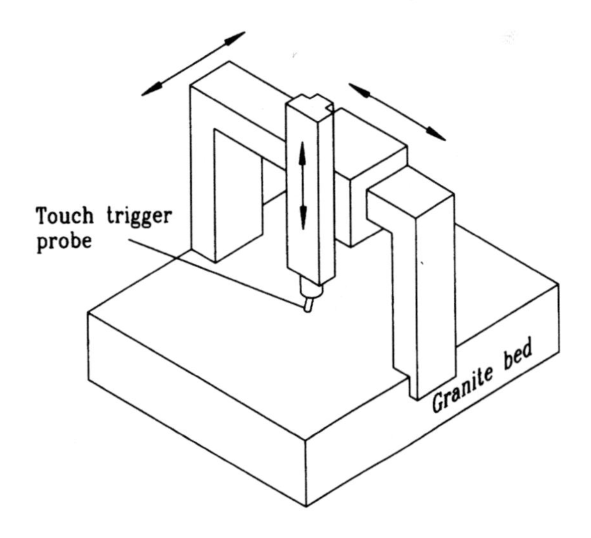

Figure 11.2 *Schematic of coordinate measuring machine*

We have so far introduced a few of the basic mechanical measuring devices, but you should also be aware that with the onward march of technology there is an increasing choice of other sophisticated devices: optical devices using lasers, and vision systems, to name but two.

Fixed gauges

The system of limits and fits (Chapter 3) is designed to ensure that components fit together so that assembly has the properties (easy running, slide, interference, etc.) needed to function correctly. When inspecting components produced in high volumes, we are often more interested in whether the component will fit with its mating part than in the absolute values of certain of its dimensions. In such instances, limit (GO or NO-GO) gauges can be used.

A simple example would be a gauge to check that the diameter of a hole is within tolerance. A GO gauge, in this case a cylindrical shaft, would be offered to the hole and would pass through it, provided the hole was large enough (i.e. above its minimum allowable size). The NO-GO gauge, another cylindrical shaft similar to the GO gauge but with a different diameter, when offered to the hole should not be able to enter it. The NO-GO gauge is checking that the hole is not too large (i.e. above its maximum allowable size).

Unfortunately, as with other components, these gauges cannot be manufactured to absolute dimensions, so need their own tolerances. As a general rule the tolerance on the gauge should be at least ten times tighter than the tolerance on the item being measured. The purpose of these gauges is to avoid accepting defective components, so the tolerances on both gauges must be within the tolerance zone of the work. The GO gauge will engage the work each time it is used, so over time will wear. A wear allowance is made when setting the tolerance for the GO gauge. The tolerance zone is normally set in from the maximum material limit by this allowance (see BS 969 for further details).

Taylor's principles for gauging lay down the fundamental functions of such gauges in two statements:

- The GO gauge checks the maximum material condition and should check for as many dimensions as possible.

- The NO-GO gauge checks the minimum material condition and should only check one dimension.

In effect this means that several NO-GO gauges may be needed.

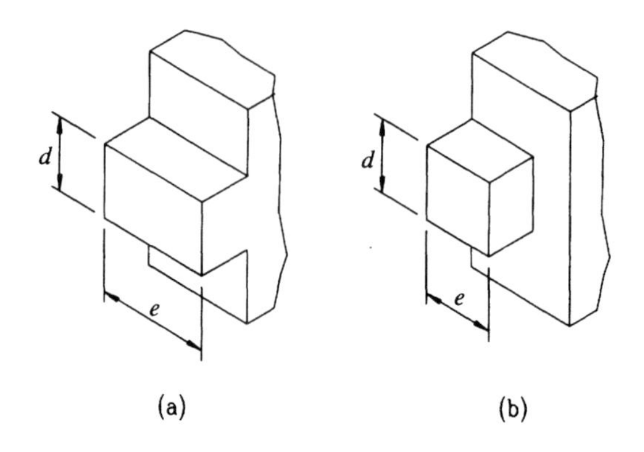

Figure 11.3

This is best explained by an example. We are to check the tab on the item in Figure 11.3, specifically that dimensions *d* and *e* are within tolerance. Assume we have two gauges, each featuring a rectangular hole. If the component is as (a) i.e. within tolerance, the GO gauge will fit over the tab, indicating the maximum material conditions of the dimensions have been met. The tab will be unable to enter the NO-GO gauge, so is correctly passed as satisfactory. Now, consider component (b), where dimension *d* is within tolerance, but an error has been made with *e*. The GO gauge will indicate no problems: the maximum material conditions have not been exceeded. As dimension *d* is correct, it will prevent the tab from entering the NO-GO gauge, which will therefore also indicate an acceptable component. This is obviously incorrect. We need a separate NO-GO gauge for each of the dimensions *d* and *e*. Applying Taylor's principles, we would have used two such gauges.

The example given is a very simple one; fixed gauges of this type come in many shapes and sizes and are commonly used to check tapers, threads and splines.

11.3 Inspecting non-geometric features

So far we have explored ways of ensuring that the geometry of the component meets the specification requirements, but there are often a number of other characteristics affected by the manufacturing process that need to be met. For example, how do we know that a weld does not contain hidden faults or how do we know that the correct high-strength steel has been used, rather than an unsuitable alternative? One way would be to section the weld, or perform some chemical analysis on the suspect steel. However, such tests would damage or destroy the component. These sorts of problems can be covered by non-destructive testing (NDT) and/or following certain procedures.

11.3.1 Proof testing

This involves loading the product to a predetermined value, normally the design load (which will be above its working loading). Deflections at various critical points are measured during the test to ensure that no damage has occurred. The test generates a level of confidence that the product will perform safely under working conditions, even if subjected to a degree of overloading. This method is often used for pressure vessels, and can be performed after manufacture, or often on site when the installation is complete. This is all very well for confirming that the item has been correctly designed and manufactured, but is of little use during the manufacturing process.

11.3.2 Visual inspection

One of the simplest inspection methods is just to examine it – preferably in well-lit conditions. Provided the inspector is experienced in looking for particular faults this is a suitable method. Investment casting for use in food processing equipment needs to be free of surface inclusions and have an acceptable level of surface finish. We have already mentioned methods of measuring surface finish (section 3.3.2), but these are normally only suitable for a fairly flat surface rather than for instance a complex casting. The experienced inspector's eye can be a formidable inspection device. Another example is assessing the quality of a sprayed paint finish on, say, a motor vehicle. Here again the inspection will be mainly visual under good lighting. The inspector will often also use a tactile check with his/her fingers to check the smoothness of the finish. A right-handed inspector will always use the left hand for this check as it will be more sensitive. Such procedures are simple and do not require expensive equipment, but have the disadvantage of relying on the skill of the inspector, and are not able to discover defects below the surface.

11.3.3 Dye penetrant

This test is suitable for non-porous components. The surface in question needs to be cleaned then the penetrant liquid applied, often from an aerosol can. The liquid is drawn by capillary action into any cracks or flaws. After a short wait to allow this to take place, excess penetrant is wiped from the surface, and a developer applied. This is an absorbent material which draws the penetrant from the cracks so that they become visible on the surface. Contrasting colours are used to make the traces more visible. For example, some kits use red for the penetrant and white for the developer. Any red lines appearing against the white background will indicate flaws. Sometimes the penetrant contains additives that fluoresce under ultraviolet light so that they can be seen more easily. This process is suitable for identifying both surface defects and through cracks.

11.3.4 Magnetic particle

Imagine a bar of ferrous material that has no flaws. If we magnetise the bar by placing it within an energised coil, lines of magnetic flux will be created along the axis of the bar. Now, if the bar was replaced with one having defects, such as cracks running perpendicular to the axis of the bar, the lines of magnetic flux will have a discontinuity in the area of the crack. This is the principle employed in this method of testing. The component concerned is cleaned, placed within a magnetising coil and coated with a powder of magnetic particles, which align themselves along the flux lines. Any faults at or near the surface running perpendicular to the magnetisation axis should become visible. The axis of magnetisation is then changed, by moving the coil or the component, and the test repeated to check for flaws running in different directions. The process can identify cracks both on and slightly below (up to about 5mm) the surface.

11.3.5 Radiographic

This process uses the same principle as with medical X-rays. Essentially, the component in question is placed between a source of radiation (X-rays) and a photo-sensitive film. The transmission or absorption of the X-rays is affected by the thickness and density of the component. Any internal cracks, air pockets, etc. will therefore also affect the transmission of the radiation, and will be visible on the developed photographic film. The process is particularly useful in identifying faults in castings or welds. It is expensive and time-consuming and tends to be used for coded work (pressure vessels), for process development, or on a statistical sampling basis for higher volume production. As radiation is a safety hazard, appropriate steps have to be taken to avoid subjecting personnel to the radiation.

11.3.6 Ultrasonic, acoustic

Ultrasonic inspection involves sending high-frequency vibrations through the item being tested. Any flaw within the item will reflect part of the transmitted beam. By careful interpretation of reflected vibrations, flaws such as internal cracks, or density changes, can be identified. In fact, the process can even be used to measure thickness. This is particularly useful when access is only available to one side of the component, such as, for example, the leg of a North Sea oil drilling rig.

Acoustic testing uses lower-frequency vibrations, but in a passive manner. As cracks develop in components under stress, sound waves are emitted. Ice cracking under the loads imposed by skaters can emit quite loud audible warnings of impending failure. The process involves continuously monitoring the sounds emanating from the component or structure and interpreting any increased changes in the emissions. It is particularly suited to monitoring large structures such as nuclear power reactors, bridges, or even the earth (for earthquake detection).

11.3.7 Leak detection

A variety of tests can be used to detect leaks. Probably the simplest is to fill the component in question with air, place it under water and watch for any rising bubbles, just as you would check a bicycle tube for punctures. Later, in Chapter 14, we will discuss a case study where this process was superseded by the use of dye penetrant.

A completely different type of leak test will be used by the mass-production motor manufacturer to ensure that the final product is able to withstand wet weather without suffering water ingress. The car will simply be placed in a water test booth where it will be subjected to spraying by a variety of water jets. A simple visual examination will be used to check for faults.

At the other end of the scale, products produced by the vacuum industry have to be airtight. One of the tests used is to fill the component with helium and use a detector to check for helium emitting from any joints or suspect areas.

11.4 Traceability

Sometimes non-destructive tests are not available, and destructive tests are obviously not suitable. Take as an example a pressure vessel which has been designed with the assumption that a particular strength steel is to be used in its construction. The manufacturing facility may be using a number of different varieties of steel.

Frequently there is no visual distinction between a specialist high strength steel and a low-cost general purpose steel. The problem is addressed by using a system of traceability. In this case the material supplied by the steel mill would be stamped with an identification number and have an accompanying certificate which confirmed its composition specification. Once received, the material needs to be stored in a quarantine area (an area where it is separated from uncertified material).

In all probability the pressure vessel will be designed to a code and the end user will require certification to prove to his insurance company that the vessel has been manufactured to the code. If this is the case an independent inspector will be used to ensure that Lloyds, ASME or other requirements have been met. The inspector will need to witness any cutting of the raw material so that the identification stamp can be transferred to the pieces cut for individual components.

The individual components will probably need to be welded to form the vessel. The specifications for the welds will have been defined at the design stage when weld procedure was drawn up. These define exactly how the weld should be performed. Assuming the weld is made manually, the welder will need to be a 'coded welder'. This means that within the previous six months he will have successfully performed a trial weld using the appropriate procedure. This trial weld will have been witnessed by the independent inspector, who will also examine the weld after it has been sectioned, before passing the welder as competent. The weld procedure will define any weld preparation. These preparations again have to be witnessed by the independent inspector before the welding takes place. Records are kept covering all the operations involved in the vessel manufacture.

This example involving the use of an outside inspector may be an extreme example with an obvious safety reason. However, there are many instances when it can be employed to the direct benefit of the company. It is often very easy on a casting to include a batch number identification. Imagine there was an error in the alloy mix on one batch, for example, that reduced the strength of the end product, and for one reason or another this error was missed during the manufacturing process. Subsequently the product fails in the customer's hands. The failed component is analysed and the problem identified. Without the batch number the company may have to replace perhaps thousands of products in the field.

11.5 Process control and feedback

So far our consideration of quality control has been restricted to inspection, i.e. looking for faults or errors in the products after they have passed through one or more stages of the manufacturing process. If this is very thorough it will prevent faulty products reaching the customer, but at the expense of a lot of rectification work, or, even worse, high levels of scrap. Two actions are needed. First, care needs to be taken to ensure that the process selected for each operation has the ability to perform the operation to the required standard on a regular repeatable basis. Second, there needs to be a feedback system so that errors can be corrected as soon as they occur, or, ideally, *before* they occur!

11.5.1 Process capability

No manufacturing process will consistently produce totally identical effects on its input material. If this were possible there would be no need for tolerances to be applied to the component dimensions. There are many reasons for the variations in the process, some of which can be eliminated (or at least minimised) and some which are inherent in the process. Errors that may be reduced can be due to factors such as operator error due to poor training, faulty materials or inappropriate tooling. Errors that cannot be reduced could be due to variations in the input material (within the material specification), wear in the moving components of the machine tool, vibrations due to lack of rigidity in

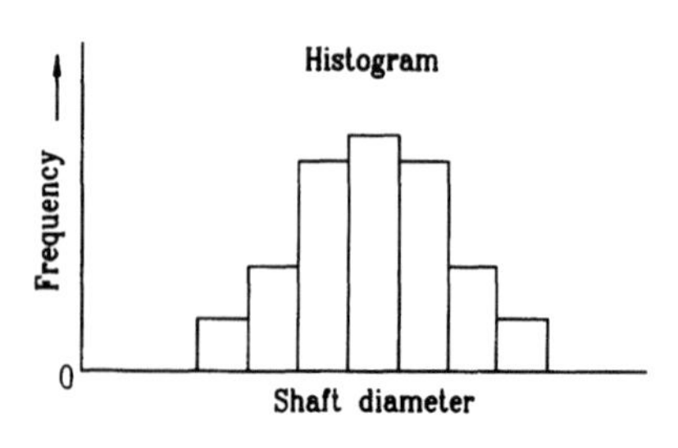

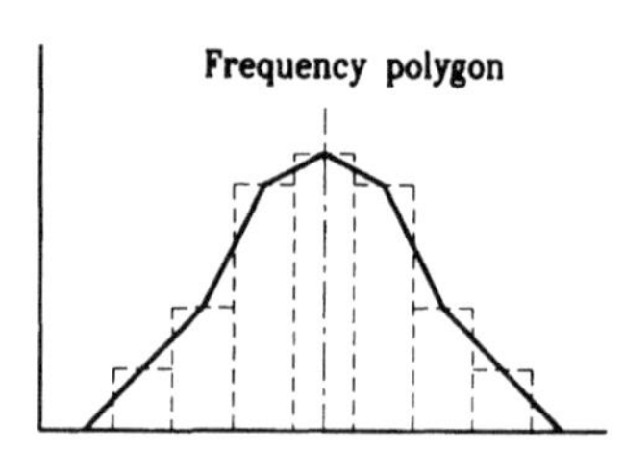

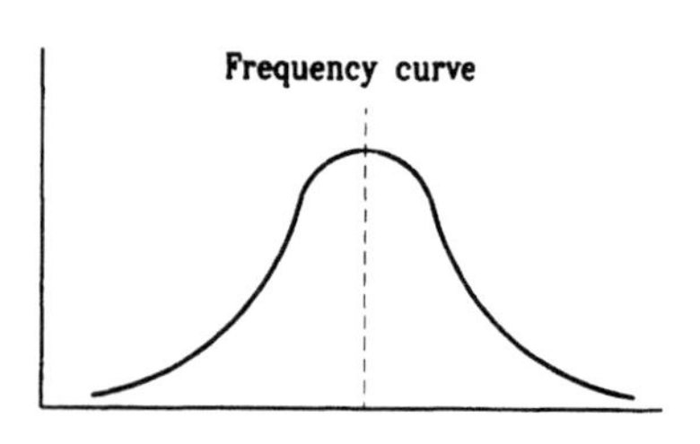

Figure 11.4 *Evolution of the frequency curve*

the machine tool, etc. These latter errors are inherent in the process and cannot be reduced. When selecting a manufacturing process these errors (representing the capability of the process) must be taken into account.

Process capability can vary over a period of time. It may be improved by more consistent input material quality, minor changes or adjustments to the process, or perhaps the addition of more modern control equipment. It can also deteriorate with time. Imagine two nominally identical machine tools, both supplied by the same manufacturer to the same design. Both are to be used in a production environment; one was commissioned six months ago while the other has been in service for the last ten years. The newer machine is almost certain to be capable of holding closer tolerances than its older cousin, which is likely to be suffering wear to its sliding parts, free play in its bearings, and no doubt some backlash in its feed drives. We need a procedure for measuring this capability.

Process capability is measured by first ensuring that no correctable errors are present, then analysing the output of a batch. (The procedure used should, of course, be fully documented so that at a later date the capability can again be measured under similar conditions.) At this point a brief reminder of statistical analysis seems appropriate. Take a simple example. Imagine that we are checking the capability of a lathe to produce a particular diameter on a shaft. During the test we produce a number of shafts and measure the achieved diameters. These are then plotted on a frequency histogram, the width of each block representing a narrow diameter range and the height representing the number of occurrences of shafts within that range. Joining the mid-points of the tops of the blocks will create a frequency polygon. The area of the frequency polygon will be the same as that of the histogram, which in turn represents the total number of occurrences. If we then 'smooth' the frequency polygon we will create a frequency curve, as in Figure 11.4.

If the number of samples is sufficiently large and the diameter classes (width of the histogram blocks) are sufficiently narrow, in many cases the final frequency curve will approximate to the *normal distribution curve*. The area beneath this curve represents the total number of occurrences. The curve is symmetrical about its centre line, which represents the mean of the readings. (The mean is simply the sum of the readings divided by the number of readings.) We are particularly interested in the spread of the readings away from the mean. A wide spread would tend to indicate a wide variation in shaft diameters produced. A measure of this spread is the *standard deviation*

$$\text{standard deviation} = \sqrt{\frac{\sum(x - \bar{x})^2}{n}}$$

where $\bar{x}$ is the mean of a set of n values of x.

Now the normal distribution curve can be represented by an equation. This can be shown to demonstrate a relationship between the standard deviation, the mean and the normal curve. In Figure 11.5 we see that 68.26% of our samples will be within one standard deviation of the mean and that 95.46% will be within two standard deviations from the mean.

Imagine that our capability test produced 100 shafts, which when inspected had a mean diameter of 25.21mm and a standard deviation of 0.07. From the relationship between mean and standard deviation, this tells us that 68.26% (i.e. about two-thirds) of the shafts produced will have diameters between 25.21 + 0.07 and 25.21 − 0.07 (i.e. between 25.28 and 25.14) Or, to put it another way, one in every three shafts produced will be outside this range. The right hand graph in Figure 11.5 indicates that for a normal distribution almost all (99.73%) of the samples will be within ± 3 standard deviations from the mean. Using the figures above this tells us that almost all the shafts produced would lie between 25.42 and 25.00 (i.e. 25.21 ± 3 × 0.07). This, then, is the limit of the capability of this process.

Figure 11.6 illustrates the importance of matching the process capability with the required tolerance. If the component specification calls for a tolerance smaller than the process capability (a), then scrap items are certain to be produced. Figure (b) shows a situation where the tolerance and process capability are very closely matched. This will produce good components, provided the process is kept exactly centred. If not, situation (c) will occur, with creation of some scrap components. Finally, situation (d) shows a safe condition where the process has a greater capability than required by the component. Obviously care needs to be taken to ensure that this margin of safety is not too great if costs are to be kept under control. Using a sledgehammer to crack a nut would be effective but not very efficient!

11.5.2 Statistical process control

Even if we carefully match process capability with component requirements, errors will still occur: humans make mistakes, machines break down, raw materials can have flaws, etc. Some form of inspection system is needed if quality

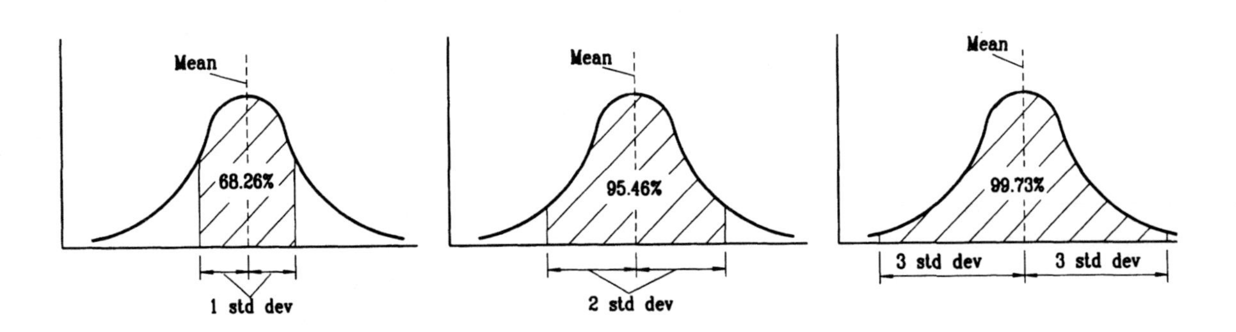

Figure 11.5 *Relationship between standard deviation and normal distribution*

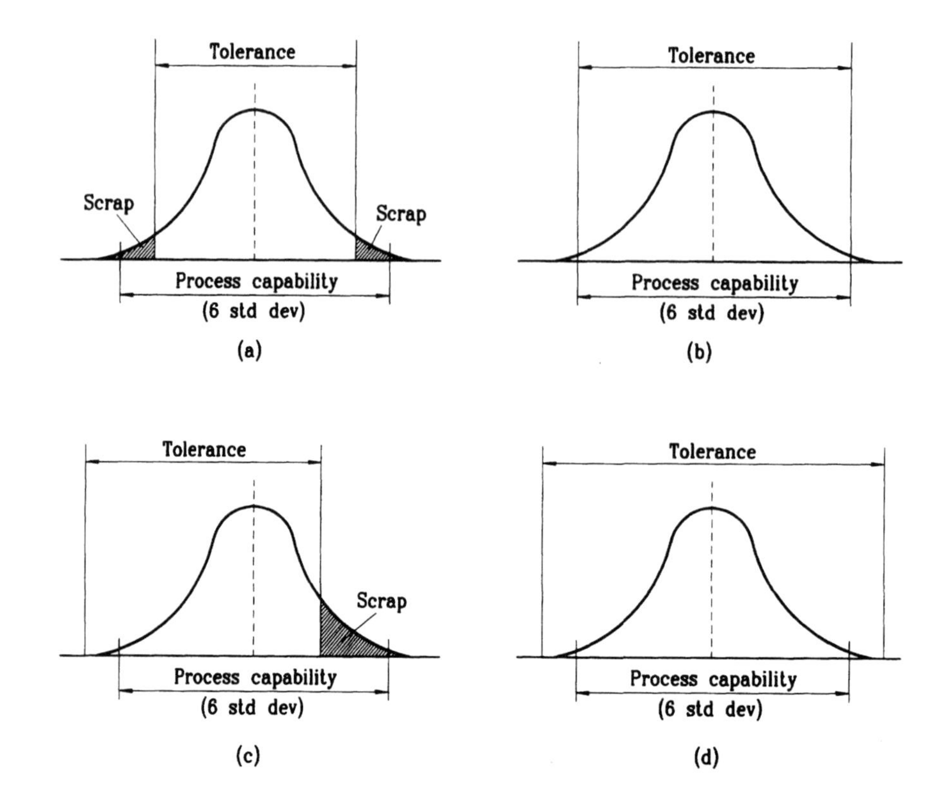

Figure 11.6 *Component tolerance and process capability*

control of the manufacturing output is to be kept to an acceptable standard. How frequently should we inspect? If we inspect every item as soon as it finishes each process, promptly feeding back information on errors detected, and ensuring corrective action is taken, scrap levels will be minimum and no faulty goods will leave the factory. Unfortunately, our costs will be so high that we are unlikely to stay in business long! The alternative is to inspect samples from each of the items produced and use the information gained to make decisions about the quality of the product. This is known as statistical process control (SPC).

The problem we now have is knowing how large the inspection sample should be and how frequently it should be inspected. Statistical analysis can lead us to the answers. The sample size and frequency of inspection should depend logically on the number of faulty components found in each inspection. If several consecutive inspections fail to reveal any faults, it is reasonable to assume that the process is well under control and that the frequency of inspection can be reduced. If, however, the inspections reveal increasing numbers of faults, the process needs to be examined more frequently. For details of how this is applied in practice the reader should refer to appropriate standards such as BS 6000/1/2.

11.5.3 Quality control charts

There are two reasons for measuring the quality of components leaving a process. The first is to ensure that only acceptable components are passed to the next process (which may be the customer). The second is to indicate that the process needs some adjustment or modification. Ideally we need to monitor the process continuously so that we can detect trends in quality and make adjustments before any out-of-tolerance items are produced. If the data from the sample inspections is plotted on a control chart any drift in a particular dimension will soon become apparent. This does not necessarily mean that there is a fault in the process. For example, on a machine tool the cutting tip will gradually wear with use. Dimensions cut by the worn tip will differ from those generated by a new tip. The control chart will indicate the drift in the component dimension, so that the tip can be changed before any out-of-tolerance pieces are produced.

11.6 Summary

By the end of this chapter you should appreciate the importance of quality in the manufacturing process. You should have an understanding of some of the commonly used inspection equipment and techniques, and appreciate that the reason for inspecting is not simply to weed out faulty components, but also to feed back the information and remove the reasons for the faults.

11.7 Questions

Now check your understanding of this chapter by trying to answer the following questions.

1. Describe Taylor's principle in relation to fixed gauges. (11.2.1)
2. Some inspection techniques involve the use of dye penetrant. What does this method check for and are there any limitations in its use? (11.3.3)
3. Why would you undertake a process capability study on a manufacturing process, and how would you perform one? (11.5.1)
4. What do the initials SPC stand for? Why would SPC be used? (11.5.2)
5. What does the term 'traceability' mean, and why is it needed? (11.4)
6. What do the initials CMM stand for? Describe the concept of such a machine. (11.2.1)
7. What do you understand by the term 'proof testing'? (11.3.1)
8. Name three methods of NDT. (11.3)
9. Give an example of the use of quality control charts. (11.5.3)
10. What techniques could you employ to test for flaws in a casting? (11.3)

Part III
Effective Integration of Design and Manufacture

12 Designing for Manufacturing Processes and Materials

12.1 Introduction

The chapters so far have introduced the basic concepts of the design and manufacturing processes. To maximise our chances of achieving a successful product, we need to integrate the skills and techniques from these areas. The remainder of the book will concentrate on this integration.

We will start, in this chapter, by considering how components can be designed to simplify the manufacturing process. Always remember, the aim is to be able to produce the components reliably in the simplest possible manner, without compromising any of the design functional requirements. The word 'simple' has been used rather than 'cheapest' because almost without exception the simpler the production technique, the cheaper will be the component. In this chapter we will concentrate on individual components, and move to assemblies later.

Initially we will discuss a number of manufacturing processes (mainly, but not exclusively, for metals), covering the benefits available to the designer, and some pointers as to how the designer can help the process. The latter sections of the chapter will comment on design considerations for non-metals.

12.2 Casting

12.2.1 When to use castings?

The casting process is appropriate for a wide variety of applications, from the very large one-off iron castings used in bridge structures by the Victorians to the small intricate die castings produced in great numbers and used in today's computer disc drives. As with any process, it should only be selected if it offers significant benefits over the alternatives.

The primary benefit of using a casting is to progress the raw material to a form nearer to that of the finished component than is available from stock material. This will minimise the cost of subsequent finishing operations. The following component characteristics will influence the advisability of using a casting process.

Shape of component:
Internal cavities that would be difficult to machine. External features that are complex in shape and hence difficult to machine.

Required surface finish:
This will in part define the type of casting process chosen. Remember that the casting should limit the amount of machining required. At the very least it will reduce the amount of metal to be removed and at best it should eliminate totally the need for machining.

Material cost:
Casting should reduce material wastage. This will be more important for the more expensive materials.

12.2.2 Which casting process?

As introduced in Chapter 8, there are a variety of casting processes for metal components: sand, shell, investment, die casting etc. The processes are all common in that molten metal is poured into a mould and allowed to solidify to form the required shape. The more sophisticated the process the better will be the emulation of the final shape. A die or investment casting will, for example, be capable of producing a more intricate shape to closer tolerances than will a sand casting. As with most choices in life, nothing is free. The die or investment casting will each require a much greater initial invest-

Table 12.1

	Sand	*Shell*	*Investment*	*Die*
Detail	Approximate shape only	Limited detail	Intricate detail	Intricate detail
Surface finish	Crude finish typically 50μm	Slightly better than sand typically 25μm	Smooth finish typically 6.3μm	Smooth finish typically 3.2μm
Typical tolerance per 25mm	± 2.5mm	± 0.5mm	± 0.1mm	± 0.05 to 0.2mm ± 0.2mm
Component size and weight	Wide range, determined by handling equipment available; typically 25g to 3 tonne	Wide range; typically 25g to 150 kg	Very small to medium size; normally < 10kg	Very small to small/medium; normally < 5kg
Quantity *Typical minimum*	Low to moderate 1	Medium 500	Medium 50	High 10 000
Suitable materials	Ferrous and non-ferrous	Ferrous and non-ferrous	Ferrous and non-ferrous	Al, Zn, Mg, Cu Alloys
Tooling cost	Low	Medium	High	Very high
Labour content	High	Low	High	Low
Typical components	Garden furniture; heating boilers; crankshafts	Valve bodies	Public telephone dialling plate; disc drive components	Carburettor components; toy cars

ment in tooling before the first casting is produced. Table 12.1 summarises points to consider when selecting a particular process.

The selection of a particular casting process will be based largely on the range of characteristics in the table. However, it is important that the designer exploits to the full the advantages of the process. For example, a simple component that is to be machined over most of its surfaces would be a poor contender for an investment casting; a sand casting would probably be a more suitable choice. However, where a component is to be made from an investment casting, the designer should check for any features that could be available from the process at very little extra cost. For example, could the company logo be included on the product? There would be an increased tooling cost but the production cost might not even be affected.

It is also important that the designer understands the limitations of the process. For example, although investment casting is often used by artists to generate large bronze sculptures, it is more commonly used for smaller components. As described in Chapter 8, the wax impressions of the component will be joined together to form a large 'Christmas tree', which will be used to create a large multi-cavity ceramic mould. The maximum size and weight of this mould will be determined by the handling facilities in the foundry. The size of component will obviously affect the number of cavities possible within the 'Christmas tree' mould. Generally, the greater the number of cavities, the more cost-effective the process becomes. So although a weighty component with simple features could be produced by this process, it is unlikely to prove to be the most economic choice.

12.2.3 General design points

The designer of a cast component must always bear in mind the mechanics of the casting process in order to achieve a successful product. All castings involve the flow of molten metal into a mould, which is then allowed to cool. An ideal casting allows the solidification of the metal to occur in a uniform fashion, starting from the points in the mould furthest from where the molten metal is fed in. Changes in section will hinder this process. Thick sections will contain more heat energy than thin sections and will therefore cool more slowly. The subsequent differences in temperature will generate thermal stresses in the component which may in turn lead to shrinkage cavities being formed. The design should aim for uniform thickness in all sections. Where this is impossible, the section change should be as gradual as possible.

Although the supply of molten metal via the runners and risers should in part compensate for contraction that occurs during the solidification process, problems may occur if a thin section is used to feed a thicker section. Premature solidification in the thin section may starve the thicker section of metal and result in shrinkage cavities.

The following points should also be kept in mind:

1. Care needs to be taken where sections join and at corners. Avoid sharp edges and potential hot spots, such as the insides of sharp corners, by ensuring that the internal radius is no less than the section thickness (see Figure 12.1).

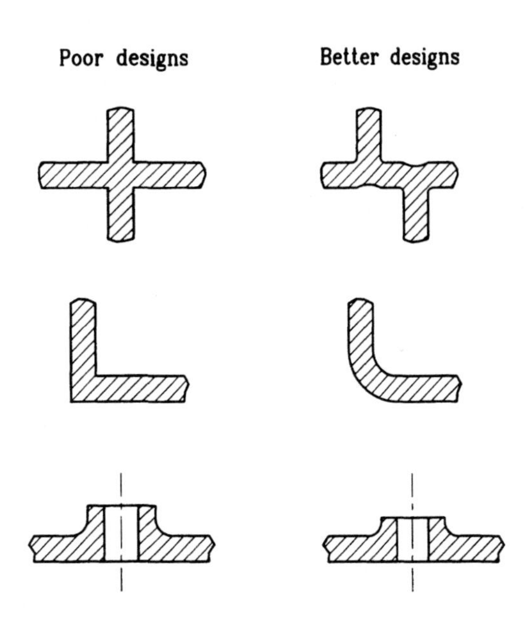

Figure 12.1 *Designs to avoid hot spots*

2. Large, flat areas may warp as they cool, so should be avoided.

3. The design must allow the pattern to be removed from the mould or, in the case of a permanent mould, the casting from the die. A joint line must be selected and a taper or draft (approx. 3°) included on those surfaces normal to the joint line.

4. If possible, keep the joint line on a single plane. This will simplify the moulding operation, particularly for sand castings. This also applies to other casting processes but is not so critical for investment or die castings, where the joint line may be cranked, or even curved, to facilitate removal of the casting. A simple planar joint line will help minimise the tooling costs, but a complex one will not necessarily increase the cost of the production process. The good surface finishes available from die and investment casting means subsequent manufacturing

operations on the component are often not needed. The joint line will always leave a line on the final component, but care in selecting its location can minimise its aesthetic impact.

5. Avoid re-entrant shapes as these will increase the difficulty of removing the pattern or the casting. They may result in the need for more components to be used in the pattern, hence increasing its cost. If internal cavities are required with a sand casting, it may be necessary to create a cored hole in the casting. This hole will allow the core to be located prior to casting, and will help during the fettling process as the sand is removed from the casting. The final design may incorporate a core plug to fill the hole if that is a functional requirement. A cast-iron car engine block will incorporate water passages which will have a number of such core plugs (see Figure 12.2).

6. Location of holes should be chosen with care. Those whose axes lie in the direction of the die opening-movement can easily be accommodated in the main die but holes in other directions will require retractable cores to allow the casting to be removed. The designer should consider whether some of the holes could be produced more economically by a later drilling operation.

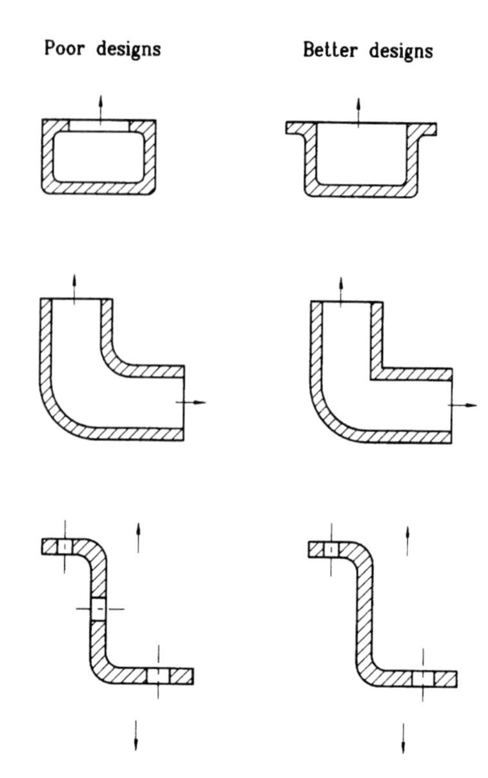

Figure 12.2 *Designs to avoid complex cores*

12.3 Forging

12.3.1 When to use forgings?

As with casting, the primary objective is to obtain a raw-material format closer to the final shape than is available from stock material. Although much greater detail will always be possible with processes such as investment casting, forging can be a worthy competitor for simpler components suitable for sand casting, provided they have no internal cavities. For example, a car engine block could be cast but is almost certainly too complex a shape to be forged. However, the crankshaft could either be cast or forged.

A casting may have surface and internal defects arising from impurities in the melt and uneven cooling. A forging, on the other hand, is far less likely to suffer from such internal defects. The often considerable degree of hot working of the material will tend to close up any existing porosity and modify the grain structure. The process results in some alignment of any voids or inclusions in the direction of material flow as it is worked. This can be thought of as creating fibres in the material, which in turn create a directionality to certain mechanical properties. A forging will tend to be stronger and more ductile in the direction of the fibering than across the fibres. This is an important characteristic that the engineer needs to exploit when selecting an appropriate manufacturing process. For example, the connecting rod of a car engine is a shape that would lend itself to production from either a casting or a forging. However, since most of the loading it will receive in service will be along its

Table 12.2

	Open die	*Closed die*	*Upset*	*Cold headed*
Detail	Approximate shape only	Closer control of detail than open die	As closed die	Can produce bolt heads in final form
Accuracy *Typical tolerance*	Poor	Depends on size 5 kg − 0.5mm + 1.5mm		Good e.g. 25mm dia ± 0.1mm
Quantity *Typical minimum*	Low volume 1	High 10 000	Medium to high	High 10 000
Tooling cost	Low	High	High	Medium
Labour content	High degree of skill required	Medium	Low	Low
Typical components	Horseshoe	Connecting rod	Cylindrical, e.g. engine valve	Bolt

length, the forging could be seen to offer certain advantages through the creation of longitudinal fibres.

12.3.2 Which forging process?

As introduced in Chapter 8, there are a variety of forging processes, including open die, closed die, upset, cold heading, etc. The Table 12.2 summarises some of the differences to be considered at the design stage.

12.3.3 General design points

The design of forging dies is fairly specialised and hence best left to the expert, who is able to draw on experience with his/her particular forging equipment as well as the basic engineering rules. The component engineer considering the use of a forging is recommended to consult and involve the forger at the earliest possible stage in the design process. However, there are a number of general points that should be borne in mind:

1. The high forces involved will tend to wedge the workpiece into the die, so a suitable draft angle should be used to ease its release. Allow 5 to 7° on external surfaces and 7 to 10° on internal surfaces.

2. Avoid deep recesses and tall bosses. If possible, the boss should be no higher than two-thirds of its diameter.

3. Remember that you are expecting the material to flow into the die halves. Make the path as easy as possible for the material by allowing generous radii. The radius leading to a rise should be of an order equivalent to the height of the rise. Sharper radii are possible but may require additional stages to the process, each adding to the cost. Normally

radii should not be less than 3mm, possibly down to 1.5mm on small components.

4. The location of the parting line is important. It will influence the grain flow and die costs. When choosing the parting line location the following should be considered:

   - Where possible it should be kept to a single plane as this will minimise the forging costs.

   - Approximately equal volumes of metal should come on either side of the parting line and ideally the height of rise across the workpiece should be uniform.

   - When the flash is trimmed the forging fibre will be cut at the parting line. Ideally this should not be at a location on the component where it will be subjected to its highest stresses when in service.

   Obviously it will not be possible to meet all these requirements on some components and an appropriate compromise will be needed. The last point may only be relevant in highly stressed components but remember that the fibering is an advantage that the forging has over the casting, so do not lose this benefit by a poor choice of parting line.

5. The aim of the process is to form the metal into a shape close to that required, with an ideal requirement of no further finishing operations. However, the geometric limitations of the process, errors arising from die wear, corrections for warping, and build-up of surface scale mean that some subsequent machining is nearly always required. A machining allowance of about 3mm is normally required.

12.4 Powder metallurgy

12.4.1 When to use powder metallurgy?

Powder metallurgy (PM) can be used to produce parts which are very close to their final shape and require no further machining. Whilst this is a potential advantage over forgings and most forms of casting, the end product is certain to have a higher degree of internal porosity than an equivalent casting and is therefore not suitable for applications requiring a high degree of mechanical strength.

Another advantage of PM is that metals that would not normally be combined can be with this process. For example, copper can be combined with carbon to produce brushes for electrical machines, and cobalt and tungsten carbide can be combined for cutting tools. The porosity can also be used to advantage, e.g. self-lubricating bearings.

The process is normally used for small (less than 500g) components required in high volumes, avoiding the need for subsequent machining and hence giving the process a cost advantage.

12.4.2 General design points

The process, described in Chapter 8, involves compression of the alloy, which is fed into a die in powder form. Initially, the powder has a relative density of about 38%, which will rise to over 90% after processing. When designing for PM manufacture the following points should be kept in mind.

1. The compression is in one direction only. Dimensions in the direction of the compression axis will be affected by the mass of material present and the compression process, enabling control to about +2% of the dimension. Dimensions normal to this axis will be more closely controlled by the die and can often be held to within +0.5% of the dimension.

2. This uni-axial compression process means that features such as cross holes, screw threads or reverse tapers cannot be accommodated. If the die is lubricated, straight-sided components can be produced.

3. Normally it is good design practice to avoid sharp edges on components. This is particularly important for PM parts. Avoid features that would require sharp edges on the punch, which would be vulnerable to damage at the high pressures involved. The example in Figure 12.3 (b) shows how the design of a chamfer can be improved for PM. The combination of the shallow angle and the small flat (about 0.2mm) should result in a satisfactory punch life.

4. Deep components will incorporate a density gradient as the compacting pressure will not be transmitted uniformly throughout the depth. As a general rule the depth of a component should not exceed 2.5 times its diameter. Figure 12.3 (c) shows a method of effectively reducing the component depth to minimise problems through varying density. The cost of modification to the lower part of the tooling in this instance would probably be offset by the savings in component weight and raw material.

5. Weight for weight, components produced by PM tend to be more expensive than forged or cast items. Only select this process if the item cannot be produced from a casting or forging, or if the efficiency of material usage and elimination of subsequent machining operations give the PM process a clear advantage.

6. Tooling for PM production tends to be expensive, so the process is normally only economically viable for high volumes (10 000+). However, HIP (hot isostatic pressing), when used to form large parts in materials difficult to produce in alternative ways, can be economic with very low volumes (20–30).

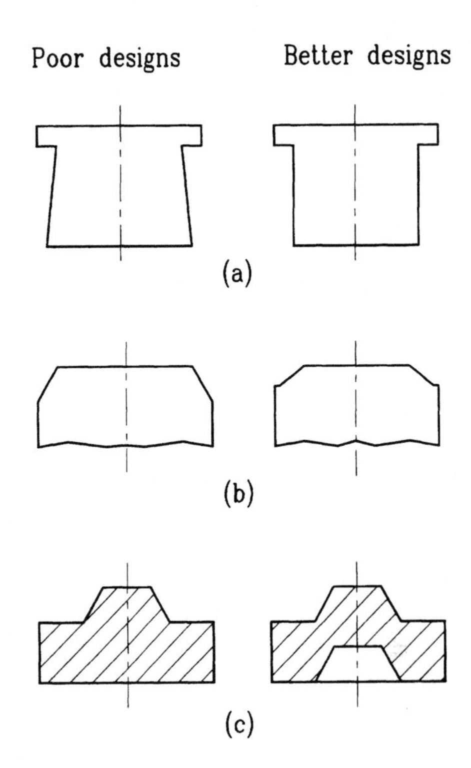

Figure 12.3 *Designing for powder metallurgy*

12.5 Material removal

12.5.1 When to machine?

Machining simply removes material from the raw item in order to create the dimensions and surface finishes within the tolerances specified on the component drawing. The raw material could be in one of a number of alternative forms: casting, forging, stock material, etc. As we have seen, the casting or forging try to replicate the completed component but often need to be finish machined for those features that are either technically or economically not possible with the original process. Many components are machined directly from stock material, i.e. standard forms readily available from stockholders: bar, tube, plate, etc. Such stock material offers a number of advantages. It is readily available on short delivery times, so helps minimise inventory costs, and weight for weight it will be cheaper than buying a forging or casting although there will most certainly be more wastage. However, the biggest single advantage is that stock material is not dedicated to a particular component in the way a casting or forging is. In the event of a change in demand, the stock material is much more likely to be suitable for producing other components than would be the dedicated material forms.

12.5.2 Which machining process?

Metal cutting processes all depend on relative movement between the tool and the workpiece. The machining process is normally defined by the type of relative movement. Table 12.3 summarises some of the more commonly used processes. In practice, the variety of machines under each of the headings is so wide that the table can provide only the most basic of comparisons.

Table 12.3

	Shaping	*Milling*	*Turning*	*Drilling*	*Grinding*
Tool movement	Linear	Rotating	Linear	Rotating	Rotating
Workpiece movement	Linear	Linear	Rotating	Static	Rotating or linear
Component detail	Flat surface	Flat surface; profiled surface; holes and pockets	External diameters; internal diameters; axial holes; radial flat surface	Holes	Diameters or flat surfaces
Surface finish	Medium to crude finish; typically 1.6 to 25μm	Moderate to good; typically 1.6 to 6.3μm	Moderate to very good; 0.4 to 6.3μm	Moderate; 3.2μm would be typical	Excellent; typical 0.4μm
Typical tolerance (mm)	± 0.03 to ± 0.06	± 0.01 possible; ± 0.03 typical	± 0.005 possible; ± 0.03 typical	± 0.03 possible; ± 0.1 typical	± 0.003 and better
Component size	Small to quite large 1m × 2m. For larger components use planer	Depends on machine; up to 1m cube is common but up to 2m cube is possible	Depends on machine; 0.5mm to 300mm dia is common, but up to 2m is possible	Up to 25mm dia. is common, over about 60mm normally turned	From 0.5mm dia. to approx. 1m × 150mm
Relative cost	Low	Medium to high depending on machine and tolerances	Low to very high depending on volume, machine and tolerances	Normally low	Normally high

Turning, for example, can be performed on a simple lathe, with the linear tool movement being controlled manually by the operator. Such a lathe might be used for low-volume components. Higher-volume components might be produced on a multi-spindle lathe, where several components could be turned simultaneously, or on a bar machine which feeds the raw material automatically as bar to produce the components. Many lathes are now computer controlled; some are equipped with driven (rotating) tools and have the ability to perform milling operations. The capital cost of machines at each end of the scale can vary enormously, by factors of 1:10 and more.

12.5.3 General design points

Any machining process will require the workpiece and the tool to be held rigidly in their correct relative positions. This 'setting up' of the machine needs to be done each time a new component is loaded onto a machine for a new operation. The use of fixtures will simplify the loading of subsequent components, but the loading and unloading time is wasted time as far as material removal is concerned. If the component has to be transported from one machine to another, there will be an increase in lead time and part completed stock, i.e. WIP (work in progress). The lower the number of machines and operations involved, the lower the machining costs will tend to be. The ideal target is a design requiring only one operation on one machine.

The following further points should be kept in mind:

1. The way that dimensions are given on the drawing can affect the manufacturing process. See the tolerance examples in Chapter 3 for ways of reducing the number of operations by careful allocation of tolerances.

2. Obviously the design should avoid tolerances being any tighter than they need be. Where tight tolerances are needed, try to avoid spreading them over a number of components so that, for example, only one item in an assembly of, say, three components needs to be ground as well as turned.

3. The workpiece will need to be held in a fixture or lathe jaws during the machining operation. This means that part of the component will be hidden and unavailable to be machined. Try to ensure that such hidden areas do not need to be machined, to avoid an additional operation.

4. Avoid blind tapped holes. If this is not possible, ensure that the thread does not continue to the bottom of the hole, to allow clearance for the tap. Insufficient clearance will simply result in a broken tap (see Chapter 13).

5. Make sure that the selected hole size is available from a standard drill; preferably one that is stocked by the manufacturing department. (These are often specified in company design manuals.) A blind hole should normally have a conical end, as would be produced by a standard drill bit. Any variance from this will add operations and cost.

6. Deep holes are more expensive to produce than shallow ones. Try to avoid holes deeper than three times the diameter.

7. Entry and exit points for holes should be normal to the component surface. This helps to avoid uneven loads on the drill bit (Figure 12.4).

8. If a component is to be turned, try to ensure that all diameters are concentric and that any plane surfaces are normal to the axis of the diameters (see Figure 12.4).

9. Small internal radii on a turned item should be equal to the tip radius of the cutting tool. Try to ensure that standard cutting tools can be used when selecting internal radii.

10. A turned component will normally be held in chuck jaws. Make sure that there is a suitable diameter (can be internal or external) for the jaws to grip. The required length for gripping will depend on the component size, but for small items would normally be at least 6mm.

11. If possible, provide a datum that can locate against the chuck jaws. This will be a fixed datum for all similar components as far as the cutting tool is concerned. If the datum is at the other end of the part, the cutting tool will need to be set to this datum each time a new part has been loaded. (This does not apply to a bar machine, where the bar is fed forwards to a datum point.)

12. Avoid components that are very long relative to their diameters. They may need additional support during turning and may well deflect under the cutting forces. This deflection will result in poor dimensional control.

13. Internal features on long components are likely to need cutting tools of a shape that will preclude sufficient rigidity. Surface finish and dimensional control are likely to suffer.

14. Components requiring machining operations other than turning will normally need to be held in some form of fixture. The part design should take account of this by providing a suitable base and reference points.

15. Internal corners produced by a milling cutter (in a plane parallel to the cutter axis) cannot have radii less than that dictated by the cutter diameter. Their radii should be kept as large as possible. Internal corners in a plane normal to the cutter axis cannot have a radius less than the tip radius of the cutting tool. The designer should select radii that enable standard tools to be used wherever possible (see Figure 12.5).

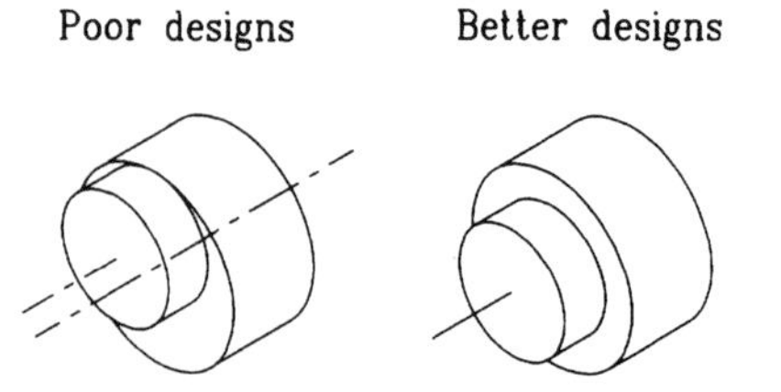

Concentric diameters are easier to machine

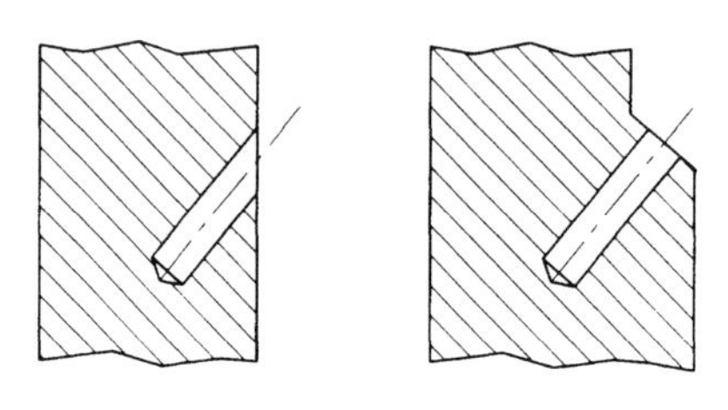

Drill entry should be normal to surface

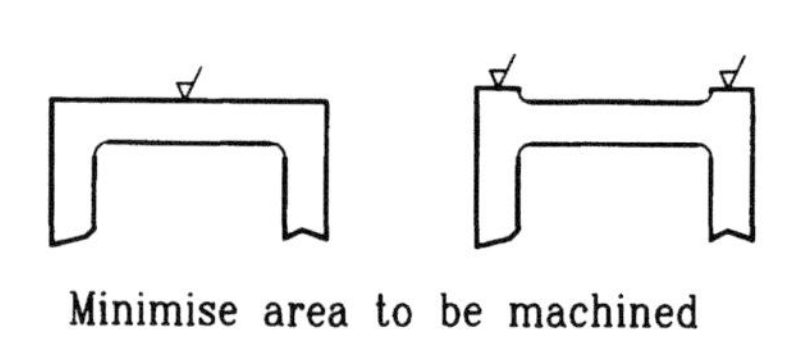

Minimise area to be machined

Figure 12.4 *Designing for machining*

16. Try to design to limit the number of cutting tools required. Each tool change incurs cost. This is obvious on a manual machine, but it also applies on the latest CNC machines with automatic tool change systems. A sophisticated turning machine may have 12 or even 16 tools in the turret but, even so, the lower the variety of tools needed, the greater can be the number of sister tools. (When a cutting tool has reached the end of its life a sister tool can take over without the need to stop the machine.) If a small range of tools can cope with a variety of different components, the change over time between different components will be reduced, thus allowing more output from the machine.

17. Avoid machining unnecessary areas. For example, rather than a whole surface,

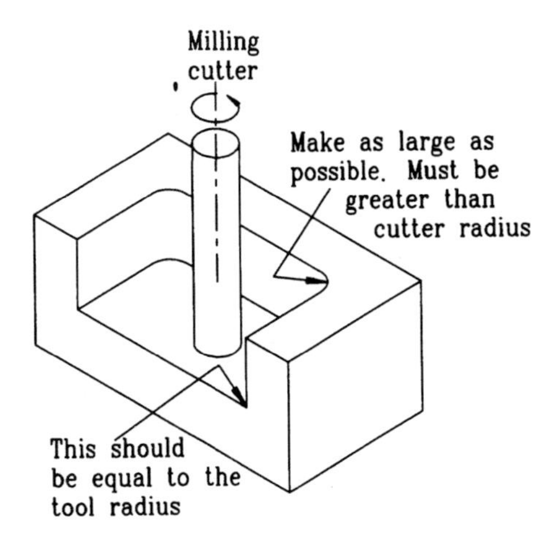

Figure 12.5 *Taking account of tool size*

perhaps two raised areas could be machined to achieve the same functional requirement (Figure 12.4).

12.6 Plastics

12.6.1 When to use plastics?

The term 'plastic' can refer to a wide and ever-increasing range of materials. In general terms plastics offer the designer advantages of low density (specific gravity normally ranges from 1.1 to 1.6), good corrosion resistance, good insulation for both heat and electricity, and they are easy to form. They are frequently self-coloured, and can normally be moulded to give a good surface finish.

However, they tend to be limited in terms of strength and dimensional stability, and are not suitable for high-temperature applications. Ultraviolet radiation can degrade many plastics, while some tend to absorb moisture, causing an adverse affect on the physical properties.

A good example of the use of plastics are the cases that are sometimes used for equipment such as video cameras or electric hand tools. Here, a case is often produced in a single, one-piece moulding including not only the top and the bottom of the case but also the hinges and the catch. Even if the tooling is costly, the elimination of any assembly operations must offer huge cost benefits.

12.6.2 Which plastic?

Plastics can be divided into two groups: thermoplastic and thermosetting. A thermoplastic material will soften as it is heated and harden as it cools. The process can be repeated many times. As one would expect, there is a quite rapid degradation in mechanical properties with increasing temperatures. Thermosetting plastics set as the result of an irreversible reaction. They offer quite good properties of strength retention with increasing temperatures but have an upper limit of around 250°C. Properties of common plastics are listed in Table 12.4.

There is a massive range of plastics materials available which is constantly growing as new variants come on to the market. The designer should liaise closely with the material supplier in order to gain the full benefits available from a particular material. If the material you are considering is very new, make sure it is fully evaluated for your particular application. See the centrifuge example in Chapter 1.

12.6.3 General design points

The wide range of plastics available and the alternative manufacturing processes (see Chapter 8) – i.e. moulding, extrusion, thermoforming, casting, etc. render it impossible to provide a comprehensive list of design points within the space available. Again, you are recommended to contact the materials supplier. The points listed relate to the potential benefits that should be borne in mind when considering the use of a plastic:

1. Many plastics are self-coloured and require virtually no further finishing process after leaving the mould. It is amazing that some plastic hub caps for car wheels are still painted. Not only does this process add cost but it can produce an inferior product if the paint starts to peel in later life.

Table 12.4

	Tensile strength MN/m^2	*Max working temperature °C*	*Specific gravity*	*Comments*
Thermoplastic				
ABS	28 to 55	120	1.02 to 1.06	Good weather resistance, low weight, good strength and abrasion resistance, flammable, poor chemical resistance
Acrylic	40 to 75	95	1.12 to 1.19	Good optical clarity, wide colour range, weather resistant
Nylon	55 to 80	120	1.1 to 1.2	Good strength, low coefficient of friction, excellent dimensional stability, but absorbs water and discolours at high temperature
Polycarbonate	66	120	1.2	High strength and impact resistance, used for safety glasses and shields
Thermosetting				
Epoxy	28 to 90	165	1.1 to 1.7	Good strength, elasticity, chemical resistant. used as adhesives and in composite materials
Melamine	35 to 55	175	1.76 to 1.98	Highly resistant to heat and water, commonly used for table mats

2. Avoid trying to match a self-coloured plastic component with one that is painted. Many plastic car bumpers are produced in a standard dark grey with no attempt to match the body colour. This makes good engineering sense in simplifying both the manufacturing process and the subsequent supply of spare parts. One company used self-coloured bumpers, attempting to match the painted body colour. When new, the match was good but with time the colours faded differently, spoiling the effect. Now most cars with colour-keyed bumpers have the plastic components painted.

3. Make use of the flexibility available from plastics by incorporating hinges and snap fit clips into a single moulding. Remember, reducing the number of components invariably reduces the component cost.

4. The moulded surface of many plastics is frequently of very high quality. Use this to incorporate operating instructions, warnings, etc. Once the mould has been produced, the additional information appears for no extra cost.

5. Any moulded component will have some indication of the manufacturing process,

such as flash lines at the mould joints or ejector pin marks. Obviously the position of these will be largely determined by the practicalities of the process, but attention to detail can reduce the cost of removing the marks. For example, a round knob may have flash lines running axially on opposite side of its outer surface. The provision of axial serrations on this surface might serve both to enhance grip and hide the flash lines.

6. Flash lines or joints between components can often be disguised by varying the surface finish so that the feature needing to be hidden lies along the joint between the different finishes. Products such as electric razors or hair dryers provide good examples.

7. Check the working temperature of the chosen plastic. Components on the top of a car fascia must be capable of withstanding the high temperatures generated when the car is left standing in the sun. At the other end of the scale, the choice by some manufacturers of a bumper material proved poor when, in cold weather, a minor knock would demonstrate the lack of resilience at low temperatures by cracking the plastic.

12.7 Ceramics

Ceramic materials are compounds of metallic and non-metallic elements. They are often in the form of oxides, carbides or nitrides. Advantages that they offer the designer include high compressive strength, high hardness, good wear resistance and the ability to withstand high temperatures. They have fairly low densities and are dimensionally stable with good corrosion resistance. The main disadvantage is their low ductility, failure often occurring as a brittle fracture. At present, the main areas of application are for seals, bearings, turbocharger rotors and cutting tools.

12.8 Composites

12.8.1 When to use composites?

Composites are normally selected for one of two reasons – the manufacturing process or the resultant material properties. At the low-tech end of the composite scale, hand-laying-up layers of glass fibre and resin over a mould is a common way of producing items such as replacement car body panels, complete car bodyshells (in low volumes), and boat hulls. As in Chapter 14, section 4, where car bodyshell production is discussed, the manufacturing method avoids high capital outlay in press tooling, and if a self-coloured gel coat is used, the need for subsequent painting is eliminated. The composite material offers improved corrosion resistance when compared to a steel alternative.

Where composites are used for sports or military equipment, this is because the material used offers an attractive combination of mechanical properties. A Formula One racing car tub (body), for example, makes use of these properties. The production method differs from the manual low-cost process described in Chapter 14 (example 14.6). After laying up, the uncured assembly is placed in a plastic bag from which the air is then evacuated. The system is placed in an autoclave where the pressure and temperature are raised to specified levels for a specified time to cure the composite. The pressurising is to eliminate any air gaps inside the composite. The greater control of the process at this high-tech end increases costs, but also increases the levels of consistency in mechanical properties of the final result. Consequently, lower safety factors can be used to create a structure with dramatically better stiffness-to-lightness ratios than are possible without composites. Minimising the weight on a racing car is obviously important, but high stiffness is also needed to provide acceptable cornering qualities.

As Table 12.5 shows, a carbon fibre composite has the potential to offer an almost fourfold increase in the stiffness-to-lightness ratios possible from more conventional metals! Other materials, such as silicone carbide, offer even

Table 12.5

Material	*Young's modulus (E) GN/m²*	*Density kg/m³*	*E/density × 1000*
Steel	210	7800	27
Titanium	120	4500	27
Aluminium	70	2700	26
Silicone carbide	510	3200	159
Carbon fibres	410	2200	186
Carbon fibre composite	170	1700	100

better ratios, but not without a penalty – rapid crack propagation. Composites do not have this problem, as demonstrated by military aircraft with composite wing structures surviving projectiles passing through the wings without catastrophic cracking.

Finally, to emphasise the fact that composites are not restricted to exterior body components, they are commonly used for brake discs on high-performance racing cars and aircraft, and in some cases are even used for springs. The rear leaf spring of the Chevrolet Corvette is a single spring made from a unidirectional glass-reinforced epoxy composite. This replaced a 10-leaf steel spring for only 20% of the weight of the original.

12.8.2 General design points

1. The mould should avoid sharp corners or any features where it may prove difficult to ensure close contact between the layup and the mould. Failure to attend to this may result in air inclusions in the composite which will dramatically reduce the strength of the final component.

2. Avoid undercuts. They are possible but will complicate the mould design – it may need to be split.

3. The layup should be tailored in terms of thickness and direction of the fibres to provide additional strength in the areas where it is most needed. The fibres are strongest when loaded in tension along their length. If loaded transversely, the strength-to-weight ratio will be reduced dramatically, probably to well below that of alternative materials. It is essential that the designer ensures that the fibre alignment is used to best advantage.

4. This ability to tailor the material can apply theoretically to other properties. For example, carbon fibres have a negative coefficient of thermal expansion, so when combined with an epoxy which has a positive coefficient, a component that is dimensionally stable over a range of temperatures is possible. Some components in the space industry use this property, but you should be aware that this is a very specialised area and unless the temperature range is very

limited, problems are likely with the bonding between the resin and the fibres.

5. Reinforcing ribs can be incorporated by including fillers of other materials (wood, foam, metals) in the layup. Care must be taken to ensure that the material chosen does not react with the resin or it may be subject to later corrosion. Sometimes composites are used for their non-magnetic properties. If so, take account of the effect of the rib former.

6. Take particular account of areas of high compressive stress, especially if the stresses could create local bending in a direction that might cause the fibres to buckle and delaminate.

12.9 Summary

You should at this point have a greater appreciation of the need for the designer to be aware of the manufacturing process and material characteristics in order to be able to take full advantage of the available potential benefits. Remember that, in a competitive world, if you do not fully utilise these benefits and one of your competitors does, he/she may be in a position to steal your customers and perhaps your livelihood!

In particular, the reader should understand when to use castings, forgings, powder metallurgy or machining, and how the design can take advantage of the chosen process. He/she should also be aware of when to consider non-metallic materials – plastics, ceramics or composites.

12.10 Questions

Now test your understanding of the chapter by trying the following questions.

1. As a designer, what factors would influence your choice between the use of an investment casting and a sand casting? (12.2)

2. Consider a simple butterfly valve that is to be used in the food processing industry, the main components of which are to be manufactured in stainless steel. The design features a fixed seal in a two-piece valve body. The flap (which rotates 90 degrees) to open or close the valve is produced integral with its stem. What raw-material form would you recommend and why? (12.2 to 12.4)

3. What is the difference between a thermosetting plastic and a thermoplastic material? (12.6)

4. If a hole is to be created as part of a machining operation how can the designer simplify the manufacturing task? (12.5)

5. Weight for weight, components produced by PM tend to be more expensive than forged or cast items. Why is PM ever used? (12.4)

6. What problems can arise if a casting is designed to have widely differing section thicknesses? (12.2)

7. When should the use of a ceramic be considered? (12.7)

8. A fully dimensioned drawing should specify the radii at changes of section. Assuming the component is to be machined, what factors should the designer consider when specifying these radii? (12.5)

9. Why might you decide to produce a component from a composite material? (12.8)

10. Name some of the reasons that might be used for deciding to produce a component in a plastic material. (12.6)

13 Designing for Joining and Assembly

13.1 Introduction

The aim of the engineer is to generate products that meet the customer's requirements of function, quality, reliability, etc. at the lowest practical cost. In general terms, the simpler the product, the easier this will be to achieve. One criterion for simplicity is the number of individual components required. More components mean more joints, more potential tolerance problems, and more variety for the manufacturing system to control, and in the end they will add more cost to the product. When considering an assembly or joining process the engineer should first question the need for the joint.

The need can arise for a variety of reasons, including

- the physical size of the assembly
- the materials used (e.g. a combination of metals and non-metals)
- the manufacturing process employed for the components (a casting and a pressing may need to be assembled)
- the need to dismantle for servicing during the product life.

Once the need for the joint has been established, the next stage is to select an appropriate technique and develop the product to take full advantage of the chosen method. Chapter 9 introduced a number of common techniques, so we will now concentrate on designing to gain full benefit from the process.

Table 13.1

	Heat application		*Chemical*	*Mechanical methods*			
	Welding	*Braze, solder*	*Adhesives*	*Screw*	*Rivet*	*Clip*	*Interference*
Ability to dismantle	Permanent	Semi-permanent	Permanent	Yes	Semi-permanent	Depends on design	May be possible
Distortion during process	Yes, with most types	Some	No	No	No	No	Some
Suitable materials	Fusible	High melting point metals	Most	All	All	All	All
Ability to hide joint	Excellent	Good	Good	Poor	Poor	Poor	Possible

13.2 Choice of process

Joining processes can be split into a number of broad categories, summarised in Table 13.1.

The characteristics compared in the table are simply the first points that the engineer needs to consider. A range of other factors may be relevant, such as the type of load applied, the operating environment, the accessibility of the joint, finish considerations and so on. These will be examined in the following sections.

13.3 Heat application

13.3.1 Welding

The welding process involves the application of heat to melt the surfaces being joined. They blend together and, on cooling, form a continuous structure. The result is a permanent joint whose strength is as good as the parent metal. There is a very wide range of welding techniques. These can in general terms be split into two categories which we will call continuous and localised.

Continuous welding

This category will include processes such as gas welding and variants of arc welding, MIG, TIG, etc. Heat is applied either by gas or an electric arc, the process often being shielded by an inert gas to prevent oxidation and the formation of scale on the joint. The process is sometimes called fusion welding, as the parts to be joined are fused together. There are essentially two types of weld used, the fillet and the butt weld.

The fillet weld consists largely of deposited weld metal into which some of the surrounding parent metal has melted (Figure 13.1(a)). The effective size of the weld is based on the height (t) of the enclosed triangle (Figure 13.1(b)). Any weld metal deposited outside the triangle is surplus material.

The butt weld is similar, but the deposited metal standing proud of the surfaces of the parent metal is called overfill and does not necessarily add to the strength of the weld (Figure 13.1(c)). For maximum strength the butt weld will penetrate the full depth of the material. If stressing is not critical, a partial penetration butt weld is often used(Figure 13.1(d)).

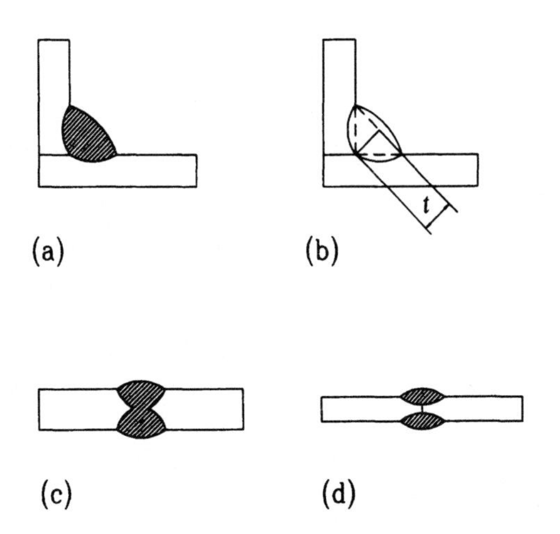

Figure 13.1 *Fillet and butt welds*

As the weld cools, cracks can form. The proportions of the weld can influence the degree of cracking that occurs. Ideally, the ratio of the width of the weld to its depth should be close to unity. Narrow joints with deep welds are prone to cracking, so the edges of the materials to be joined are often chamfered to allow the weld to have better proportions. Material thicker than about 3mm usually has some form of edge preparation to enable the weld to achieve full penetration. This involves making a chamfer on the edges to be joined, thus providing a V- shaped space that will be filled during the welding process. A single V is normally only suitable for up to 20mm-thick material. Above this a double-sided preparation is normally needed.

Assuming that no defects are present in the weld, it should be as strong as the parent metal, so in the case of a full depth butt weld, the static strength of the joint is determined by the cross-sectional area of the plates being joined.

Sizing a fillet weld is similar except that it is based on the size of the weld throat. In many cases the weld loading may be more complex than simple loading in tension or in shear. In the following example, a simplified approach of calculating the resultant force and applying this to the ultimate shear stress criteria is used to generate a conservative design. Figure 13.2 shows a yoke attached to a plate with twin fillet welds at each foot. Imagine a load F has been applied to the yoke. This will result in reactions at each of the joints between the feet of the yoke and the plate, which can be resolved into V vertically and H horizontally. Assuming the plate is part of a larger structure, and taking moments about the lower foot of the yoke,

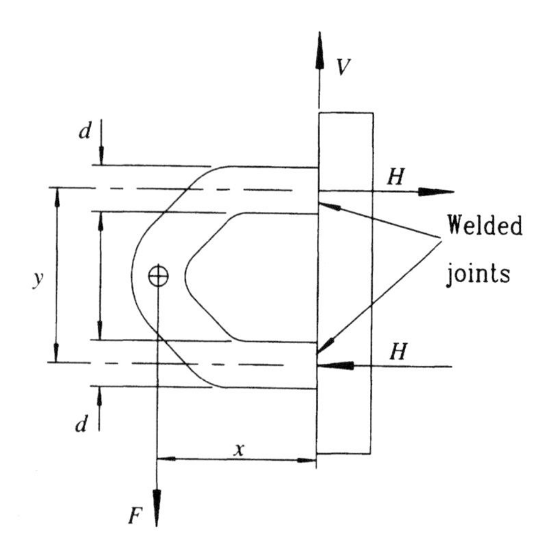

Figure 13.2

$$Fx = Hy$$

Hence

$$H = F\frac{x}{y}$$

The horizontal load on each welded joint is H. The vertical load on each welded joint is $F/2$. The resultant load at each joint is the combination of these, i.e.

$$\text{Total load} = \sqrt{\left(H^2 + \left(\frac{F}{2}\right)^2\right)}$$

Substituting for H

$$\text{Total load} = \sqrt{\left(\frac{Fx}{y}\right)^2 + \left(\frac{F}{2}\right)^2}$$

Now if the weld size is t (Figure 13.1(b)) and is a twin fillet weld, i.e., covering both sides of joint, the effective weld throat area at the foot of each leg of the yoke will be $2dt$ (Figure 13.2). Hence the shear stress in each weld is given by

$$\frac{F}{2dt}\sqrt{\frac{1}{4} + \left(\frac{x}{y}\right)^2}$$

Note that this example is an oversimplification (the lower joint may in fact take the full horizontal load by direct contact alone), but should result in a conservative design.

The loads considered above are static loads. If the component concerned is subjected to continuously fluctuating loads, failure may occur by fatigue. The welding process will impart significant thermal gradient on the component, which in turn will suffer a degree of internal stress. Examining a welded joint under the microscope will almost certainly reveal very small cracks at the toe of the weld. Under a fluctuating load these cracks can grow. A welded component will therefore tend to have a lower fatigue life than a similar component machined from solid. Shot peening can sometimes be used to increase the fatigue life of a welded joint.

Distortion

The high thermal stresses incurred frequently result in some degree of distortion. Figure 13.3(a) shows a typical type of distortion that can result from a single V butt weld. In this case there is a greater heat input at the top of the weld than at the bottom, the subsequent greater shrinkage at the top causing the distortion as in the diagram. The distortion could be reduced by using a double V butt weld, particularly if access permitted both welds to be done simultaneously. If this is not possible it may help to use an unequal double V preparation, the V for the first weld being smaller than that for the second.

The distortion on T joints is more difficult to avoid. If the two fillets (Figure 13.3(b)) are produced simultaneously, at least the distortion will be symmetrical as both the welds contract on cooling.

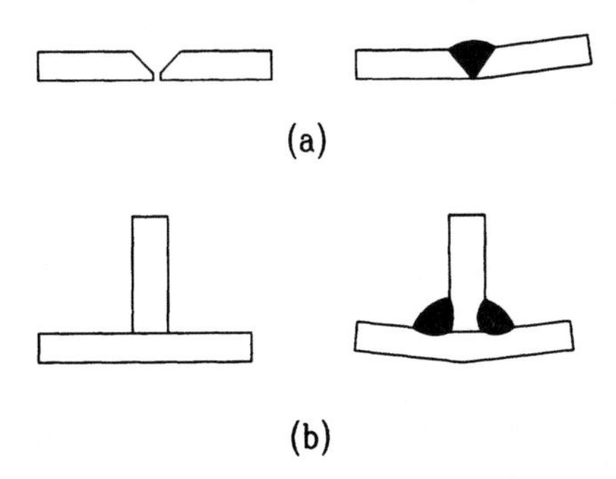

Figure 13.3 *Distortion from welding*

Although such internal stresses in the components can be reduced by heat-treating after welding, the designer needs to be aware that welding will induce stresses that can lead to distortion. Any machining that may be required should therefore be performed after all the welding has been completed. Even then there are dangers. If the component has not been stress relieved (i.e. heat-treated, see Chapter 10), machining may remove a surface containing some stresses, and as a result the component could then distort.

Electron beam welding (EBW) can be used for critical components where distortion is to be minimised. As the process takes place under vacuum, the size of the vacuum chamber will be a determining factor in assessing the viability of the process. (Large chambers are expensive and tend to take a long time to evacuate.) No filler material is used and the weld is very narrow, so the design must allow only small gaps at the joint (0.02–0.2mm). As the process is very rapid and the applied heat energy is very localised, distortion is kept to a minimum. Ideally, a clear line of sight is required in the plane of the joint to allow access for the electron beam. Figure 13.4(a) shows a recessed joint where access is limited. Figure 13.4(b) would be a much easier design for electron beam welding. Although the process tends to have a high-tech image it can also be used for fairly mundane products. A bi-metal bandsaw blade, for example, is almost certain to use EBW to join the expensive tool steel for the teeth to the lower-cost backing area. The welding process in this case reduces the cost of the raw material and improves the product (the high ductility of the backing strip reduces the instances of the band breaking).

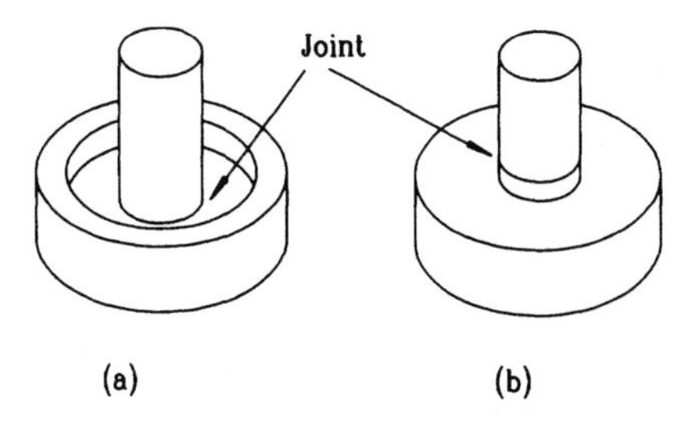

Figure 13.4 *Modifying for electron beam welding*

General design points

Figure 13.5 shows some typical joints suitable for welding. Note that for the thicker materials a form of edge preparation is shown.

1. Material thickness is probably the most important design parameter. Materials to be welded should be of a similar thickness.

2. Continuous welding will seal gaps – use this to your advantage.

3. In all cases sufficient access must be provided for the welding equipment. Ideally, welds should be in the horizontal plane, welded from above, not below. Remember, if access is possible but restricted, the quality of the weld may suffer.

4. Avoid placing welds at areas in the design where stresses or deflections are critical.

5. Consider the sequence in which the welds should be made and make this clear on the drawings.

6. Avoid welding items after they have had their final machining operations. The distortion inevitable after welding will affect the tolerances on the machined areas.

7. Be aware that two components welded together effectively become one. There will be no joint line between them to restrict fracture propagation.

8. Avoid making structures too rigid by welding as this will restrict their ability to redistribute high stresses and may result in premature failure.

Figure 13.5 *Joints suitable for welding*

Localised welding

An alternative method to continuous welding is to apply pressure and an electric current between the items to be joined. The electrical resistance in the joint will melt the metals locally, so forming the weld. This is called spot welding because it results in a weld spot typically 8mm in diameter. The process is commonly used in the mass production of assemblies made from sheet metal (car bodies and a range of white goods).

An advantage to the designer is that the weld does not require a skilled operator, so a high degree of reproducibility is possible. Unlike the continuous weld, it does not provide a sealed joint. In fact, the gaps between the spots are often a source of future corrosion in the product's later life. A variant of the process, seam welding, where the spots effectively are overlapped, can overcome this problem. In this instance, the items to be joined pass between a pair of rollers through which the pressure and a pulsed current are applied. However, this method can only be used if there is sufficient access for the bulky welding equipment. Conventional spot welding can also only take place on those welds that allow access for an electrode at each side of the joint.

The main points for the designer's consideration are:

1. Fast, and hence low-cost, form of joining.
2. Suitable for robot or other automated operation and hence high volume production.
3. Heat input is localised, so distortion is not generally a problem.
4. Joints need to be sealed against moisture ingress after welding to reduce corrosion.
5. Suitable for sheet metal assemblies.
6. High degree of reliability and reproducibility without the need for skilled operators.
7. No filler material is needed.
8. Material has to be overlapped to form the joints.
9. The equipment involved has a high capital cost.

13.3.2 Brazing and soldering

Unlike welding, brazing and soldering do not involve melting the components being joined. Molten metal alloy fillers are used to make the joint. Soldering uses alloys with melting points up to about 430°C, whereas brazing takes place at temperatures above 430°C but below the melting points of the materials being joined.

The lower temperatures involved should be considered by the designer. Compared with welding, they can allow the joint to be made more quickly, and the problems of distortion, warping, etc. are very much reduced. In fact, thin complex assemblies that would be difficult to weld can be brazed successfully. However, the disadvantage is that subsequent heating can destroy the joint by melting the braze material.

The strength of the brazed joint is dictated by that of the brazing alloy, which will be significantly lower than the base metal. Most metals can be brazed (a variety of suitable alloys are available) – in fact, the method can be used for joining dissimilar metals.

The process involves capillary action, so the joint must be designed with this in mind. Figure 13.6 shows some typical joints suitable for brazing. Fluxes are used to aid wetting. This means that any surplus must be removed to avoid future corrosion problems.

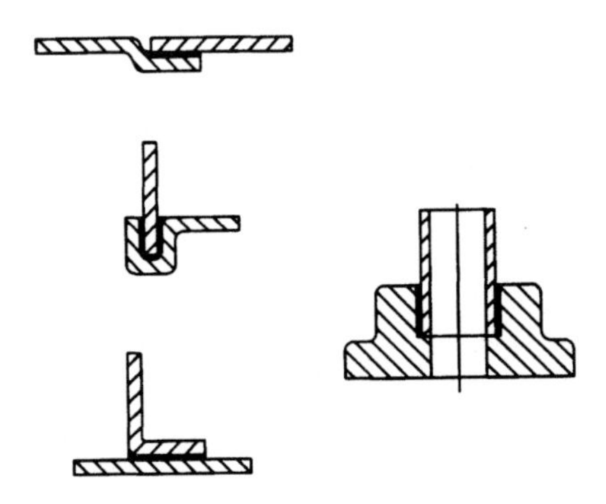

Figure 13.6 *Joints suitable for brazing*

13.4 Chemical application

13.4.1 Adhesives

The performance of joints achieved through the use of adhesives depends on the type of loading they are subjected to. They tend to be strongest in compression and shear, less strong in tension and generally poor under peal and cleavage loadings (Figure 13.7). Ideally, they should be designed from the outset as a bonded joint and not arise simply as a substitution for another method.

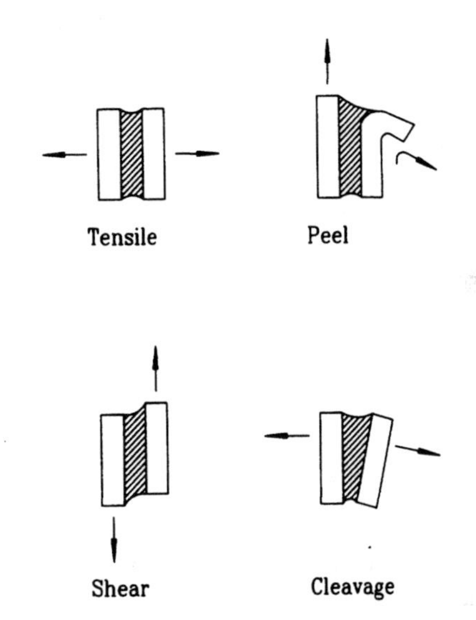

Figure 13.7 *Joint loadings*

The process offers a number of advantages:

1. Curing temperature need not be high. For some, room temperature is adequate and few require more than 180°C.
2. Almost all engineering materials can be joined to one another.
3. Materials of widely differing thickness can be joined.
4. The low temperatures involved mean distortion is not normally a problem.
5. As with continuous welding, joints can be both sealed and joined structurally at the same time. This can help corrosion resistance.

6. Smooth contours are possible on the final product as no through holes are needed.
7. Stresses in bonded joints tend to be more evenly distributed than in an equivalent riveted or spot-welded joint.

Against these need to be set a number of drawbacks:

1. Preparation and cleanliness are important for consistent joint performance.
2. Generally not suitable for high temperature applications. Most adhesives are not stable above about 180°C.
3. It is difficult to predict the life expectancy of a joint, especially in hostile environments.
4. Curing time may be a restriction at the manufacturing stage.
5. Some adhesive chemicals require precautions to prevent harm to the assembly workers.
6. Joints are normally permanent (although this may not be a disadvantage).

The bonded-joint designer should attempt to minimise stress concentrations and spread the load over the maximum possible area. The shape of the joint should also be such that the main loading is either compression or shear (Figure 13.8). Remember the shear strengths of common adhesives is in the range 14 to 40 MPa, whereas the capabilities in tension are only in the range 4 to 8 MPa at room temperatures.

Attention must also be paid to the method of assembly. When components are brought together, the surfaces to be joined should move towards each other in a direction perpendicular to the joint surface in order to avoid any shearing movement that might redistribute the adhesive. If possible, the design should incorporate features that enable the items to be located in their correct positions. Any additional cost of such features has to be offset against the need for assembly clamps, jigs and additional assembly time that may otherwise be needed.

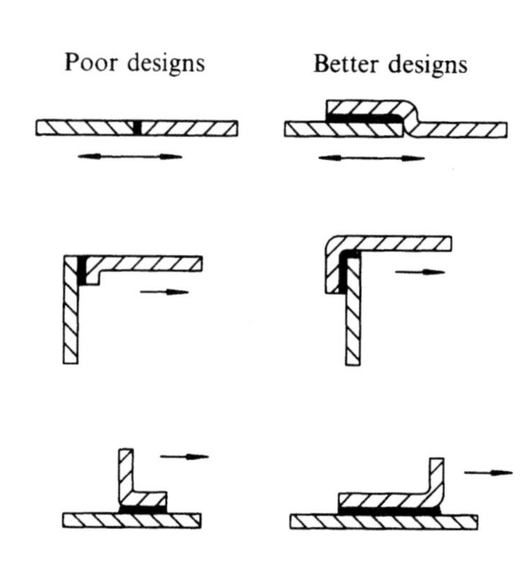

Figure 13.8 *Design for joints with adhesives*

13.5 Mechanical joints

Mechanical methods of joining components can be split into three categories – integral design features, threaded and non-threaded fasteners.

13.5.1 Integral features

Here we are considering means of securing joints mechanically, but without the use of third party devices such as nuts and bolts.

Permanent deformation

Sheet metal components can often incorporate integral fasteners, such as tabs or crimped edges. Figure 13.9(a) shows a seam joint produced by a sequence of fairly tight bends. The bend radius would normally be about half the material thickness. If there is a requirement to seal the joint, this can be done with fillers such as polymer sealants, adhesives or even solder. Applications range from car-radiator edges to drinks cans.

Electrical components often use the type of joint shown in Figure 13.9(b), which is a crimped joint, where external pressure has been applied to create a mechanical interference by permanent deformation of the components. The lanced joint (c) is a similar process except that the deformation is restricted to the tabs.

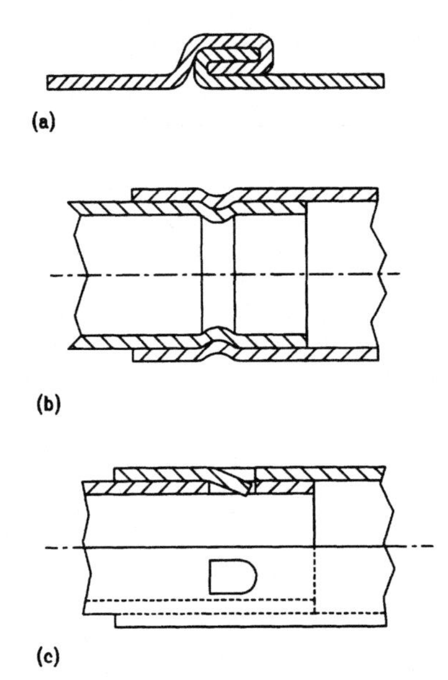

Figure 13.9 *Integral mechanical joints*

Joints of this type are not suitable for components that need to be dismanted later.

Interference

In Chapter 3 we introduced different types of fit, one of which was called interference. As an example, imagine a disc that is to be connected to a shaft. The principle of interference is that prior to making the joint the shaft has a slightly larger diameter than the hole into which it is to be fitted. Once it has been forced into the hole, the resulting distortions of the two components create a pressure at the joint which holds the parts together.

The choice of materials and the degree of interference determine the permanency of the joint. For example, a bronze bush may have an interference fit with its housing to prevent the bush rotating in service. At some stage in the product's life the bush may need replacing, so the degree of interference will be chosen such that it can be removed without causing damage to the housing. The material's thermal properties are often used to advantage during assembly. The shaft may be cooled or the housing heated to reduce the degree of interference during the actual assembly process. The designer should bear in mind the effects that in-service temperatures may have on the joint.

Interference fits can be used for assemblies which are required to transmit torque across the joint. For example, a pulley could be fitted to a solid shaft without a keyway, by using an interference fit. In such a case the designer has to ensure that the degree of interference is sufficient to avoid any slippage between the components. This can be done by considering:

(a) the pressure needed to transmit the torque

(b) the change in shaft outside diameter.

(c) the change in pulley inside diameter.

(d) the required interference.

Each step is dealt with as follows:

(a) Calculate the required pressure, p, between the surfaces to transmit the torque. This pressure will be the radial stress in the components at the joint.

(b) Consider the shaft. The pressure as calculated in step (a) will enable us to calculate the hoop stress on the shaft. Now Young's Modulus (E) gives us the ratio of stress to strain for a given material, so if we know the values for E and the stress, we can calculate the strain. Remembering that strain is the change in length compared with the original length, we should be able to calculate the change in shaft diameter required to create the necessary pressure at the joint.

$$\frac{\text{Stress}}{\text{Strain}} = E$$

Hence

$$\text{Stress} = E\frac{\delta}{L}$$

L, the length, is in this case the circumference of the shaft $= \pi d$ (if the shaft is diameter d), δ is the change in length.

Now assume the change in diameter is i

$$\delta = \pi(d + i) - \pi d$$

That is,

$$\delta = \pi i$$

$$\text{Strain} = \frac{\delta}{L} = \frac{\pi i}{\pi d} = \frac{i}{d}$$

From this we see that the strain can be expressed as terms of the diameter and the required change in diameter.

The diameter change as the components are forced together will create hoop and radial stresses. Remembering from stressing theory that strains in orthogonal directions are related by Poisson's Ratio, v:

$$\text{Hoop strain} = \frac{(\sigma_\theta - \sigma_r)}{E} \qquad \text{(A)}$$

Where σ_θ = hoop stress, and
σ_r = radial stress.

The radial stress will be the same as the pressure p at the joint as calculated in step (a). The shaft is solid. This means that the hoop stress will equal the radial stress and can also be represented by p. We are now able to calculate the change in shaft diameter that will occur at the joint.

The change in shaft diameter,

$$\delta d_{\text{shaft}} = \text{hoop strain} \times \text{diameter}$$

$$\delta d_{\text{shaft}} = \frac{(p - vp)}{E_{\text{shaft}}} d \qquad \text{(B)}$$

Using the convention that compressive stresses are negative, p in Equation (B) will be negative. As Poisson's Ratio will be less than unity, the calculated value for δd_{shaft} will be negative. This indicates a reduction in shaft diameter, as we would expect.

(c) Next consider the pulley, assuming that it has a solid hub with an inside diameter of d and an outside diameter of D. The material has a Young's Modulus of E_{hub}. The hoop stress will be at its maximum at the inside diameter

$$\text{Max hoop stress} = \delta_{\theta\,\text{max}} = \frac{(D^2 + d^2)}{(D^2 - d^2)} \times p \qquad \text{(C)}$$

Note that Equation C only gives the numerical value of the stress, and not its sense. In this case the hoop stress will be positive. This should be obvious if you imagine increasing the stress to the point at which something fails. The pulley would fly apart.

The change in hub diameter, δd_{hub} is given by

$$\delta d_{\text{hub}} = \text{hoop strain} \times \text{diameter}$$

Knowing the radial stress is p, the hoop stress is σ_θ and using Poisson's Ratio we have

$$\delta d_{\text{hub}} = \frac{(\sigma_{\theta\,\text{max}} - vp)}{E_{\text{hub}}} \times d \qquad \text{(D)}$$

Note that p is a compressive stress and is therefore negative, so the term $(-vp)$ will be positive. As we have already shown that $\sigma_{\theta\text{max}}$ is positive, so the value for δd_{hub} will be positive. In other words, the diameter of the hub will increase, as we would expect.

(d) The total allowance for the interference will be the sum of the required increase in hub diameter and decrease in shaft diameter. That is,

$$\text{the required interference} = \delta d_{\text{hub}} + \delta d_{\text{shaft}}$$

(Note that the numeric values are used, ignoring the signs, since we are interested in the total interference.)

The following example will demonstrate the above process:

A solid steel shaft, 50mm diameter (d), is driven by a cast-iron hub (outside diameter (D) 150mm and length 100 mm). The connection between the shaft and hub is via a press fit. If the joint has to be capable of transmitting a torque of 5000 Nm without any slippage, calculate the necessary degree of interference.

$$E_{steel} = 207 \text{ GPa}$$
$$E_{cast\ iron} = 103 \text{ GPa}$$
$$\text{Poisson's Ratio, } \nu = 0.25$$
$$\text{Coefficient of friction, } \mu = 0.3$$

1. Calculate the required radial pressure, p

   $$\text{Torque} = (\mu p \times \text{joint contact area}) \times \text{radius}$$

   $$5000 = \left(0.3p \times \pi\frac{50}{1000} \times \frac{100}{1000}\right) \times \frac{25}{1000}$$

   Hence

   $$p = 42.44 \text{ MN/m}^2$$

2. Now calculate the change in shaft diameter needed to create the pressure from Equation (B).

   $$\delta d_{shaft} = \frac{(p - \nu p)}{E_{shaft}} d$$

   Hence the decrease in shaft diameter is 0.0077mm.

3. Now for the hub; from Equation (C)

   $$\text{Max hoop stress} = \sigma_{\theta\ max} = \frac{(D^2 + d^2)}{(D^2 - d^2)} \times p$$

   $$\sigma_{\theta\ max} = \left(\frac{150^2 + 50^2}{150^2 - 50^2}\right) \times 42.44$$

   Hence

   $$\sigma_{\theta\ max} = 53.05 \text{ MPa}$$

   Using Equation (D)

   $$\delta d_{hub} = \frac{(\sigma_{\theta\ max} + \nu p)}{E_{hub} \times d}$$

   $$\delta d_{hub} = \frac{53.05 + 0.25 \times 42.44}{103 \times 1000} \times 50$$

   the required hub deflection is 0.0309mm.

4. The total allowance for the interference will be the sum of the required increase in hub diameter and the decrease in shaft diameter. That is,

   $$\text{total interference} = 0.0077 + 0.0309$$
   $$= 0.0386\text{mm}$$

   Note that this is the calculated interference that should allow sufficient pressure to be generated at the joint contact surface to prevent slippage when the torque is applied. When selecting suitable tolerances make sure that this is the minimum interference.

Although, as we have seen, it is quite acceptable to design an interference fit capable of transmitting torque, probably a more common use of the interference fit is for the location of the inner and outer races of rolling element bearings. In this case, the objective is not to transmit torque via the fit but simply to ensure that the races do not rotate with respect to the shaft or housing respectively. Bearing catalogues will define the fit and necessary surface finish. As with all interference fits, the designer should make sure that the components only need to be forced over the minimum distance, partly to ease the assembler's task and also to minimise damage to the components being joined. If a separate seal is used to protect the bearing, select a seal diameter such that the bearing can pass over the seal area without causing any damage to its surface.

Threaded

The final use of integral features involves creating a thread on each of the items to be joined. Screw threads, as we will see in the next section, are more commonly found on standard (often re-useable) fasteners such as nuts and bolts. These components are produced in very high volumes on specialist machinery, giving a low-cost fastener. However, there are a number of occasions when there are benefits from producing integral threads.

- If the joint requires adjustment, as with the track rod ends on a car steering. An integral thread is provided to allow the tracking (parallelism) of the front wheels to be set.
- Compared with other integral methods of joining, this will allow the joint to be disassembled easily at a later date.
- There is high functional efficiency (see Chapter 15) when compared with the use of nuts and bolts. As the fastening is integral, the number of component parts in the overall assembly will be less, thus simplifying the whole manufacturing process.
- Specialist fine threads may be used that are not available with standard products. For example, this might apply in the production of accurate measuring equipment.

There are also some limiting factors:

- The most common way of producing external threads is by turning the component. This means they are normally only economic to produce on a component that can be rotated and held in a chuck on a diameter sharing its axis with the surface on which the thread is to be cut.
- Internal threads (below about 25mm) are normally tapped so do not need to be capable of being rotated. Care must be taken at the design stage to ensure that sufficient clearance for the tap is provided at the base of any blind holes (see section 13.5.2).
- If the thread becomes damaged in service, the customer may be obliged to purchase a replacement part. In the short term this may be an advantage, but care needs to be taken not to turn the customer against future purchases of your products.

There are many examples of the use of integral fasteners, ranging from the screw caps on soft-drinks bottles to compression fittings used by plumbers. Most of these are produced in very high volumes and could be considered as a variant on standard fastenings. They will be produced on specialist equipment, which may overcome some of the limitations above. For example, the very simple threads such as those in the screw caps of bottles are produced by other techniques, such as moulding. Overall, an integral thread will provide a reusable means of joining components with the minimum number of parts, but unless the volumes are very high, you should first consider the benefits of using standard fasteners.

13.5.2 Threaded fasteners

Chapter 4 introduced the concept of the standard fastener. Here we will concentrate on its application.

Bolt loading

In principle bolts should only take loads in tension. The bolt tension provides a frictional force between the components being bolted. This friction, in a correctly designed and assembled joint, is the only resistance the bolted joint has in shear. If the design requires greater loads to be accommodated in shear, other features should be added, such as dowels or spigots.

The designed frictional force will only be achieved if the bolt is tightened sufficiently. If this does not happen a shear load, may appear on the bolt. A prudent designer should take account of the likelihood of this happening by considering the effect of loose bolts.

Joint loads

We will now consider a bolted joint under a number of loading conditions, examining the joint line between the bolted components.

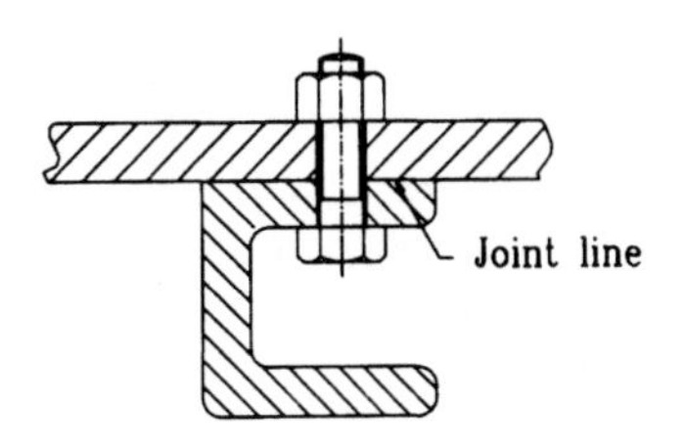

Figure 13.10 *No external load*

(a) *No external load* (Figure 13.10). On initial tightening the bolt will extend by δb and the component by δc. In the absence of any other forces, the initial tension in the bolt F_b must equal the compression on the component, F_c. That is, the joint will remain safe if $F_b = F_c$.

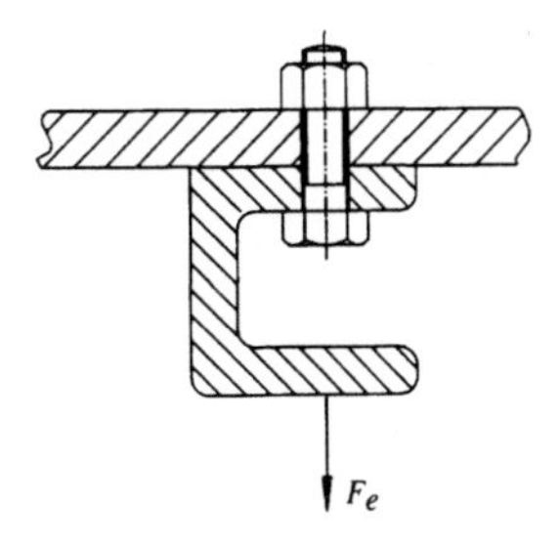

Figure 13.11 *External tensile load*

(b) *External tensile load* (Figure 13.11). Ignoring any distortion of the components, adding an external tensile load F_e, will increase the tension in the bolt but will reduce the compression between the two components at the joint face. It will not make the joint part as long as the increased bolt tension F_i is able to take the external load and still hold the components together by applying a compressive load between them. That is, the joint remains safe providing $F_i > F_e$.

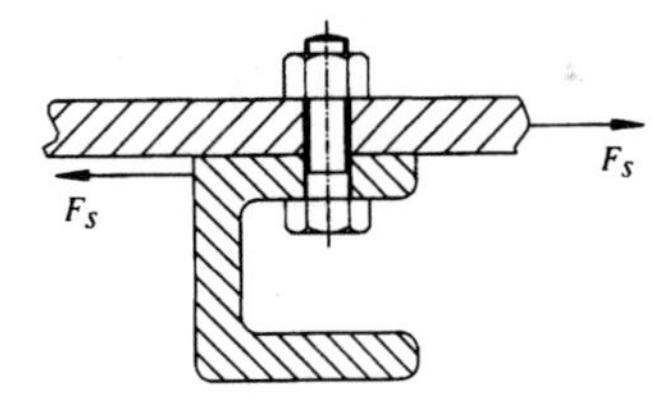

Figure 13.12 *External shear load*

(c) *External shear load* (Figure 13.12). Now consider the above F_e to be replaced with a shear load F_s. In this case the resistance to movement between the components will be provided by the frictional force between them. The joint will remain safe provided $\mu F_b > F_s$.

(d) *Combined tensile and shear loads* (Figure 13.13). In this instance, as the shear load does not affect the tensile condition, (b) must still apply, i.e. $F_i > F_e$. However, as the external tensile load has altered the compressive forces at the joint line, the frictional force available to resist the shear loading will be changed. For the joint to remain safe the following must also apply: $\mu(F_i - F_e) > F_s$.

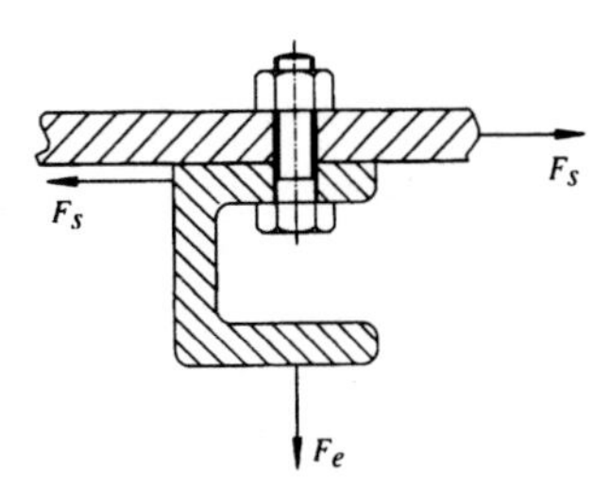

Figure 13.13 *Combined loads*

Bolt loads

Next we will examine the loads in the bolt. Both the bolt and the components are elastic and can be considered as springs of stiffness k_b and k_c respectively, where

$$k_b = \frac{A_b.E_b}{l_b} \qquad k_c = \frac{A_c.E_c}{l_c}$$

Note this is derived from

$$\frac{\text{stress}}{\text{strain}} = E = \frac{F}{A}.\frac{l}{x}$$

$$\text{i.e. stiffness} = \frac{F}{x} = \frac{A.E}{l}$$

where E = Young's Modulus
F = force
A = cross sectional area
l = length
x = change in length

Now examine the force acting beneath the head of the bolt. If the external force F_e is applied, the bolt and the component will both extend by Δ. It then follows that

$$F_e = k_b\Delta + k_c\Delta$$

That is,

$$\Delta = \frac{F_e}{k_b + k_c}$$

The initial bolt load, F_b, will be increased by the force needed to extend the bolt by Δ (i.e. $k_b\Delta$). The total tension in the bolt F_{bt} is therefore

$$F_{bt} = F_b + k_b\Delta$$

Substituting from above for Δ we have

$$F_{bt} = F_b + \frac{F_e.k_b}{k_b + k_c}$$

Examining this equation, we see that if the stiffness of the bolt k_b is large relative to k_c, the total bolt load approaches $F_b + F_e$. If k_c is large relative to k_b, the total load approaches F_b. Hence, provided that the joint does not part, the actual bolt load is always between the initial tension and the sum of the initial tension and the external load. In practice, the stiffness of the component is often unknown. In this case, design on the safe side, taking

$$F_bt = F_b + F_e$$

Note that if the external load fluctuates, this will result in a fluctuating load on the bolt, which could give rise to fatigue problems.

The maximum allowable loads in a bolt can be estimated from

$$\text{Max load} = \frac{0.8.\sigma_y.A}{K}$$

where σ_y = yield stress for the bolt material, A = tensile stress area, and K is a factor that depends on the material and method of manufacture of the thread, normally in the range 1.5 to 3.5.

Tightening torque

As we have seen, it is often necessary to ensure that bolts are tightened sufficiently to generate the required initial tension. This is normally achieved by measuring the tightening torque with a torque wrench. The required torque can be estimated by:

$$T = C\ D\ P$$

where T = torque (Nm)
D = nominal thread diameter (mm)
P = bolt tension (kN)

C is a factor which depends on the friction in the threads. This is affected by the materials, level of lubrication, surface condition, etc. (For steel, take C to be between 0.19 and 0.25.)

Although this method only has an accuracy of about 30% it is easy to apply. Note that care should be taken if the bolt is clean and well lubricated; the reduced thread friction can easily result in an overtightened bolt.

An alternative but costly method is to use special washers. These have raised portions that collapse, giving a visual indication that the required tension has been reached. Necked bolts can also be used. The bolt has a double head designed so that the top head, used for tightening, breaks away when the maximum designed torque is applied.

In practice, it is not uncommon for the bolt tension to decrease over time for a number of reasons, such as bedding-in of the contact surfaces, possible yielding of the materials, or simply the nuts becoming loose. Where it is important for joints to remain tight, the tightness of bolts should be checked later.

General application considerations

(a) Allow sufficient space between adjacent bolts and also between bolts and the edges of the bolted components (Figure 13.14).

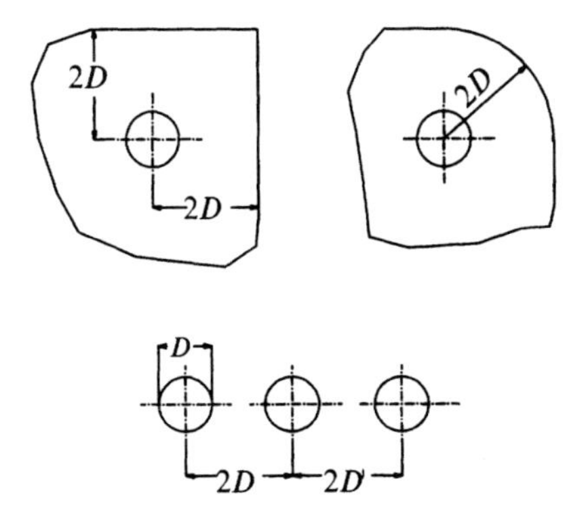

Figure 13.14 *Bolt spacings*

(b) Ensure blind holes are drilled sufficiently deep to allow clearance for the tap (Figure 13.15). A tap that encounters the end of a blind hole is almost guaranteed to break level with the top surface of the hole, thus making its removal difficult.

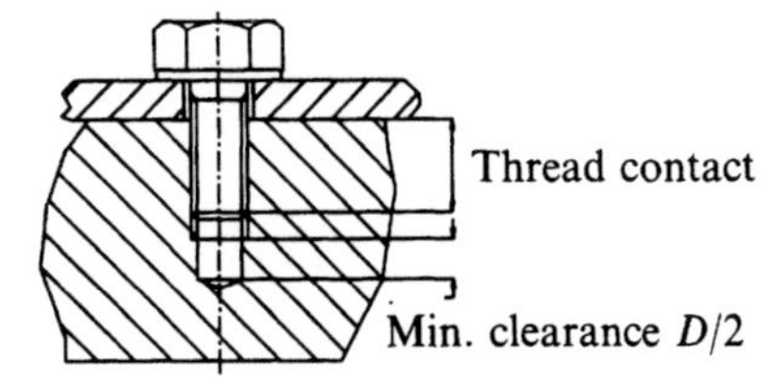

Figure 13.15 *Blind hole depth*

(c) Always ensure adequate access for assembly, both for positioning the components and for allowing sufficient room for spanners or sockets.

(d) Where possible, position vertical bolts with the head uppermost. This should ease assembly and may help if the nut becomes loose.

(e) Plain washers should be used They act as a bearing surface and help spread the load.

Figure 13.16 shows the design used to illustrate the joint loads and a modified version taking the practical considerations into account.

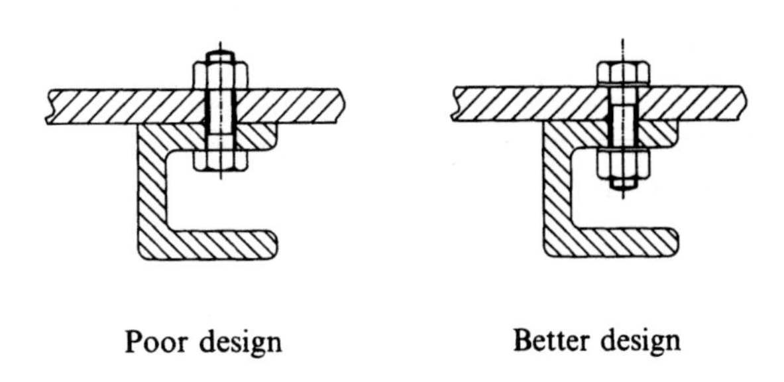

Figure 13.16 *Alternative bolt installations*

13.5.3 Non-threaded fasteners

Threaded fasteners are very useful for those joints that may need to be taken apart during the product's life. However, they tend to be more time-consuming to assemble than some of the non-threaded alternatives that were discussed in Chapter 4. These alternatives, although cheaper to assemble, are not so accommodating to the future disassembler.

A riveted joint differs from a bolted joint in a number of ways:

- It is semi-permanent. It can be drilled out but this is much more difficult than releasing a nut and bolt.
- It is much faster and easier to apply. There is no possibility of cross-threading, or not tightening sufficiently.
- It is highly reliable and should not come loose in service.
- The loading is different from bolts. A solid rivet is driven into place when hot. As it cools it shrinks, pulling the components together. Some of the load will, as in the case of the bolted joint, be taken by the friction forces between the joined components. However, as the levels of shrinkage, interference and friction are difficult to control, riveted joints are sized assuming that the rivet will take the full transverse load. The designer has therefore to check that first, there is sufficient cross-sectional area on the rivet to take the load and, secondly, that there is sufficient material around the rivet to prevent it tearing out.

Many other varieties of mechanical clips are now used, particularly in high-volume industries. Trim panels in cars, which years ago would have been secured by self-tapping screws, are now held in place by metal or plastic clips that are simply pushed into place. Not only does this make the assembly operation simpler, but eliminating the need for a tool to contact the clip allows the designer to hide the fastener completely (often to the distress of a future dis-assembler).

13.6 Automated assembly

Designing for automated assembly implies that there needs to be something different about components that are to be assembled by machines. This is not strictly true. The main difference is that machines are far less adaptable than their human counterparts, who are normally only too willing to accommodate a slightly awkward assembly operation. Although people will adapt and cope with a less than ideal operation, there will be a penalty in terms of time and cost. If the designer simplifies the assembly task so that it can be performed automatically, it is almost certain that the manual assembly operation will also have been simplified and quickened. Attention to most of the points raised in this section will therefore frequently be of benefit to both types of assembly.

Consideration to automated assembly must be given very early in the design process. The following pointers should help:

- One component should be singled out as the primary item in the assembly and there should be a logical sequence for assembling the components. This primary component will in effect become the jig to which the other components are attached. Ideally, it should posses a low centre of gravity. This is first to help with orientation of the part as it is fed via some form of hopper, and secondly to ensure stability of the assembly.
- Wherever possible try to avoid (or at least minimise) the need for component orientation. Figure 13.17(a) shows a component that needs to be oriented head up. This can be done easily in a feed mechanism, but to orient the flat would be difficult. Assuming the flat is used to lock the item in place it would be far better to add a groove to the component which will provide the same function without needing to be the correct way round.
- It may be necessary to add some features whose only function is to help with the component orientation during assembly. Fig-

ure 13.17(b) shows how the addition of a flat can identify the orientation of the small hole. Take care how such features are added, to avoid over-high costs. In this case, for instance, the flat could be added for very little cost if the component was a plastic moulding or die casting.

- Each of the above components will probably be sorted and fed in a head-up attitude. This is easier if the head is as large as practicable and has a minimum radius beneath the head (Figure 13.17(c)).
- Parts when fed in a hopper will have a tendency to lock together unless the designer has deliberately taken action to minimise this effect. A component such as the one shown in Figure 13.17(d) may become entangled with its siblings unless the designer has chosen a slit width of less than the material thickness. Open-ended springs may also tangle if the wire diameter is less than about half the coil pitch. Serrated washers will cause similar problems – use plain washers if possible.
- Bear in mind that the surface of a component that is used by gripper jaws effectively becomes a datum for the assembly process. It may be necessary to provide a gripping surface that relates to a functional feature even if this is not needed in the final product. The position of the hole in Figure 13.18(a) is functionally independent of the external surface. However, by adding a machined surface concentric with the hole, we have added a datum that can be used by the grippers.
- Some components may differ only in their internal features. Adding an external feature which has a fixed relationship to the internals can help.
- Keep the variety of items such as fasteners to a minimum. This will reduce the number of assembly feed mechanisms. As a general rule, minimising the number of different bolt diameters will help to contain manufacturing costs by minimising the numbers of drill bits stocked and the changeover times on machines (as well as simplifying the stocking and purchasing functions).

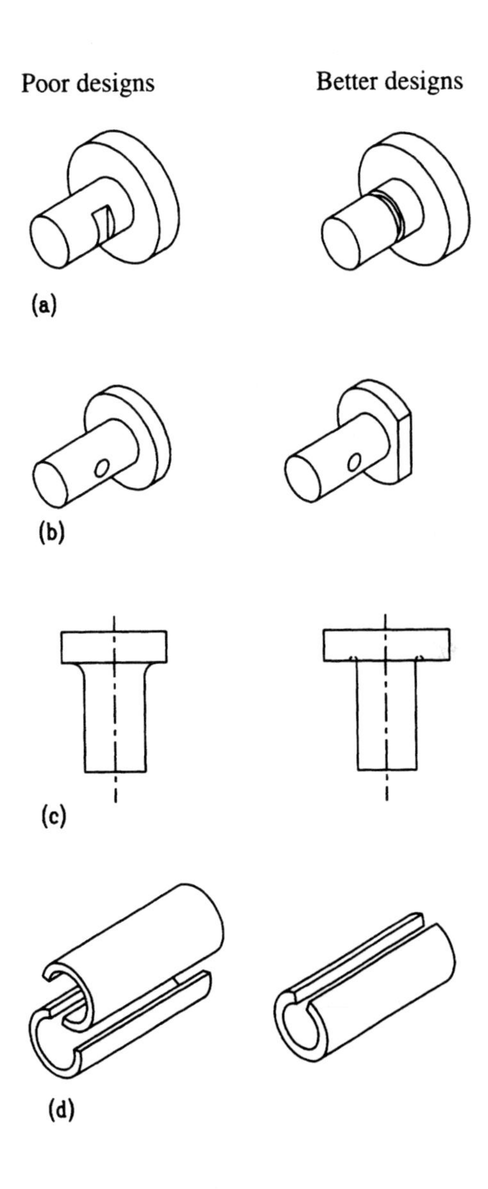

Figure 13.17 *Features to ease automatic assembly*

- Sometimes the need for a separate fastener can be eliminated. Figure 13.18(b) shows how a rivet can become an integral part of one of the components.

- O-rings are easier to locate on a shaft than on an internal groove in a bore: Figure 13.18(c).

- Use generous lead in chamfers to ease the location of components: Figure 13.18(d).

- If one of the components is a die casting, investment casting or plastic moulding, consider providing hexagonal pockets for nuts. This will help locate the nut radially, so easing the bolt's start and also prevent the nut rotating during tightening. There will be an increased cost for the die in the first place but in many cases this can be recovered many times over during the assembly process.

- When selecting a head type for a screwed fastener, the hexagonal or socket types will allow a more even application of torque than the slotted variant. If these are not possible, use a cross head rather than a single slot.

- Make sure adequate access is available for nut runners etc.

As these points show, it is vitally important to consider at the earliest possible stage how the product is to be assembled. For example, a small modification to a casting can ease the assembly process greatly.

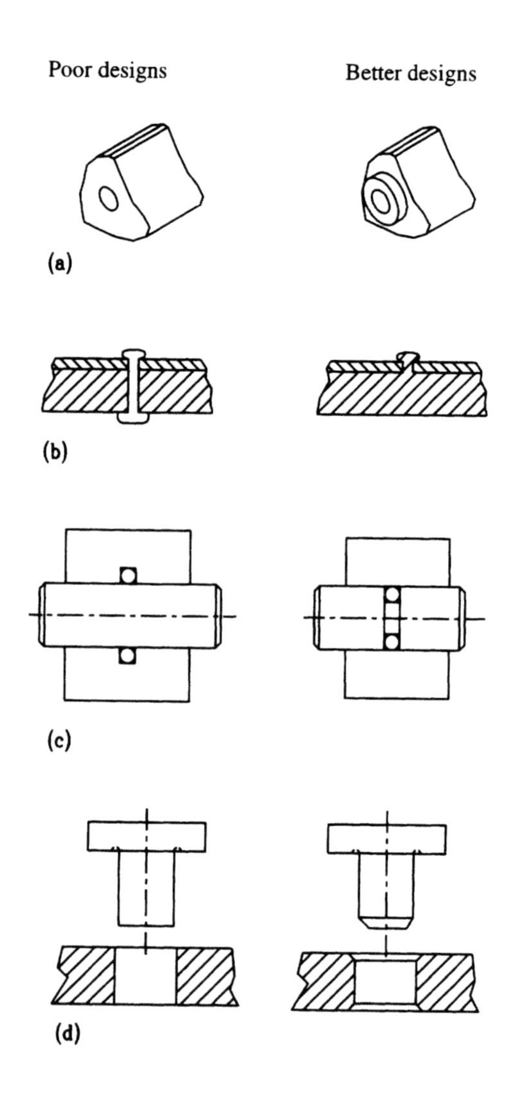

Figure 13.18 *Details to ease assembly*

13.7 Summary

This chapter has built on the basics of the joining processes, as introduced in Chapter 9. These include welding, brazing and use of adhesives, plus a variety of mechanical methods. At this stage you should understand:

- the principles of the processes

- how to select a particular technique

- some of the detail considerations that the designer must make in order to take full advantage of the system chosen

- some of the aspects that need to be considered if the assembly operation is to be automated.

13.8. Questions

Now test your understanding of this chapter by trying the following questions.

1. What is the essential difference between soldering and brazing? (13.3.2)
2. A joint made with adhesives is stronger under some types of loading than others. Would a given joint be stronger in tension than shear? (13.4.1)
3. When does a welded joint normally need an edge preparation? (13.3.1)
4. Sketch two joints that would be suitable for brazing. (13.3.2)
5. What type of load should a bolt be subjected to? (13.5)
6. Describe the essential difference in the design of a bolted joint and one that is riveted. (13.5.3)
7. Why would a plain washer be used under the head of a bolt? (13.5)
8. Name two features that could be included on a casting that would help with automated assembly. (13.6)
9. What are the limitations of using tightening torque as a measure of ensuring that a bolted joint is correctly assembled? (13.5)
10. Why would an interference fit be used between the outer race of a rolling element bearing and its housing? (13.5)

14 Influences on Design and Manufacturing Choices

14.1 Introduction

Part 3 has, so far, discussed many of the details that both design and manufacturing engineers need to take into account when generating a product capable of both being made and meeting its functional requirements.

There are, however, many additional demands that must be met for the product to be acceptable both to its potential customers and to the producing company shareholders. These demands, although frequently non-specific to either design or manufacturing engineering skills, can have a significant effect on many of the engineering choices made.

As an example, consider the many alternative methods of construction used for dwelling houses. Within the developed world the basic requirements for family accommodation must be broadly similar, so it might not be unreasonable to expect a trend towards a standard method of dwelling-house construction. After all, this has happened in the car industry, with the same basic model, a world design, being sold in many different markets. There are external influences at work – the availability of materials, local skills, planning requirements, aesthetic considerations, etc. – that all combine to affect the design and manufacture of the final product. The end result is a considerable variation in construction methods, from prefabricated timber buildings, to brick cladding with timber frames for internal walls, to stone structures, and so on.

This chapter will examine a number of external influences that are likely to sway the engineering decisions taken when introducing a new product. The approach will differ from the previous two chapters by concentrating on examples to demonstrate how the compromise nature of engineering requires the designer and manufacturer jointly to address these influences.

14.2 Aesthetics

The appearance of a product is often not given the priority it deserves, largely as a result of two popular misconceptions:

1. Aesthetics is the province of the artist, stylist, architect or industrial engineer, rather than the 'true' design or manufacturing engineer.

2. Aesthetics is really only of importance to consumer products.

Let us deal with each of these in turn.

The artist or stylist obviously has a great deal of influence on the final shape of an artefact, but unless it is an original painting or sculpture, the final product will have been produced according to the manufacturing engineer's instructions. The overall shape of a plastic moulding will be 'styled', but the choice of materials, surface finish, closeness and location of the joint lines, suitability for automatic assembly, structural strength, etc., will be the result of close cooperation between the skills of the design and manufacturing engineers. Aesthetics are an influence that has to be accounted for by the producing engineers.

When preparing for a job interview we all take particular care of our personal appearance, in the belief that the first impression created tends to be lasting and is likely to influence the outcome of the interview. The same principle applies to both industrial and consumer products. The first contact with a new product is frequently visual, an immediate (and often lasting) impression being created. This applies not only to the first sight of a new car in the showroom, but equally to a new engineering product at a demonstration or at an exhibition. The appearance, the feel, the fit and finish all build to form an impression of the product. This first impression will, of course, not be the sole basis for choosing the product, but will be the first stage in the selection process. If the impression gained is poor, that is the end of the process, and the product is rejected. Aesthetics frequently form a link, a vital link, in the acceptance chain of events.

Example 14.1

An industry that considers appearance an important quality in the selection of its technical equipment is the food process industry, in particular dairies and breweries. Obviously the many pumps, valves and associated assemblies of pipework have to be designed to avoid internal crevices, to have hygienic joints and smooth, easily cleanable internal surfaces, and to be resistant to corrosion. To achieve this, pump and valve bodies are often fabricated in stainless steel. Any weld joints must be polished so that the surface in contact with the product (milk, beer, etc.) remains smooth and crevice-free. However, it is also frequently a requirement that the outside of the joint is also polished! This only affects the appearance in terms of the product's technical functionality, and it adds to the component cost, so why is it needed? Breweries and dairies often take parties of visitors around their plants and consider it important for the plant not only to be hygienic, but to be seen to be hygienic.

Now consider the effect of this aesthetic requirement on the design and manufacturing decisions. Polishing on these kinds of product is generally a hand operation, the success of which requires a degree of subjective judgement. This means that it is difficult to control in terms of time, quality and cost, and so tends to be expensive. Inevitably, the food processor will foot the bill for this extra work, at least until an alternative supplier can provide a suitable item at a lower cost. Pressure is on the design and manufacturing engineers to provide the required end product at reduced cost.

Alternatives examined are reducing the number of joints (by the use of more complex pressings or castings), creating a more visually acceptable weld, automating some of the polishing operations, attempting bead blasting as an alternative finish, etc. The integrated nature of product engineering means that some of these changes will almost certainly have a knock-on effect on other aspects (see Example 14.9 later in this chapter).

Example 14.2

In some cases, despite sound engineering reasons for the original introduction of a concept, aesthetics can develop to become its eventual raison d'être. From the mid-1930s onwards the most common method of vehicle wheel manufacture within the motor industry has been a pressed steel fabrication. This has proved durable, reasonably resistant to corrosion, and able to form a satisfactory seal when fitted with a tubeless tyre. During the 1950s and 1960s an alternative type of wheel design became popular for those involved with motor sport: the light alloy wheel. This offered two major advantages of light weight and high strength. In the interests of gaining a high power to weight ratio, competition cars tend to be as light as possible. The alloys wheels helped in this objective in a particularly useful way by reducing the unsprung weight. In simple terms, the unsprung elements of a car are those that do not move when a vehicle at rest is rocked. The sprung elements are those that should ride smoothly as the vehicle proceeds over bumps in the road. The lower the unsprung weight (and especially the ratio of unsprung to sprung weights) the more effective the suspension will be in reducing the influence of irregularities in the road surface. A racing driver may not be interested in obtaining a smooth ride, but he is very interested in ensuring continuing close contact between the wheels and the ground, as without it he has no means of controlling the vehicle's progress.

Today, alloy wheels are frequently offered as options or standard equipment, particularly on the more expensive and powerful variants of motor cars. They are sometimes called 'sports wheels', presumably in recognition of their origins. However, their role today is simply an aesthetic one: they offer manufacturers a fairly inexpensive means of differentiating between high and low cost variants of a particular model.

Although undoubtedly visually attractive, these wheels are vulnerable to damage (from kerbs, or careless tyre fitters), more prone to corrosion than their steel counterparts, and often have problems retaining an adequate seal with tubeless tyres in later life. Much engineering effort, involving both design and manufacturing skills, has been expended in attempting to reduce these problems. The aesthetic appeal of steel wheels has been enhanced by more complex pressings to produce the actual wheel (starting with the Rostyle wheel from Rubery Owen), and complex metal pressings to cover the steel wheel cosmetically, which today have been superseded by plastic covers, some of which imitate quite successfully the appearance of the alloy product. Some of the problems with the alloy wheel arise from corrosion affecting the seal at the tyre/wheel interface. Solutions tried include various finishes or coatings to minimise the corrosion, and at least one company offers a wheel with a steel rim and alloy centre. Unlike the pressed steel wheel, the alloy product is usually cast. This has the advantage of being able to incorporate a wide variety of visually attractive shapes, but has the potential disadvantage of a lack of homogeneity. It is very easy to produce a porous casting. The casting surface is normally finished with a paint or lacquer, which will disguise the porosity when new. If the coating degrades, the tubeless tyre will deflate. Some manufacturers now produce the wheels from forgings in an attempt to avoid the porosity problems.

These two examples demonstrate the importance of appearance for both industrial and consumer products, to the extent that the fundamental functionality seems to be compromised (costs of the hygienic pumps and valves are raised by the 'unnecessary' finish, and the alloy wheels can have problems retaining air pressure). Always remember that the product is produced to satisfy a need and that this need should be defined in the PDS. Any specific aesthetic requirement must be included in this document. Meeting the PDS is the product's fundamental function (see Chapter 5 regarding PDS).

The previous examples were chosen to demonstrate the 'power' of aesthetics, and the subsequent impact on engineering decisions. The next example shows that even when appearance is not considered to be important it can affect the product.

Example 14.3

The Mini, introduced in 1959, was probably one of the last motor cars to be designed and engineered under the close control of a single individual. Alec Issigonis introduced an impressive number of new concepts – small wheels at the corners of the car, transverse engine with gearbox beneath it, front wheel drive, rubber cone suspension, etc. It was from the start intended to be a low-cost car that could be manufactured in various parts of the world. To simplify body manufacture, the various panels were joined by spot welded flanges on the outside of the vehicle. This helped access and was thought to require less skill on the part of the welder. Aesthetics at this stage were not considered important.

In some countries where there were no available metal-pressing facilities the only way to provide the necessary local content was to organise manufacture of the body shell in glass fibre. The Mini was produced this way in Chile in the 1960s, without the external flanges. In 1991 glass-fibre-bodied Minis were introduced in Venezuela. External seams were included to improve the appearance!

14.3 Ergonomics

In general, engineering artefacts are designed and built to assist human activity of one form or another. Some of these have little direct human contact, but the majority have at least some form of human interface. Ergonomics, the relationship between the machine and its operator and the working environment, must be considered when developing most products.

Example 14.4

As with most influences on product design and development, compromise plays a large part. The current QWERTY keyboard layout was largely determined by the method of mechanical operation of the early typewriter. Each key was connected via a system of levers to a small block with a raised impression of the letter. Depressing the relevant key resulted in the block being propelled towards a ribbon impregnated with carbon. The impression block briefly pressed the ribbon onto the paper, thereby 'printing' the letter. The action was similar to that in a piano (with hammers striking the wires), with one essential difference – each letter had to strike a precise point on the ribbon and paper. The impression blocks were arranged around the arc of a circle, such that the mechanism would result in their striking the paper at the centre of the circle. An indexing system ensured that the paper moved the correct amount after each letter was printed. This system had one drawback – certain combinations of keys pressed in rapid succession would result in the impression blocks and their support levers becoming entangled. The solution was to evolve a keyboard layout that minimised this problem, by spacing out those letters likely to be used in succession. The result was the QWERTY keyboard. Although a more ergonomically acceptable layout may have been considered for the early typewriter, a compromise with the available technology had to be made.

Today, almost all keyboards are electronic and do not need to use the QWERTY layout. Various attempts have been made to introduce a more ergonomic keyboard involving different key positions and heights. In individual tests, although these have proved easier to use than the flat QWERTY system, its wide usage and acceptance have prevented the alternatives being commercially successful. In this instance a compromise is being made with customer acceptability.

Example 14.5

In some cases, ergonomic changes to the accepted standard are introduced. As technology advanced within the telephone industry to make the introduction of push-button instruments a possibility, British Telecom carried out an involved research programme to arrive at the most logical layout of the push buttons. The apparent ideal was three rows, each with three buttons, and 1, 2 and 3 being in the top row (Figure 14.1).

The electronic calculator preceded the introduction of the push button telephone by a number of years, with that industry standardising on a similar button layout. There was, however, one essential difference: 1, 2 and 3 formed the bottom row. The existence of this standard did not prevent BT introducing its alternative layout.

14.4 Quantity

Although aesthetics and ergonomics affect the overall product introduction process, their largest influence is probably at the design stage. The number, or volume of production, needed affects *all* stages of the design to production process and

Figure 14.1 *Comparison of keyboard layouts*

is probably the single most important influence to be accounted for.

In general, the number required will have a significant effect on the resources likely to be available for both development and production. The greater the quantity, the greater the resources. Obviously this only applies within an industry – clothes pegs are produced in somewhat higher numbers than space shuttles, but do not require the same level of investment.

Example 14.6

Within the motor industry most high volume motor vehicle bodies are manufactured from pressed steel panels that are spot welded together and a finish is then applied to delay the onset of corrosion, while giving an attractive appearance. The panels are produced on large and expensive presses. Over the years, component cost savings have been introduced by reducing the number of panels, and hence the number of spot welds needed. This has been made possible partly by the introduction of ever larger and more expensive presses. Much of the subsequent manufacturing process is automated, particularly the welding and painting. This automation gives many advantages, including regular production rates, consistent quality, and controlled costs. However, it requires enormous capital investment, which can only be recovered over production runs of many tens of thousands of vehicle bodies.

As the bodies are made in such large numbers a small cost-saving feature introduced at the design stage could result in very significant overall savings. It therefore makes sense to employ substantial resources during the design process to catch as many of these potential savings as possible. This in turn results in large investments in computer systems, development and testing programmes, etc.

The introduction of a new body starts with a stylist creating the shape as a clay model, which will be a full size replica of the final product.

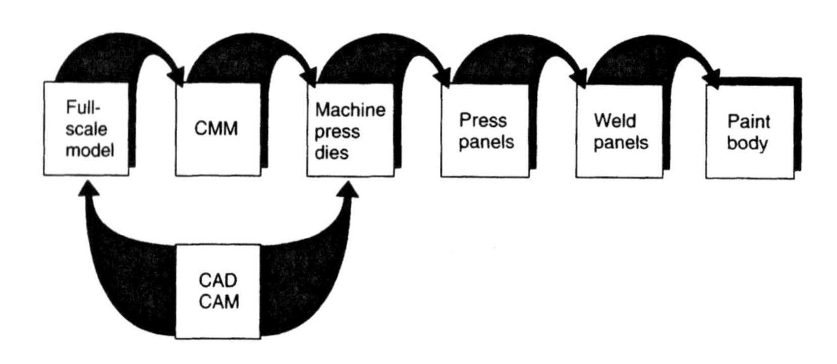

(a) Stages in the production of a steel body

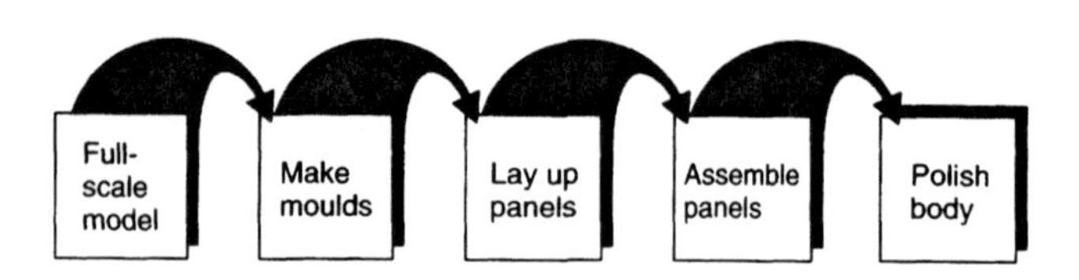

(b) Stages in the production of a composite body

Figure 14.2 *Alternative methods of vehicle body manufacture*

Usually a number of variants will be produced before a final design is chosen. (The selection process may involve customer clinics where members of the public are shown the clay models, usually without knowing which company's future products they are seeing.) A CMM (coordinate measuring machine) is then employed to record the shape of the model (or buck) numerically. This data will be passed to a machine tool for producing the press dies. Many large companies now use CAD to generate the initial shape, and use this data to machine a full-size styling buck from high density foam. The buck is still needed to help visualise the final shape at as early a stage as possible – an unattractive styling feature may jeopardise sales and prevent economic production.

The small-volume motor manufacturer will start the process in a similar way, but will normally choose an entirely different method of body manufacture. The equivalent of the styling buck will be produced, and as with the high volume manufacturer, this will be used as the master from which dies will be made. In this case the 'dies' or mould will be taken directly from the master, which will be retained for further sets of moulds if required. The mould will be used to lay up panels in composite material. This will normally start with a coloured gel coat followed by layers of fibre impregnated with resin. This is a labour-intensive task, the quality of the final product being dependent on the quality of the initial workmanship employed. After curing, the panels are removed from the mould and assembled in a jig. Finally, the joint marks are removed and the gel coat polished to give a finish similar to the painted metal product. This is very labour intensive and difficult to control in terms of cost and quality. (See Chapter 12, section 12.8, and Chapter 8, section 8.5.3, for more information on composite construction.)

The steel and composite end products are broadly similar in terms of functionality, despite dramatically different production processes. The low-volume producer has avoided the high-capital-cost elements, the press tools, welding robots, and the painting process. However, these have been traded for a high labour content, with its attendant drawbacks.

Which process is the more economic? The answer depends on the quantity, high volume steel fabrication, low volume composite structure. In terms of actual cost a painted steel bodyshell produced in high volumes will be about 40% of the cost, of an equivalent low volume composite product.

The implications of investment on quantity and product cost will be explored further in Chapter 17.

14.5 Safety

Today, almost all new products will have been developed with safety in mind, from the manufacturing process to final use by the customer. This is largely the result of combinations of common sense, competition, and pressure from consumer organisations, reinforced by increasing levels of legislation.

The Health and Safety at Work Act was introduced in the UK in 1974. This placed a responsibility on individuals to ensure that procedures within the work place were safe. These included appropriate diligence over design calculations, testing, and selection of materials, etc. as well as safe practices in the manufacturing environment. If a faulty product resulted in death or injury, any subsequent claim for damages would normally require proof of negligence on the part of the producer. Product liability legislation, as already exists in the USA, is almost certain to be introduced within Europe. The concept was introduced to the UK by Part 1 of the Consumer Credit Act 1988, which implements the European Directive on strict liability. With product liability, any person injured by a defective product will need only to prove that the product was defective and that the defect caused the injury. Safety as a significant factor in design and manufacture is here to stay.

Codes of Practice

These are procedures for performing tasks that have proven over a period of time to result in a safe outcome. Standard procedures are well documented for products with obvious potential for becoming a hazard. Although there is often not a direct legal requirement to follow these procedures, it would be very difficult to provide a viable defence, following a product failure resulting in injury, if the relevant code of practice had been ignored. This applies to manufacturers' recommendations regarding applications of standard components as well as relevant national standards.

In many cases there is a direct legal requirement (either from UK legislation or from EEC directives) to comply with particular standards. Pressure vessels must, for example, be designed to BS5500. The codes do not simply cover the design, they apply equally to manufacturing methods. Material has to be identified (often with a casting melt number) and remain traceable throughout the manufacturing process. Operations such as welding may only be performed by a coded welder who has passed an inspection test within the previous six months. Any critical welds in the assembly are normally inspected by an independent authority such as Lloyds or Veritas.

The move towards standardising methods is continuing, with many companies attempting compliance with BS5750. In essence, this standard requires organisations to introduce defined procedures (with regular reviews of the procedures) into all aspects of the organisation.

Overall, increasing levels of legislation will ensure that safety is set to play an ever more important role in product development.

Fail-safe

The safety characteristics of a product can be enhanced by taking great care with the relevant calculation methods, testing, and controlling the manufacturing process. However, no matter how much care, is taken there will be occasions when components fail. If potential modes of failure are examined at an early stage the design can ensure that such failures do not cause dangerous situations. For example, railway signals used to be angled down to indicate 'go', and would be pulled to a horizontal position for 'stop'. If the operating cable or rod failed, gravity would automatically set the signal to 'go'. This is obviously an unsafe design! It did not take many accidents before the fail-safe version was introduced, with the cable having to pull the signal to the 'go' position against gravity. This is a fail-safe design as far as the signal is concerned, but of course does not guarantee that the driver will not miss the signal. Over the years a number of devices have been tried, ranging from audible warnings in the cab, to systems that apply the brakes automatically if a signal is disobeyed. The dead man's handle is another attempt at fail-safe design, where the train driver has to maintain pressure against a spring-loaded switch for the train to proceed. If he or she fails to keep the switch closed, through illness or lack of concentration, the power to the train will be cut and the brakes applied.

Generally the most reliable fail-safe systems are based on simple mechanisms involving natural forces such as those resulting from pressure, gravity, springs, heat, etc. For example, imagine an interruption in the gas supply to a house during winter. As soon as the supply is cut off the gas-fires and boilers would simply stop working. No problem so far, but what happens when the supply is reconnected and the gas taps have been left in the 'on' position? Gas is fed to various appliances, but cannot reach the burners because of the presence of a simple fail-safe device. The pilot light in a boiler keeps the supply on by heating a bimetal strip. If the supply is cut, the pilot light goes out, the strip cools and automatically closes the gas valve.

Redundancy

Sometimes it is not possible to introduce a simple fail-safe feature. Back-up or redundant systems are often used. Imagine a single-engined aircraft suffering engine failure: it has no back-up system to keep it airborne other than its ability to glide for a short period. This may be adequate for private flights over terrain offering a variety of landing opportunities, but totally unacceptable for a large commercial aircraft. These all have several engines and are able to remain airborne after at least one engine failure, albeit with reduced performance. Redundancy will apply to most of the aircraft operating systems, with at least two independent mechanisms for operating vital systems such as the navigation, the flaps, the ailerons, lowering the undercarriage, etc.

Redundancy as a principle is not new and does not only apply to high-tech products such as aircraft – it is often used to increase the reliability of products. Early motor-car engines were started by rotating manually a handle connected to the engine crankshaft. Electric starter motors were in common use by 1930, but starting handles were still supplied as a back-up on most cars until the late 1950s, when the reliability of the electric starting systems rendered the need for a back-up obsolete.

14.6 Strength, fatigue

One of the fundamental criteria in the PDS is the performance envelope within which the product must operate. It must be strong enough to survive in these conditions. The approach taken to estimate the product's capabilities in this area depends very much on the industry and the envisaged production volumes.

The high volumes in the car industry will result in extensive stress analysis work (almost certainly computer aided) prior to generating any prototypes. However, despite the sophistication of the calculations, they still represent only an estimate of the product's performance, so a number of prototypes will be produced and tested to determine exactly where the design weaknesses are. This is not to denigrate the importance of the calculations, for they are vital in predicting the likely effect of design modifications, and thus can dramatically reduce the time taken to introduce a new product.

Sometimes it is not possible to build a selection of prototypes and test them to destruction. The space, nuclear and aircraft industries will all have extensive component testing programmes, but can only simulate certain conditions. In this case a suitable safety factor is added to the calculations to ensure that the design is conservative. The same principle will apply to less sophisticated products built in low numbers. The safety factors used will be affected by custom and practice in a particular industry. When using standard components such as gears, belts, rolling element bearings, etc., the manufacturer's applications data sheets often contain a wealth of information, including appropriate safety factors, based on a very wide range of experience. In the absence of any such information, Table 14.1 could be used as an approximate guide, with simple calculations based on the material's ultimate strength.

Table 14.1

Type of load	*Ductile metals (steel)*	*Brittle metals (cast iron)*	*Timber*
Dead load	3–4	5–6	7
Mild shock (unidirectional)	6	7–8	10
Mild shock (reversed)	8	10–12	15
Shock	10–15	15–20	20

As is apparent from the table, any design must take account of both the choice of material and the estimated load.

Material characteristics

Unfortunately, it is not possible to list materials with a simple strength rating, as a number of separate characteristics need to be considered.

The standard stress/strain curve from a tensile test can reveal a number of useful material characteristics. From Figure 14.3, the linear portion of the curve indicates elastic behaviour (stress is proportional to strain), with the slope of this section (modulus of elasticity) giving a measure of the stiffness. The point of transition between the elastic and plastic regions is called the yield point. Note that in practice the yield stress is found by drawing a line parallel to the straight part of the curve at a percentage elongation offset. The intersection of this line with the stress/strain curve gives the yield stress. Any quoted values should include the percentage elongation.

The UTS (ultimate tensile stress) is the highest value of stress that the test piece was able to withstand. The ductility of a material is its ability to deform plastically without rupturing. A material that exhibits very limited ductility is said to be brittle. Brittle materials are therefore likely to exhibit closer values for the yield and ultimate stresses.

In service most (though not all) components will be designed to behave within the elastic part of the curve. Table 14.1 gives approximate safety factors based on the ultimate stress. If the yield strength is used for ductile materials these could be reduced, possibly halved.

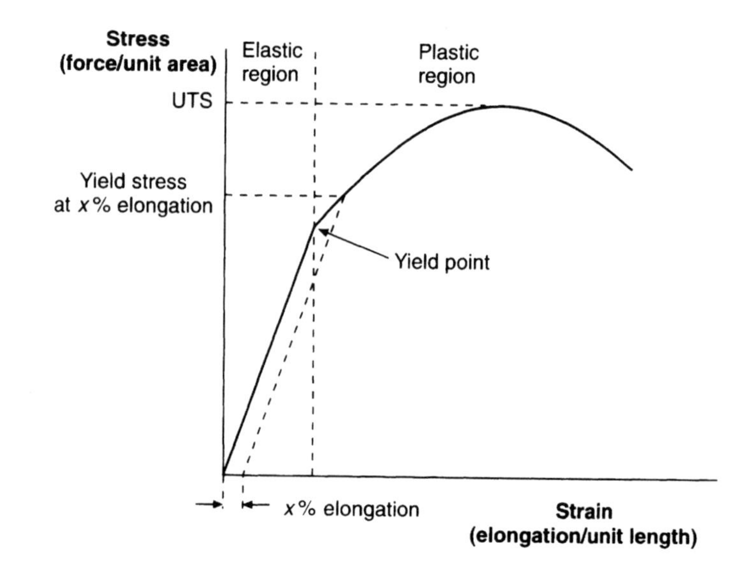

Figure 14.3 *Typical stress/strain curve*

Even if ductility is not required during the components' service life, it may be important for the manufacturing process. For example, a pressed body panel for a motor vehicle must be sufficiently ductile for the material to flow during the forming process without splitting or unduly thinning.

The stress/strain curve (Figure 14.3) provides us with test information that can be used to predict safe static loads, but it will be over-optimistic when it comes to repeated loadings. Metal fatigue is the failure of a material caused by the application of varying (often cyclic) loads. This is the most common type of failure among machine components. The failure will occur even though the stresses generated are well below the UTS and the yield stress. Figure 14.4 shows a typical endurance curve for steel where the stress versus the number of cycles to failure have been plotted. Note that below a certain stress level, known as the endurance limit, the material will not fail regardless of the number of cycles.

Fatigue fractures normally start at a discontinuity in the material's surface such as a sharp corner, a machining mark, a surface crack, etc., and gradually propagate from this stress raiser. The surface finish of the component can therefore have a significant effect on its fatigue resistance. A machined surface will have a lower endurance limit than a polished one of the same material. Treatments such as shot blasting can also increase fatigue resistance, by putting the surface of a component into compression, thereby reducing the likelihood of initiating a fatigue crack.

The design detail is vitally important in that any stress raiser will affect the fatigue resistance on an item subjected to varying loads. Shafts with features with sharp edges such as circlip grooves or keyway slots will have a reduced resistance.

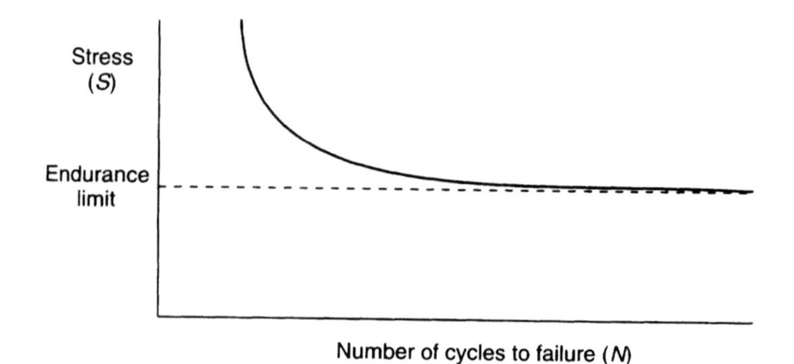

Figure 14.4 *Typical SN curve*

Wherever there is a diameter change, the step should be chamfered or have a blend radius as large as possible.

14.7 Corrosion

Corrosion is the result of a chemical reaction of a material with its environment. This is usually undesirable as the strength of the corroding material reduces gradually until failure eventually takes place. Successful products can be produced to minimise the negative effects and sometimes take advantage of the process.

Generally, metals corrode because they have a tendency to revert to a more stable state in which they are combined with other elements. In some cases this can be beneficial. Aluminium in air rapidly forms an oxide coating which is stable and hence prevents further corrosion, provided the surface coating is not damaged. Copper and lead are often used as roofing materials, because these metals receive a protective surface coating automatically: copper is covered with highly visible green copper sulphate, while lead gains a carbonate coating.

A moist atmosphere will almost certainly increase the rate of corrosion of metals, because of the electrochemical mechanism. Here, a galvanic cell is created between two materials, with the anode being the corroding metal. The cathode does not have to be another metal. For example, atmospheric corrosion is very common with many metals; a thin layer of moisture covering the surface can act as an electrolyte, enabling a reaction with chemicals in the atmosphere.

Where a design allows two dissimilar metals to be in contact, either directly or via an electrolyte, the table in Appendix B.8 can be used to predict which will be the anode and hence corrode. The higher metal in the list will always be the anode. The rate of corrosion is dependent on the separation of the metals in the table. It can also be affected by the relative areas of the exposed parts. For example, a steel rivet in a copper plate will corrode rapidly in sea water, but a copper rivet in a steel plate is likely to result in a much less catastrophic level of degradation of the steel plate. Ideally the design should avoid direct contact between dissimilar metals, by using insulating washers, bushes or gaskets.

Protective paints can be used, although in some cases they can aggravate the situation. If only the anode is painted, and there is a defect in the coating, an accelerated rate of corrosion will occur at the break in the coating! Either only the cathode, or both the anode and cathode, should be painted.

There are times when this electrochemical effect can be used to advantage, as with galvanised steel, where the zinc coating becomes the sacrificial anode, thus protecting the steel. Even if the surface coating is damaged, the exposed steel will be the cathode and hence remain protected. This is not the case with a tin coating on steel, which is only effective as long as the coating is intact. If the coating is damaged, the steel will corrode, as it will become the anode (see the relative positions of steel and tin in the table in Appendix B.8).

A crack, or joint between two components, can also provide conditions for crevice corrosion. One common example is the oxidation type concentration cell, where the most marked corrosion occurs in the part of the cell with an oxygen deficiency (i.e. the crack or crevice). In this instance the cathode reaction requires the presence of oxygen to remove electrons from the metal. These electrons are supplied by adjacent areas which do not have as much oxygen and thus serve as anodes. The cracks and crevices which have restricted access to oxygen become anodes and hence the focus of the corrosion.

As plastics are electrical insulators they do not corrode in the same way as metals. They are not immune from chemical attack, but this is very dependent on the type of plastic and the attacking chemical. A number of plastics suffer from absorption. Nylon, for example, will be unstable dimensionally if used underwater, as it will expand as water is absorbed.

The adverse effects of corrosion can be minimised by:

(a) Careful selection of materials. See Chapter 6 and Appendix B.9 for a summary of common material properties.

(b) Use of protective coatings, noting the use of the sacrificial anode (e.g. galvanising), and the potential problems of coating an anode inadequately.

(c) Designing to avoid creating sources of corrosion.

- If dissimilar metals must be in contact, use non-conducting washers or bushes to separate them.
- Avoid features which create cracks or crevices. If they cannot be avoided, ensure that they are filled with solder, weld, mastic, etc. See Figure 14.5(a).
- Avoid features that can trap dirt or moisture, or at least provide drain holes as appropriate. See Figure 14.5(c) and (d).
- Use the gaskets of the correct diameter – the wrong size may result in an unseen dead area in which corrosive material can accumulate. See Figure 14.5(b).
- Avoid fibrous gasket materials that might draw corrosive materials into a crevice by capillary action.
- If a protective paint is to be used, make sure that the design does not include inaccessible areas that will not be adequately protected.
- Provide adequate ventilation to closed areas to limit the effects of condensation increasing the rate of corrosion.

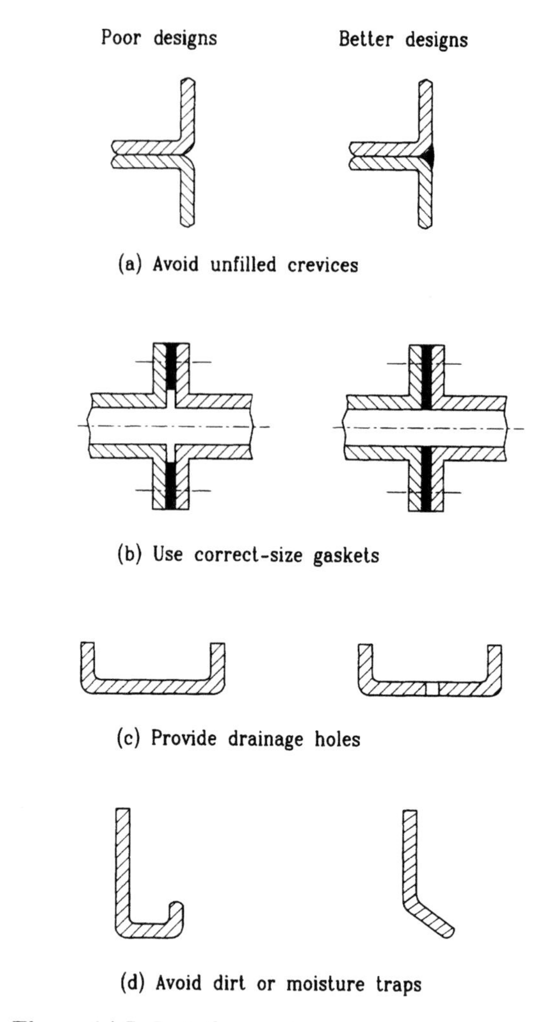

Figure 14.5 *Details to minimise corrosion*

Figure 14.5 illustrates some good and bad instances of designing to combat corrosion. The motor industry has made considerable progress in this area over the past 30 years. The accumulated effects of vehicles covering higher mileages, higher concentrations of salt used during the winter periods, and thinner steel body sections (to reduce weight and cost while increasing performance), lead to a greater vulnerability to corrosion. Today's car underbody will have a minimum of nooks and crannies to form the start of crevice corrosion, plenty of ventilation for the enclosed box sections, plastic inserts under the wheel arches, and a good chip-resistant protective coating.

14.8 Environment

The environment within which the product finally operates obviously has a significant effect on the product's gestation process. As with other influences, the effect is felt throughout the design-to-manufacture process. Three different environments are discussed briefly below, each highlighting some of the aspects needing particular attention. The third example demonstrates the iterative nature of the new product introduction process caused by the interlinking of the many skills involved.

Example 14.7

Vacuum environment: Many final operating environments contain harsh conditions caused by factors such as high pressures, wind loads, or corrosive atmospheres. High vacuum conditions have none of these, as by definition the pressure is zero (or at least fairly close to zero), and there is no atmosphere to provide wind or to be corrosive. The lack of an atmosphere is a problem when heat dissipation is considered. Convection and conduction through the atmosphere are non-existent so the design must limit unnecessary heat generation and provide adequate conduction paths through solid elements.

Now we will consider some of the special environmental aspects of which the designer of work-handling equipment for use in an electron beam welding machine must be aware. The machine operates by passing the joint to be welded beneath a narrow, high-energy beam of electrons, generated from an electron gun (see Chapter 9 for details of the process). The electron beam can only survive in a vacuum; its energy would be dissipated rapidly in air. The workpiece and its handling equipment must therefore operate within a vacuum chamber. Each time the machine is reloaded with components the chamber has to be vented to atmosphere, then pumped down again. This pump-down time can have quite an effect on the economics of the process and is influenced by a number of obvious factors such as the chamber size and the capacity of the pumps.

It is also influenced by the detail design of the work handling equipment and the components to be welded. Outgassing is a term used to describe the rate at which molecules of gas can be extracted by the vacuum pumps. Features that have a low rate of outgassing must be avoided. Certain materials should not be used: for example, a sintered bush impregnated with oil would definitely not be appropriate. The action of the vacuum pumps would simply cause the oil to evaporate, thus increasing the pump-down time and possibly damaging the pumps. Even if the sintered bush had not been impregnated with oil it would still not be suitable, as the evacuating process would be slowed by the air gradually outgassing from the pores in the bush.

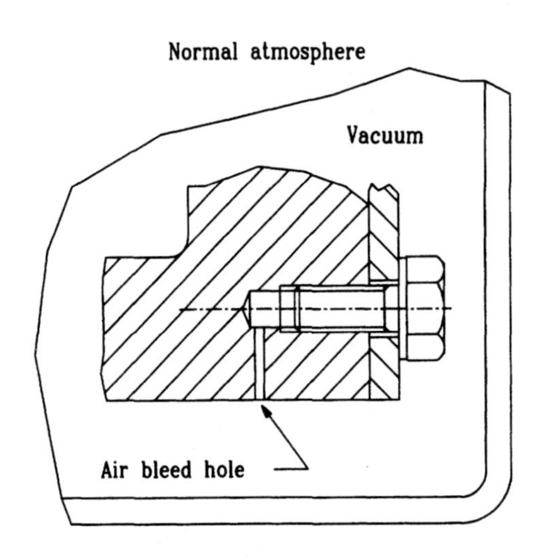

Figure 14.6 *Blind hole in vacuum environment*

The designer should avoid any feature that might cause trapped air to be released slowly. A blind hole is a good example. If it cannot be avoided, then a bleed hole should be included (Figure 14.6).

Lubrication is a problem, as most lubricants will evaporate under vacuum. Some special greases are available, but inevitably they represent a compromise between their ability to survive in the environment, and to lubricate. The need for such lubricants has to be minimised. A large EBW machine used for the repair of aircraft engine components had simple XY coordinate tables as the basis for its work-handling equipment. The tables were driven via a pair of leadscrews. To avoid the need for a gearbox and its associated lubrication problems, the leadscrews were driven directly by electric motors. The motor was built into the end of each leadscrew thus limiting the need for bearings to one set per motor screw assembly. Fewer bearings meant fewer lubrication problems!

Example 14.8

Underwater environment: Working underwater is in at least two ways directly the opposite of working under vacuum. First, ambient pressure is always greater than atmospheric, considerably so at any significant depth, and secondly the water can operate as a very effective heat sink, thus simplifying cooling aspects. It is this second aspect that plays a part in the next example.

Consider an unmanned submersible, a vehicle used to inspect underwater installations. Its prime function is to act as a stable and easily manoeuvrable platform on which sensing equipment such as sonar and cameras may be mounted. The manoeuvrability is often achieved by the use of several thrusters, strategically positioned so that the craft can be controlled by varying the thruster outputs individually. On this type of vehicle, signals are sent to and from the craft via an umbilical cable. This cable can also be used to transmit power to the vehicle. The thrusters consist of a motor and propeller assembly (Figure 14.7). During the mid-1970s, just as the potential of the North Sea oilfields was starting to be realised, the market opportunities for this type of vehicle expanded rapidly. As a result a variety of innovative solutions appeared. The budgets available to different builders varied widely, as did the solutions chosen. This is apparent when examining how one such group, with a very limited budget, tackled the problem of avoiding water ingress on the thruster unit.

The design stayed with the then common practice of using electric motors. In this case, AC induction motors were selected, mainly for their simplicity and hence expected reliability. One of the environmental problems was the incompatibility of seawater and electric motors. One solution used by a competitor was to house the motor in a special pressure-proof housing made of stainless steel. The result was a bespoke motor able to withstand water pressures to depths of around 1000 feet. This option was rejected on the grounds of cost and weight.

The chosen solution had to incorporate a commercially available motor, with a minimum of costly modifications. Of the two potential hazards – high pressure and the electrical conductivity of seawater – only one was in reality a hazard. In air there is no good reason why the motor would not operate at increased ambient pressures. Underwater the same should apply, provided contact with the seawater was avoided. The chosen solution was to fill the motor case with oil, which was in turn connected to a pressure compensating device that kept the oil pressure at a similar level to the ambient seawater pressure. The oil would prevent the normal method of motor cooling (by air passing over the windings), but would transmit the heat to the case which would be in direct contact with an enormous heat sink, the ocean.

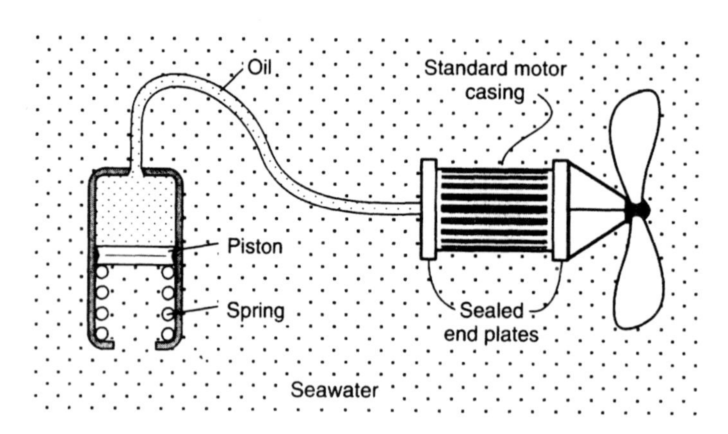

Figure 14.7 *Pressure compensation system*

The standard motor was fitted with a simple assembly at the output shaft end to take the drive and support the propeller and its associated nozzle. It used the standard motor mountings and incorporated an O ring seal. The opposite end of the motor was fitted with blanking plate with a similar seal arrangement, as was the terminal box. The finned motor case was an aluminium casting, which certainly had not been designed to be leakproof, so this had to be checked for any porosity. This was achieved by filling the motor with helium then using a sensor to locate any leaks. Any problem areas found were sealed with epoxy resin. Sealing the motor in such a low-cost way, and retaining the standard case, was only made possible by the use of a pressure compensator. This was a simple piston and cylinder arrangement, with oil on one side (connected to the oil in the motor), and ambient sea water on the other. As the craft altered its depth the piston maintained the parity between the oil and sea-water pressures.

Although the low pressure differentials reduced the likelihood of any leakage, losing a small amount of oil would be preferable to allowing a small amount of sea water into the motor. The piston was lightly spring-loaded to ensure the oil pressure was always slightly higher than that of the ambient sea water. A sensor was also added to detect movement of the piston, so that in the event of a leak the submersible pilot would be warned before the motor suffered any water ingress.

The design concepts and implementation in this example may seem rather basic, particularly for what was essentially a 'high tech' piece of equipment, but they were successful, and that is what counts in the end. Not only did the thrusters function reliably at depth, but the device was built and manufactured within a rather limited budget and timescale. The approach used took account of the low number of submersibles being built. The principles used would, of course, be appropriate for higher volumes, but a somewhat different manufacturing technique would no doubt be employed!

External influences on design and manufacturing choices rarely come singly; in this case the underwater environment was supplemented by low volume, low cost, and high reliability. In the next example, although based around the influences of a hygienic environment, a number of other influences will become apparent.

Example 14.9

Hygienic environment: Consider a valve used by the food process industry. There are some obvious features the designer must include. It must be self-draining: i.e. when the pipework is drained down none of the process fluid must remain within the valve or its associated pipework. These systems are often cleaned without dismantling by pumping a CIP (cleaning in place) fluid, such as dilute nitric acid, throughout the system. If any of this fluid remained after cleaning it would contaminate the food product. The designed shape of the valve should account for this requirement.

There must also be no cracks, crevices, or dead ends where food product may stagnate and harbour bacteria. This is a challenge that has to be addressed by use of both design and manufacturing skills. A poor surface finish will result in effective cracks and crevices. Generally, the industry strives to achieve a minimum of 1.0μm R_a *for surfaces in contact with any product.*

Now we will consider some possible engineering options that might to be available to the designers and producers of a simple valve body (Figure 14.8). The material will be 316 stainless steel, and annual sales of some 5000 bodies are anticipated.

The valve as shown is normally fitted with an actuator (not shown) which moves the stem vertically to open or close the valve. The position of the actuator relative to the body is critical to ensure, the valve seats correctly. Normally it is located by a machined spigot on the upper surface of the body which is machined at the same set-up as the valve seat, thus ensuring concentricity. Whichever method of body manufacture is considered, the machining of these two surfaces has to be included.

The required finish is readily available from a machined surface, but the final shape would need far too much material removal to consider machining from solid. Even if the shape was more suitable for machining, the chosen material is expensive, and difficult to machine.

A good finish is also easily available from pressings. However, as our valve body is too

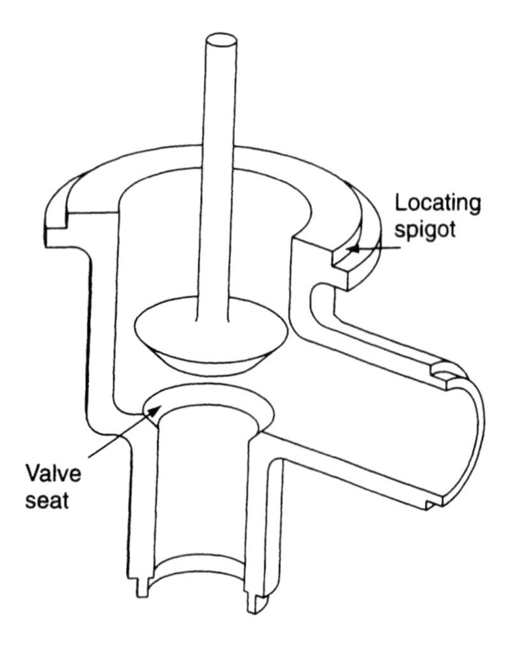

Figure 14.8 *One piece valve body*

complex to be manufactured from a single pressing, a number of joints would be required, as in Figure 14.9. These could be welded, but as explained in an earlier example, the joint would need subsequent treatment to achieve the necessary surface finish.

This treatment could be hand polishing (i.e. by a hand-held power tool). Hand operations are difficult to control in terms of time, cost and quality. Why? The time taken to polish a joint is very much dependent on the skill and attitude of the operator, as is the quality of the finish achieved. The cost (see Chapter 17) is affected because most costing systems record cost as a function of time taken. As a general rule, cost and quality are easier to control with an automated system. In this case automating the polishing would require either a special purpose machine, or possibly a sophisticated programmable robot.

An alternative might be to ensure excess weld material was applied to the joint and then machine the surface to the correct finish. This would make good sense, especially if there already existed a requirement for machining, such as the valve seat in this case, and both areas could be tackled on a single machine and at the same set-up. Adding machining operations can add considerably to the cost, but extending an existing machining operation, particularly on a relatively low volume component, can often deliver an economic solution.

Another approach to generating a smooth joint might be to examine the welding operation with a view to avoiding any subsequent finishing operations. Automatic TIG welding can generate a visually good surface. In this case the components are usually clamped together on a mandrel which rotates past a stationary TIG welding head. Electronics are used to control the speed and the current of the welder. As no additional material is added during the welding process, a fairly smooth joint can be achieved. However, while it can in some cases be considered cosmetically acceptable, it would not have a suitable finish for the inside of a valve body. Another possible process would be electron beam welding, where an even more acceptable joint could be produced. Although it is used by some valve manufacturers it is an expensive process requiring specialised equipment (see Chapter 9).

If finishing the weld is difficult, we should remind ourselves that the weld is only needed if the body is

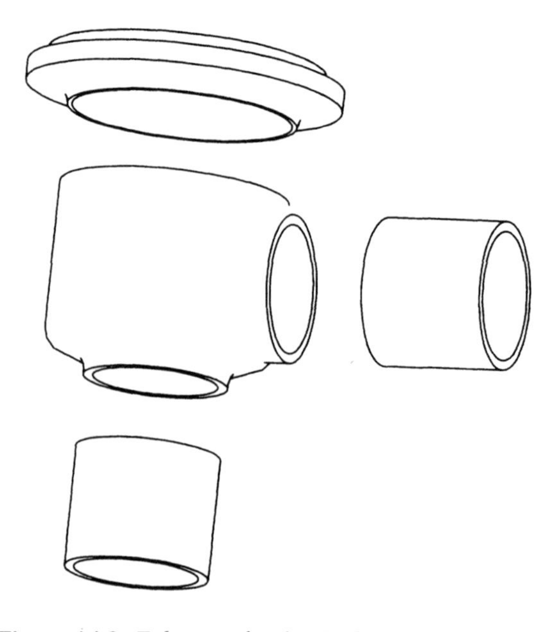

Figure 14.9 *Fabricated valve body*

to be fabricated from pressings. Could a casting be used? An approximate shape could be generated from a shell moulding or sand casting, which could then be machined over most of its surface. This would be a considerable improvement over machining from solid, and yet avoid the joining problems. An investment casting would be able to replicate more closely the required shape and could possibly limit the need for machining to the valve seat area. The surface finish on a good casting would meet the needs of the industry. Some companies do manufacture valve bodies from investment castings, but even these are not without their problems. Inclusions or porosity are quite common features in castings, but are unacceptable for a hygienic product. These problems can be minimised by careful management of the casting process (see Chapter 8) and good quality control, but this all adds to the cost. Unfortunately, some of the casting defects will be just below the surface and may only become apparent during subsequent machining operations. Sometimes the defect is difficult to repair and may result in the casting being scrapped. Even if it is possible to claim from the foundry for a faulty casting, the lost machining time and the disruption to production will be costly. One way to avoid this problem would be to X-ray-inspect 100 per cent of the castings. This is only likely to be cost-effective for very complex or critical castings such as may be used in the aircraft industry. The more likely solution for our valve manufacturer would be to assume a certain loss of castings and increase the batch size accordingly.

Although an investment casting for our valve body will contain less material than its equivalent sand casting, the higher tooling and production costs will ensure that it will be a more expensive material format.

Assuming that the manufacturer does not have its own foundry, this means the value of stock in the raw-material stores will be higher for the investment casting option. This may mean an increased average value for WIP (work in progress), although the in-house manufacturing lead time may be reduced.

Another aspect to be considered is the added value. As detailed in Chapter 16, an organisation producing goods makes its living by adding value to the raw materials it buys: turning them into more valuable products. The efficiency with which it performs this task is the main measure of its success. Much less work will be needed to complete the component if we start from the investment rather than the sand casting – i.e. the factory's ability to add value is much more limited. In boom times, when our own machine shop or manufacturing facility is likely to be full to capacity, this would be beneficial, by allowing other profitable products to be introduced. However, in other circumstances, reducing the in-house workload might be a disadvantage leading to additional waiting time and eventually to lay-offs (see Chapter 17).

The reader might wonder from the previous paragraphs why an investment casting would ever be considered for such a component. An advantage not so far discussed with this example is the additional freedom given to the designer. On the aesthetic front it may be possible to include features such as the company logo for very little extra cost. One company in this field produces valves with square flanges, by using investment castings. Although there are no particular technical advantages or disadvantages with such flanges, the products are instantly recognisable without any need for a company logo. Small features may also be included to help with assembly, such as a recessed hexagonal hole to retain a nut (see Chapter 13).

This example has demonstrated a number of important lessons.

(a) Although there is often a single best solution to a design or production problem, there can be instances where this is not the case. Valve bodies, such as the one discussed, are produced from fabrications as well as castings, the 'correct' decision varying between different manufacturers. Beware of examining a competitor's product and concluding that his or her approach is the most suitable for you.

(b) Note the iterative nature of the product introduction process. If a particular manufacturing process (such as investment casting) looks feasible, then make sure all possible benefits are gained at the design stage. Do not simply consider different methods of manufacturing the component without considering modifications to the component that would assist in its production.

(c) Manufacturing is a complex process involving more than technical decisions. In essence the aim is simple – to produce the product at the lowest cost; however, the way the costs arise is far from simple. Account must be taken of the financial aspects, such as, in this example, the levels of WIP and added value.

(d) Where one aspect of a manufacturing method is adding too much cost, or is proving difficult to control, do not simply investigate alternative ways of controlling it: find the source of the problem. Remember the hand polishing: automating the operation would have provided better control. However, polishing was only needed because the welding process left an unacceptable surface finish. In this example we needed to search for methods of eliminating the weld, or if this does not prove possible, modifying the welding technique to create a finish that did not need to be polished.

14.9 Conflict, compromise

So far we have discussed a number of examples, each concentrating on various external influences that affect the engineering decisions made. The last two of these examples emphasise that these influences rarely act alone. Sometimes it is possible to satisfy most of the requirements; however, when they are in conflict with each other the engineer will have to decide which will dominate. A properly prepared PDS should address these conflicts and thus guide the engineer.

There are many examples of such conflicts in everyday life. Take a simple sink with a single draining area, as found in many kitchens. At one time it was common for this to be made from stainless steel pressings; the width and depth of the assembly were such as to form the top surface of the kitchen unit. This maximised the working area, and minimised joint areas needing to be sealed. Nowadays it is common for the sink and draining area to be inset in the kitchen work surface. In a showroom this has a much greater aesthetic appeal, particularly with a coloured unit. However, in practice, the working area is restricted. It is less easy to clean, with more joints, and if a mixer tap is used it is all too easy to direct the water flow away from the sink area. In this case the PDS would have been quite correct to rate aesthetic appeal above ergonomic considerations.

A similar example is found on almost all motor cars. The external rear-view mirror used to be located on the front wing of the car. In this position it was viewed through a dry screen in wet weather (courtesy of the windscreen wipers), only involved a small movement of the eye from the straight ahead position, and could accommodate a reasonably wide range of driver eye locations. The current fashion is for the mirror to be located on the door. This does mean that it can be wiped clean and adjusted from within the car, but overall it is ergonomically much less successful than the wing alternative. Again, this is a triumph of aesthetics over ergonomics.

To avoid any accusations of bias, an example of aesthetics being overruled is needed. On most motor cars the pillars on either side of the windscreen sweep upwards and blend into the roof as though a single metal pressing had been used. In fact, there is normally a joint at the top of each pillar. It is not visible because it has been filled with lead and polished flat. This operation is expensive, partly because it is a hand operation, but also because of the safety requirements. Lead is recognised as being harmful if ingested, so the polishing operator generally wears a spacesuit-type helmet complete with its own fresh air supply. In addition, to protect the rest of the local workforce, the polishing operation takes place within a specially constructed tunnel with its own air extraction system. All this to disguise a very small joint at the top of each windscreen pillar! Many of today's manufacturers have deleted this operation, particularly on some of the cheaper models, probably with very few of their customers even noticing.

14.10 Summary

You should by now understand that the external influences on product development are many and varied. These influences can be split into two camps: the objective and the subjective.

The objective influences (such as strength, corrosion requirements, etc.) can be dealt with in a straightforward manner, as the processes and the means of defining their solutions are well understood. Following basic rules for calculating the physical dimensions of a pressure vessel will surely result in a design that will function adequately.

The subjective influences are more difficult to progress to a solution, as various levels of compromise will be needed. The compromise between ergonomic and aesthetic considerations is impossible to calculate coldly; it can only be obtained from a good knowledge of the market and should be defined in the PDS. Some, though not all, manufacturing decisions are almost as subjective to make. The example of the valve body manufacture should demonstrate this; the decisions may well depend as much on the perceived future for the company as on technical requirements.

15 Analysis of Existing Designs and Manufacturing Processes

15.1 Introduction

Some engineers working within an organisation designing and manufacturing products spend much of their time devising, developing and introducing new products. However, many engineers will spend their time concentrating on modifications or improvements to existing designs or manufacturing processes. The differences between the two is often very small.

Today, the pressures of the competitive world require a degree of understanding and a level of knowledge that is unlikely to be found in a single individual, for any but the simplest of products. The engineering work will be split into smaller elements and divided amongst a team. The engineer working on the new product is therefore almost certain to be concentrating on specific aspects of the new product in which he or she has a level of expertise. Similarly, the individual faced with solving a problem on an existing product will concentrate his or her efforts on the designated area. Although the constraints for each may differ, the engineering problems will be closely related.

This chapter will discuss a number of examples, thereby giving an insight into the application of some of the engineering principles covered in the previous chapters.

15.2 Value analysis

Whenever any change to a design or manufacturing process is being considered, the value of that change must be taken into account. (For the moment we will only consider value to be a relative rather than an absolute term.) Remembering the definition of the design process (Chapter 5), the value of the change must be examined from the viewpoints of both the customer and the producing organisation. The word 'value' implies cost, but we must also consider function. A change that reduces cost but fails to meet the functional requirements will be poor value to all concerned.

This point is best demonstrated by an example.

Example 15.1

When the Mini car was first introduced in 1959, it broke new ground in a number of ways, not least of which was the very compact power unit. The engine contained the gearbox in the sump below the engine, and was placed 'sideways' in the engine compartment, with the cooling radiator sandwiched between the end of the engine and the nearside inner front wheel arch. At the base of the radiator was a tap that could be opened to drain the cooling water or anti-freeze, as would be required during normal servicing every year or two. The tap was a simple device accessible from the bonnet area and operated without needing tools, in much the same way as a wing nut would be turned.

Functionally this feature was excellent, but was it good value? If competitors' products could achieve the functional needs with a lower cost solution, then they would be in a better position to offer increased value to either the producing organisation (via reduced costs) or to the customer (via reduced selling price). The pressure on producers with competitive products is relentless; they must continually evaluate and re-evaluate designs and processes simply to survive.

Before long, the tap was replaced by an even simpler brass plug with a screw thread and hexagonal head. In terms of functionality this design was not as good as the original, as a spanner was now needed to drain the coolant. However, as this was unlikely to be done more than once a year, this was not considered to be a serious problem. The brass plug was almost certainly cheaper than the tap, so offered increased value.

One of the beauties of engineering is that there are very few designs that cannot at some stage be improved upon. Realising this, the Mini producers, ever vigilant for any cost savings, examined the function of the brass plug. It enabled the coolant to be drained from the radiator, and would be required to do this possibly a dozen times in the life of the vehicle. The radiator is connected to the engine via two rubber hoses, one at the top and one at the bottom of the radiator. Disconnecting the bottom hose from the radiator would allow the coolant to escape, i.e. achieve the same function as removing the brass plug. This would be slightly more difficult to achieve than removing the brass plug, but as changing from the very convenient hand-operated tap to the slightly less convenient brass plug had probably not even been noticed by most of their customers this was not considered a problem. Today's Mini therefore has neither the plug nor the tap. The value of this component to both the producer and customer was considered to be less than its actual cost.

This example is very simple, but it does demonstrate the three main elements to consider when searching for improvements to an existing design or process:

1. *Value to the customer*
   Is the customer willing to pay extra for this feature? Or, without this feature would we lose the customer to a competitor?

2. *Value to the producer*
   Does this feature or process add cost to the final product?

3. *Functional purpose*
   Does the feature or process fulfil an important function not covered by any other feature or process?

If the answer to question 2 is yes, then it is worth asking the other two questions. If the answer to either questions 1 or 3 is also yes, then the feature or process must be included, and our efforts must concentrate on reducing the cost of providing the feature or process. If, on the other hand, the answer to all the questions is no, the feature or process should be deleted.

When considering these questions, the second two are essentially factual, but the first can be very subjective. For example, consider an electrically operated mirror fitted to the driver's door of a car. This would undoubtedly cost more than a similar manually operated item that could also be adjusted from within the car and hence meet the functional purpose. However, judging from the popularity of this option, the value to many customers is obviously high. Perhaps this is to miss the true functional purpose, which may be to provide a status symbol!

The next example demonstrates how by careful attention to detail a design can be improved.

Example 15.2

For this example we will consider the simple manual can-opener. There have been many alternative mechanisms provided over the years, possibly indicating that the door is still open for the producer of the perfect can-opener.

We will start by considering the basic functions. There are essentially two:

(i) to pierce the lid, and
(ii) to cut around the edge of the lid.

To achieve these functions the first requires

(a) a sharp point, and
(b) the means to apply pressure;

and the second requires

(a) a cutting edge, and
(b) a method of driving the cutting edge around the lid.

Probably the simplest form of can-opener is the single-piece item as in Figure 15.1.

The advantage of this design is that it is a single component, requiring no assembly, produced by simple stamping and bending. Its limitations are:

- A thin handle grip
- The same material is used for the base and handle, thus limiting the sharpness of the blade
- The method of driving the cutting edge around the edge of the lid is crude and awkward.

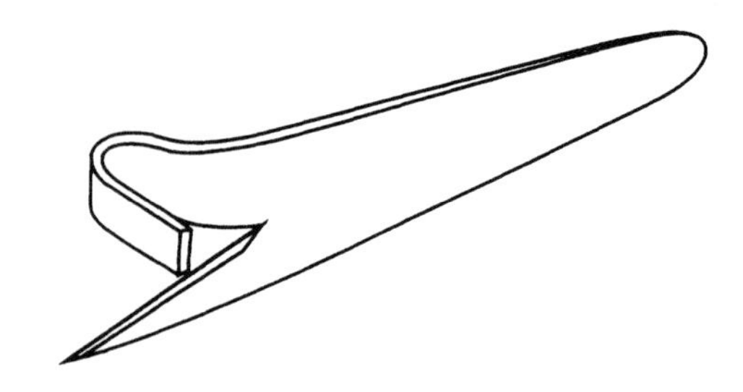

Figure 15.1 *Single-piece can-opener*

An improvement in terms of function would be the four-piece can-opener (Figure 15.2). Although this would undoubtedly be more expensive to produce, it offers a number of advantages:

- A thicker hand grip
- Separate blade material, hence the potential for longer blade life
- More scope for the designer to provide a better spreading of the load in the fulcrum area.

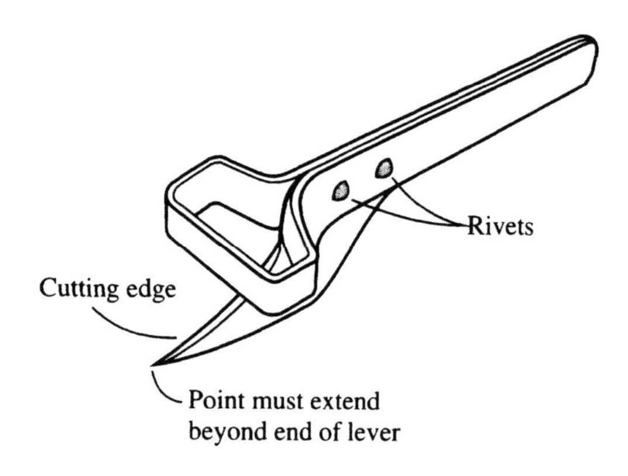

Figure 15.2 *Four-piece can-opener*

Both of the above designs are acceptable minimum-cost solutions that have been produced in very large numbers over the years. They are, however, difficult to operate in both the actions of piercing the lid, and cutting around its edge. We will now consider adding further improvements or additions to the functionality to overcome these shortcomings:

1. Use a lever to provide the pressure to pierce the lid, *and*

2. Add a means of driving the cutting edge smoothly around the lid of the tin.

A lever needs to react against a surface. Figure 15.3 shows two possible areas on the can.

Next we need to consider a means of driving the cutting edge. Friction seems the obvious choice, but the resistance to cutting will be considerable, so we need to provide increased levels of grip. A toothed wheel that could deform the surface it is running on would provide the increased grip, provided that a suitable means of keeping the toothed wheel firmly in contact with the can was present.

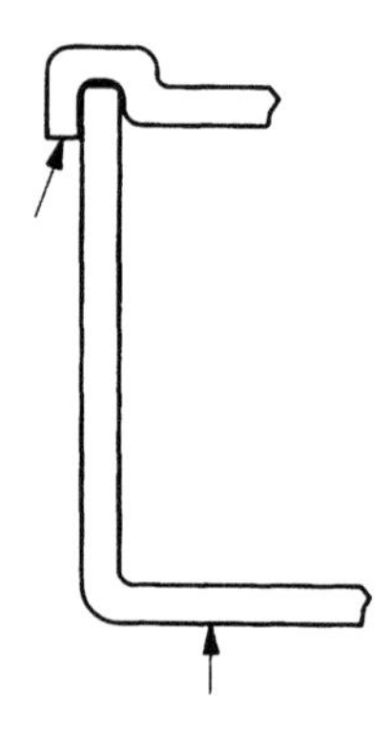

Figure 15.3 *Lever reaction points*

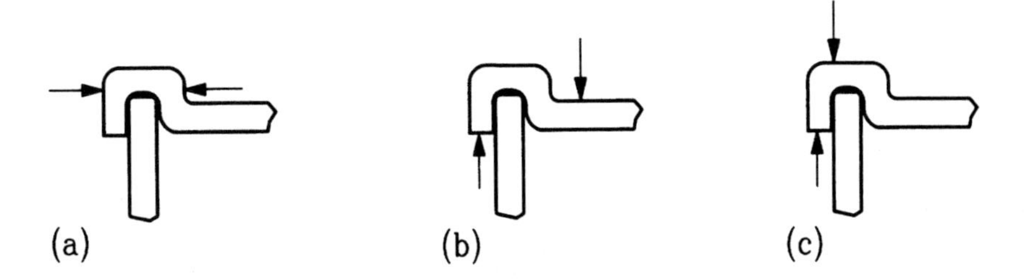

Figure 15.4 *Alternative reaction points*

Achieving the necessary levels of friction will require leverage between two opposing surfaces. Possible options are shown in Figure 15.4. Option (b) is rejected because the downwards pressure will not be on a rigid area of the can and it may tend to push the lid in, but options (a) and (c) are possibilities.

Wherever possible each component needs to fulfil as many functions as practical for maximum functional efficiency. This statement applies to most tasks that we perform. If a single action or feature can serve more than one purpose it is more efficient in terms of functionality than one that only serves a single purpose. Remember the radiator drain tap: functional efficiency was increased by deleting the tap and allowing the bottom hose to satisfy two functions – connecting the coolant supply to the engine, and providing a means of draining the coolant. We will now apply this principle to the can-opener.

The vertical upwards pressure in (Figure 15.4) (b) and (c) could possibly allow us to use the same surface for applying pressure to both the piercing and the drive systems. As we have already rejected (b), option (c) is the most promising, and the alternatives now become those shown in Figure 15.5.

How do we choose between these options? If either (a) or (b) in Figure 15.5 has no obvious advantages as regards the toothed wheel, we should turn our attention to the piercing operation. For this we need a force in the direction of P. Remembering the rule for maximising functional efficiency, we should be able to provide a single component to generate the reaction and also perform the piercing operation.

We now have two prime functional components. They need to be supported and given an interface with the operator. In functional terms, we need to achieve three objectives:

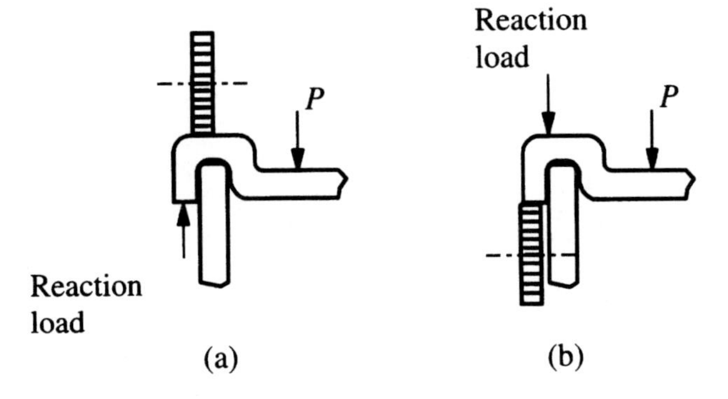

Figure 15.5 *Drive options*

(a) Drive the wheel

(b) Apply pressure, initially to pierce the lid, then to provide adequate friction for the drive

(c) Provide a force in the direction F (Figure 15.6) to ensure that the wheel stays in contact with the edge of the can.

We will take each of these in turn.

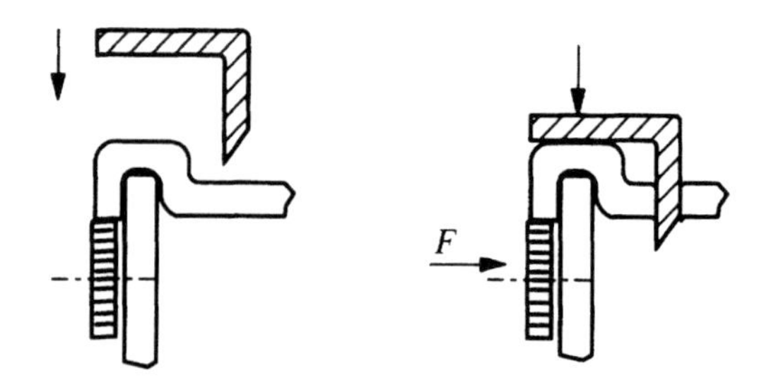

Figure 15.6 *Piercing*

(a) Driving the wheel This objective can be split into two elements:

- How to apply human load to the wheel
- How to connect the wheel to the can-opener and transmit the effects of the applied torque.

A solution would be to use a simple handle and wheel, each produced from stampings, as in Figure 15.7. The wheel would have a central rectangular slot into which the drive end of the handle would fit. The shape of the slot would allow any torque applied to the handle to be transmitted to the wheel. The handle could be retained by peening over the drive end.

Notice that as we are now moving from considering the general principles of operation of the can-opener, to the practicalities of achieving those principles, the method of manufacture must be considered. In this instance we are making the assumption that stampings are the most suitable manufacturing method. For a more complex product it would almost certainly be more appropriate to consider a variety of alternatives. Our justification for moving directly to stampings is partly lack of space (within the text) and partly that we are discussing the development of a product that is currently produced in this way. Knowledge and experience of this method is therefore likely to be to hand, thus minimising any risks.

At this point we could note that the handle shape is very easy to produce and will work efficiently, but is likely to be wasteful in material content. A modification to the shape will reduce the material content, and by allowing better nesting will reduce waste (Figure 15.8). A further

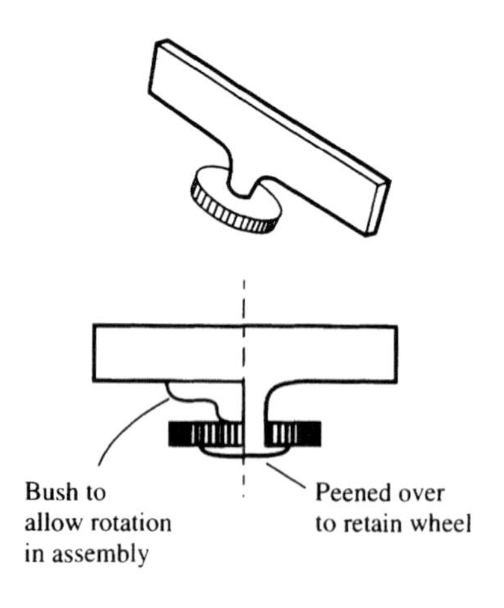

Figure 15.7 *Handle principle*

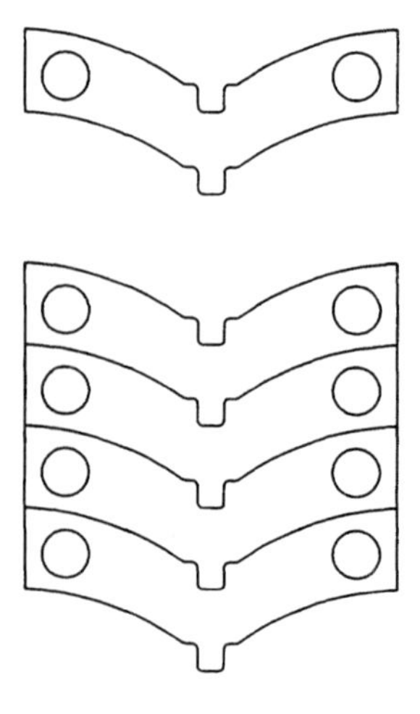

Figure 15.8 *Handle stamping*

refinement might be to introduce two holes on the handle to improve the grip between the fingers and the handle. Provided this feature can be incorporated in the single stamping operation already used to produce the handle, the production costs will not increase. (In fact, adding these features will increase the initial cost of tooling, but, provided that the production volumes are sufficiently large, the effect on the piece-part cost will be minimal.)

The assembly now needs to be held in a chassis (Figure 15.9) which will support the wheel and allow a vertical movement to be imparted to it.

We now have a means of driving the wheel, so we need to turn our attention to the next objective.

(b) Applying pressure This is needed initially to pierce the lid, then to provide adequate friction for the drive. Adding another component, the cutting handle, as in Figure 15.9, would incorporate the piercing point, the cutting edge, and the reaction load provider. This item would be connected to the chassis via a pivot, thus allowing hand pressure to generate the required loads.

(c) Providing contact For our final objective we need to ensure that the wheel stays in contact with the edge of the can. The problem arises from the fact that the can material is fairly thin, so the edge for the wheel to run against is typically only 0.5mm wide. In practice the wheel needs to run at an angle of 15° or so to the side of the can. This could be achieved by an addition to the cutting handle (Figure 15.9) when produced from a stamping or pressing this could be achieved by adding a joggle of 6 to 7mm at the front of this component below the wheel.

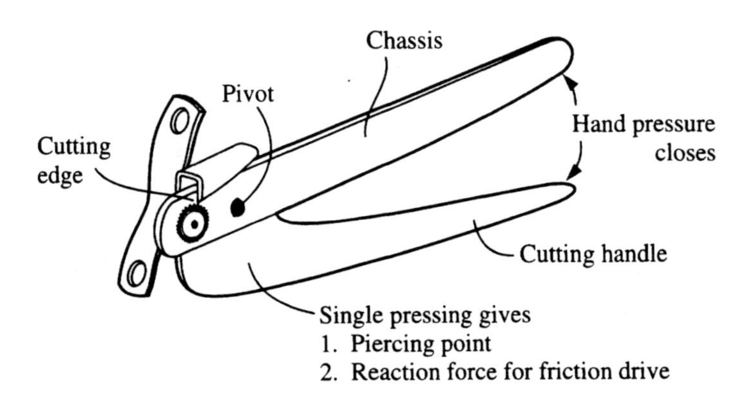

Figure 15.9 *Adding chassis and cutter*

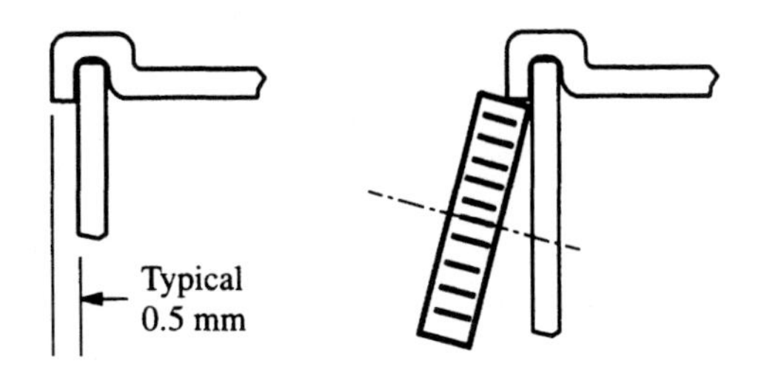

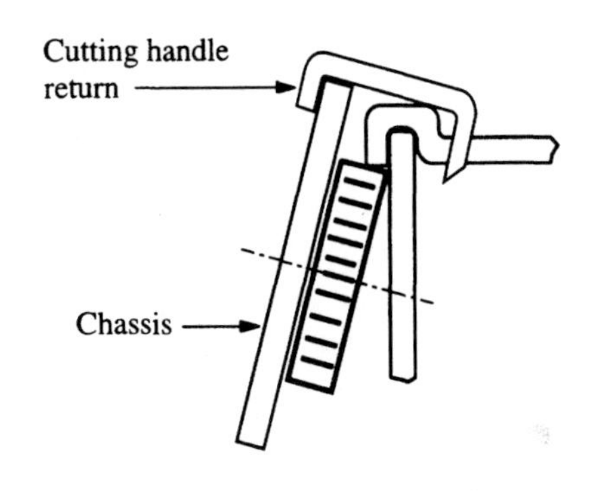

Figure 15.10

Although this will angle the wheel, we now need to ensure that the top of the wheel cannot escape from its required path. Adding a simple return to the cutting handle should achieve this (Figure 15.10).

Our revised design now contains six components:

1. Wing handle
2. Bush
3. Wheel
4. Chassis
5. Cutting handle
6. Pivot rivet.

This compares unfavourably with the number of components in the four-piece design, but we have added considerably to the functionality of the end product. Provided that these are valued by the customer (i.e. he or she is willing to pay for the improvements), we have a better product.

As with almost all designs, further improvements can always be found. The reader should consider any possible improvements either to add features or to overcome limitations in our six-piece can-opener, before referring to the end of this chapter.

The can-opener example demonstrates the importance of detail consideration of both design and manufacturing aspects related to a very simple component. The same basic principles also apply to larger, more complex components. For our next example we will examine a stage in a manufacturing process.

Example 15.3

This example covers an inspection phase in the manufacture of heat exchanger plates. A plate heat exchanger is a modular form of heat exchanger built by stacking a series of corrugated plates together (Figure 15.11). The corrugations provide room for fluid to flow between the plates in the stack. An elastomer seal prevents the fluid from escaping from the edge of the plates. A series of holes in the plates and associated gaskets ensures that a fluid can flow through the heat exchanger, passing through alternate gaps in the stack of plates. A different fluid is also able to flow through the heat exchanger using the remaining gaps in the plate stack. The two fluids never came into contact with each other, but heat would be transferred from one to the other through the plates.

Each heat exchanger tends to be a bespoke design taking account of a range of factors, including the input and output temperatures, pressures, fluid flow rates, viscosity, the corrosive nature of the fluids, and so on. Although each design may be unique, it will incorporate plates from the manufacturer's standard range of sizes and corrugation patterns. The designer can vary the number of plates, the materials used for the

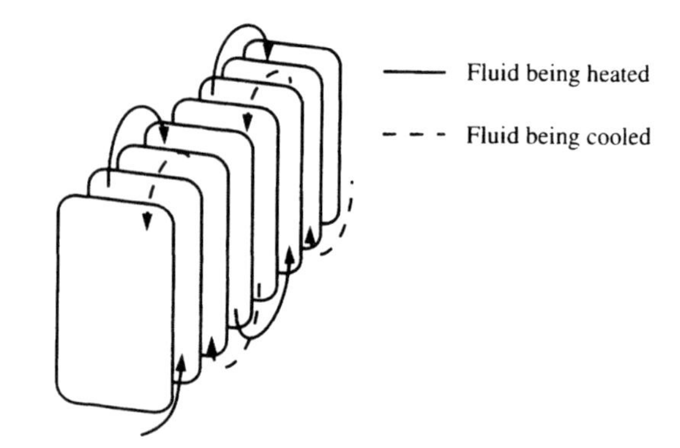

Figure 15.11 *Plate heat exchanger principle*

plates and gaskets, and the type of clamping method. The design must not only meet the functional requirements but must also be cost-effective. To remain competitive there is a continual search for ever more effective plate designs.

The efficiency of a plate design depends on a number of factors, including the design of the corrugations and the plate thickness. The corrugation design affects fluid flow (in terms of mixing, even flow across the plate, pressure drop, etc). The thickness of the plate has a direct bearing on the ease of heat transfer; the thinner the better. Over the years the drive for more efficient heat exchangers has led to an increase in the operating pressures and a requirement for ever thinner plates, with no reduction in the severity of the corrugations. As an inevitable consequence, the chances of a plate splitting during the pressing operation is forever increasing.

Even a small split in a plate can cause problems, as it would allow the two materials flowing through the heat exchanger to mix. The heat exchangers are used in a wide variety of industries, including the nuclear power and food industries. If, for example, a split plate was fitted to a machine supplied to a dairy, this could result in many thousands of gallons of milk being contaminated, which could in turn poison a large number of people. Understandably, heat exchanger suppliers take great care to ensure that machines are not supplied with split plates.

This example deals with a method of inspecting the plates for splits. One method used by a plate producer was a simple air underwater test. (This was much the same test as used when checking a bicycle tyre for punctures with a bucket of water.) It involved fitting the plate to a fixture that provided a seal around the edge of the plate, then submerging the fixture and plate in a water tank. Air would be supplied underneath the plate and the inspector would check for any bubbles rising from the plate. For a number of years this proved to be an effective method. It was generally used to check the first plates off a new run on a press. Splits could appear if, on a tool change, the press had not been properly set, or if material from a new batch with slightly different characteristics was introduced.

A number of factors conspired to generate the need for an improved inspection procedure.

- The air underwater test required each plate design to have its own test rig (since no two designs were the same physical size), so the cost of a new test fixture had to be added to the cost of each new plate design. New plate designs tended to add to the range of available plates, rather than replace previous designs. This resulted in an ever-increasing stock of test rigs. To help visualise the problem, the size of plate from one manufacturer varied from 100mm by 400mm at one end of the range to 1.3m by 5m at the other extreme.
- Increased competition provided an incentive to reduce work in progress and with it batch sizes. Reducing batch sizes focused attention on the time taken to change from producing one design of plate to the next. Testing the first off plates was becoming one of the bottlenecks in the changeover procedure.
- A small but increasing number of customers were asking for 100 per cent testing of the plates. The air underwater test was a very slow, and hence expensive, way of providing this service.

A new method of testing was required that could cater automatically for a variety of different plate sizes, with minimum set-up time when changing from one plate to another. It had to be rapid enough to cater for those instances when 100 per cent inspection was required. Finally, if at all possible it had to be capable of being automated.

A wide variety of alternatives were considered, employing many different principles. Rather than discuss the discarded alternatives we will move directly to the selected method and consider some of the aspects covered by the engineers involved.

The principle chosen was the well-proven dye penetrant test. The component concerned is covered with a liquid dye on one side. The liquid used has two important properties. First it will seek out and enter any cracks. If the component has any through cracks the dye will find them and appear on the dry side. As only very small amounts of dye will pass through small cracks, the dye's second important property, the ability to fluoresce under ultraviolet light, is used to good effect. After allowing a short time for the dye to penetrate, the component is inspected under ultraviolet light, when any through cracks can easily be identified.

This method offered at least one significant advantage over the air underwater system. An individual test fixture was not required for each size of plate. In principle, this achieves the objective of catering automatically for a variety of plate sizes, and avoids delays with set-up times when changing from one plate to another. The dye penetrant is normally applied by hand brushing and is used to inspect components on a low-volume individual basis. For the heat exchanger plates, a more appropriate method of application was to spray the plates with the dye. This method was much faster than hand brushing as well as being suitable for automating.

The chosen method was to install a small inspection conveyor line, so that the plates would pass through a spray booth, where they were coated on one side with dye penetrant. They continued along the conveyor to an inspection booth, where an inspector would check the dry side of the plates under ultraviolet light. The plates then passed through a washing and drying machine to remove all traces of the penetrant.

At first glance this may seem a complex system for checking for a very simple fault. When developing such a system the engineers involved

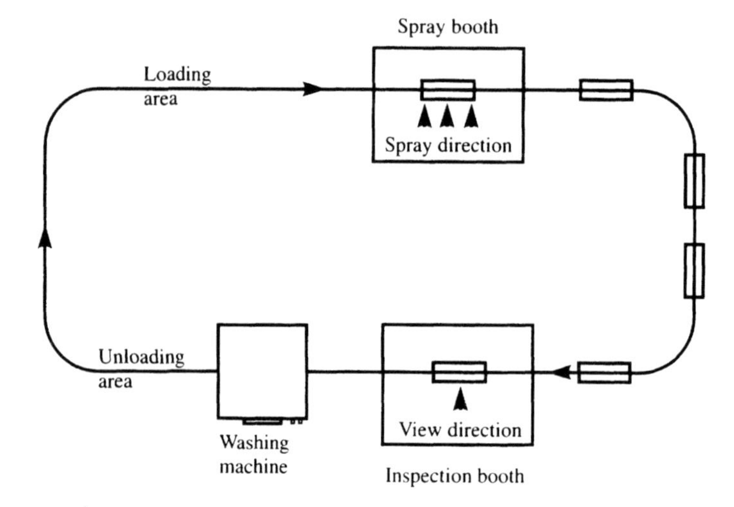

Figure 15.12 *Plate testing line*

attempt to use as many standard components as possible. The conveyor system, for example, would be supplied by a specialist and may involve components common to other conveyor systems used in the factory, to simplify future maintenance. The plates were hung on the conveyor line via special carriers. These were an in-house design with built-in adjustment so that all sizes of plate could be accommodated. The wide range of plate sizes meant that some were carried with their largest dimension vertical, and some with this dimension horizontal.

Spraying components passing along a conveyor line with paint is a very common industrial process, so finding a standard booth with spraying facilities did not prove a problem. However, some of the subsequent development work was quite interesting. As the spraying was to be automatic, and there was a wide variation in the sizes of plates, some form of component recognition was needed. All the plates were rectangular, so for each plate only two dimensions (length and height) were needed. Simple light sensors were employed at the entrance to the spray booth. Within the booth a number of spray guns were fitted one above the other on a vertically reciprocating frame. The signals from the sensors were used to switch the appropriate number of spray guns on and off at the correct times for the plates to be suitably covered with penetrant.

So far everything was very straightforward, until it was realised that there is a fundamental difference in the requirements for spraying paint and penetrant. It is important when spraying paint to ensure that the component is adequately covered, not just on the surface facing the spray gun but also around the edges. The penetrant, on the other hand, must cover all the surface facing the spray gun, but must on no account creep around the edges of the plate. (Any penetrant seen fluorescing on the reverse side would cause the plate to be rejected as faulty at the inspection stage.) Much development work has been done by the suppliers of spray painting systems to ensure good coverage around the edges with minimum wastage. There are many variables that affect this characteristic, including viscosity of fluid sprayed, the spray pressure and the spray pattern settings on the gun. Good coverage is given by equipment using the electrostatic principle. In this case, as the atomised paint particles leave the spray gun at high speed they have a negative electrostatic charge. The atomised particles act like tiny magnets and will be attracted to and deposit themselves on components that are earthed electrically. This process can result in an excellent finish, with losses through overspray being minimised. The concentration of the force field lines around sharp edges tends to result in an increased build-up of paint around these areas. It is excellent for applying paint to a motor car, but somewhat less than suitable for our dye penetrant sprayer.

A series of experiments proved that a combination of spray-gun settings and close control of the airflow across the booth (from the gun towards the component) enabled a satisfactory level of coverage to be achieved. These experiments performed on the equipment as finally installed were to identify the correct settings for the system; a separate set of preliminary trials had confirmed that the system would provide suitable coverage.

You should appreciate that engineers working on a project such as this would be expected to identify the potential problems at an early stage, particularly if they might affect the viability of the project. In this case trials were done at the paint-booth-supplier's premises prior to placement of any orders, or indeed final commitment to the process. The trials were done on similar but not identical equipment to that which was to be supplied. (This is quite normal, as spray booths are usually designed to meet individual customer requirements.) They were carried out by the potential suppliers to a close set of requirements (note the importance of the PDS yet again) provided by the engineer leading the project. The contract eventually placed with the spray-booth suppliers defined closely the required outcome in terms of penetrant coverage. Attention to the detail of the specification, preliminary trials, and the wording of the contract finally placed is vital for a successful outcome.

The conveyor line carried the now coated plates to the inspection booth. This was a fairly straightforward part of the process, but two of the points that needed to be considered were the time taken and the radius of the curves encountered. The time taken had to be sufficient to allow the dye to penetrate any cracks in the plate. The radius of curvature of the conveyor line had to be sufficiently generous to cater for the largest plate, without being unduly wasteful of space.

The inspection booth was simply a darkened area where ultraviolet light was shone on the 'dry' side of the plate. Any fluorescence would indicate a fault in the plate. Initially this inspection was done manually, but later a computer-controlled optical recognition system was investigated to enable this part of the process to be automated.

Finally, the conveyor took the plates to a conventional industrial washing and drying machine to ensure that the plates were free of any contamination from the dye penetrant.

The final result was a system that could test automatically a variety of different plate sizes for through cracks. Operators were only required to load and unload the plates at the conveyor. As a final point of interest, since this check only covered individual plates, further tests were performed on assembled machines at higher pressures. The check on the individual plates was important, partly because individual plates were sold as spares, but largely to give rapid feedback in the event of there being any problems.

15.3 Summary

The reader should by now understand the basics of value and functional analysis. An existing design or process will contain a number of features. If improvements are to be introduced, the value of each of these features must be examined.

1. *Value to the customer*
   Is the customer willing to pay extra for this feature? Or, without this feature would we lose the customer to a competitor?

2. *Value to the producer*
   Does this feature or process add cost to the final product?

3. *Functional purpose*
   Does the feature or process fulfil an important function not covered by any other feature or process?

The radiator drain tap showed how considering these three questions generated an instant cost saving by simply deleting the feature altogether!

The can-opener example demonstrated how important it is with such a product to consider the absolute detail of each feature, how it would be produced, the function it would perform and whether this would be seen as a benefit by the customer.

In a similar way, the plate inspection line demonstrated the same lessons, but applied to a manufacturing process. The attention to detail in the spray booth showed those characteristics considered beneficial when spraying paint were a disadvantage in this application.

15.4 Questions

Now check your understanding with these questions:

1. What do you understand by the term 'functional efficiency'?
2. Name the three elements that should be considered when searching for an improvement to an existing design or process.
3. Take a simple household object such as a corkscrew and perform a value analysis examination. Use the can-opener example to help you. It may also be helpful to compare different existing corkscrew designs.
4. Identify some everyday objects where the functional efficiency has been improved by reducing the number of component parts. (An example would be a single-piece moulding for a video camera case.)

Finally, did you devise any further potential improvements to the six-piece can-opener? Possible comments on the design are:

- Uncomfortable in the hands. This may be improved by folding the chassis and cutting handles to avoid hand pressure being applied to the edge of the material.

Figure 15.13 *Modified handle*

- The cutting handle uses the same material for two functions. This may result in a poor compromise between a soft, low-cost material giving a short life for the blade, or using a more expensive material to achieve good blade performance than is necessary for the handle. Should an additional component be considered, to avoid the compromise?
- If the lower handle is to be a separate component, should it be a pressing or would a forging be more appropriate? A correctly designed forging could incorporate two spigots that would locate in appropriate holes in a cutter element and be retained by peening over (Figure 15.14). In this way we would only be adding one more component and keeping the assembly operation simple. The forging from round bar is likely to be more comfortable to the hand than is the original sheet material design.
- It may be able to add further features for almost no extra cost. For example, a bottle top remover could be stamped in the end of the chassis handle during its existing stamping operation (Figure 15.13).

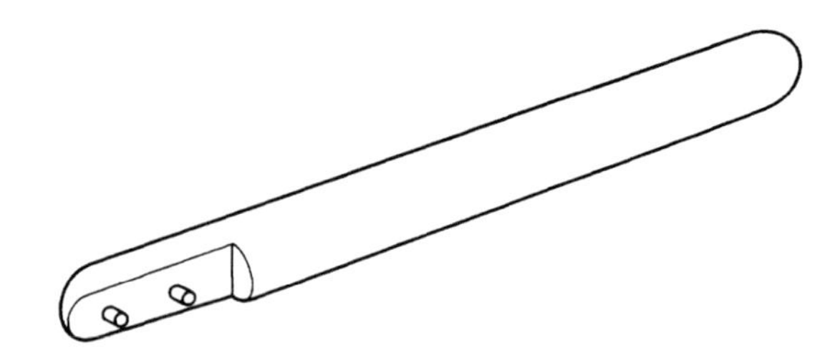

Figure 15.14 *Forged lower handle*

16 Systems for Controlling Design and Manufacture

16.1 Introduction

So far we have concentrated mainly on the detailed aspects of design and manufacture, and the ways in which these are interrelated. By now the reader should, when considering a concept or a component, have some insight into the decisions that have to be made as the concepts or components are progressed through the development to the production phase. A single component requiring a once-off design input, then produced using a set procedure, so that regular batch quantities could be supplied at the end of each week to a single supplier, would seem to be fairly straightforward. Would that life were so simple! Unfortunately, most organisations involved with design and manufacture produce a variety of products that have to be delivered to a variety of customers in varying quantities and at varying times.

Sir John Harvey Jones, a past chairman of ICI, has stated on many occasions that for a producing organisation to survive in the long term it must aim to be the best in the world at its chosen activity. If it does not, some other competing organisation in the world that has that as its aim will eventually win the customers for the product or service. The truth of this statement has been amply demonstrated many times over, particularly by a number of Japanese companies (motor cycles, televisions, calculators, motor cars, some types of machine tools, etc.). To meet this aim, organisations have continuously to review their designs, manufacturing processes, and various procedures within the organisations. So most organisations with an eye to the future are not only producing a variety of products that have to be delivered to a variety of customers in varying quantities and varying times, but at the same time are reviewing the designs and the way they are produced.

Some form of system is obviously needed to control this process. In this chapter we will introduce the main elements and information flows in a typical organisation. We will then discuss some of the systems used in an organisation largely using manual controls. Finally, we will examine how the advent of the computer can assist with the systems.

16.2 Primary tasks

In order to understand the functions of the various elements of an organisation we need to look first of all at the primary tasks that need to be performed for the organisation to thrive (or even to survive). They can be listed as follows:

1. Define aims, objectives and strategy
2. Plan future products and processes
3. Develop and implement new (or improvements to existing) products and processes
4. Obtain orders from customers
5. Manufacture or obtain products and despatch to customers
6. Recover payment from customers and pay suppliers
7. Service the system.

The order in which the tasks have been listed has no bearing on their significance to the organisation. They are all equally important in the same way that all the links in a chain are of equal importance. Any company that fails to achieve in only one of these areas is certain to reduce its chances of long-term survival. However, it does not follow that all companies will put in the same level of resources to act in each of these areas. The resources required in each area will depend upon the shape or profile of the company. A drug-producing organisation may, for example, put vast resources into planning and

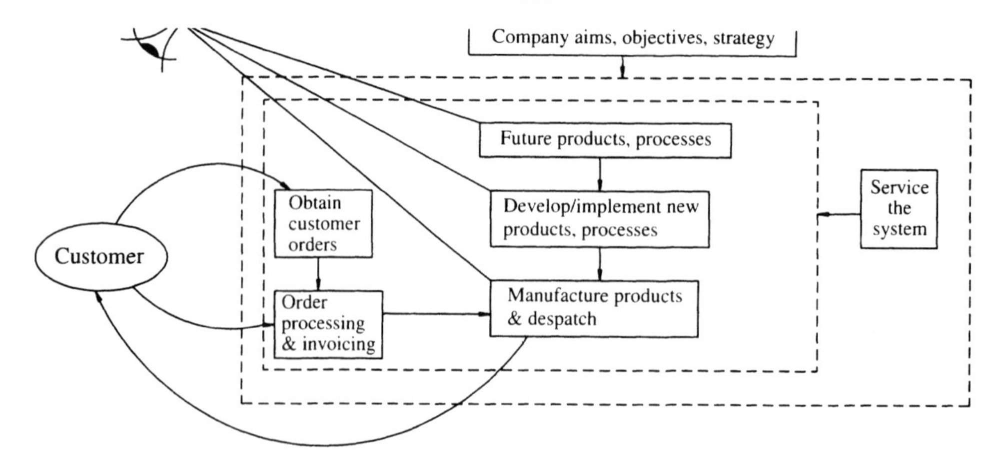

Figure 16.1 *Primary tasks of a company*

developing future products and processes, whereas a record-producing company may feel that a very high sales promotions budget is needed to generate sufficient orders from its potential customers. For the purposes of our discussions we will assume that 'our' organisation is a company designing and producing a range of products. As it has its own in-house manufacturing facility the task consuming the largest share of the available resources is almost certainly '*manufacture or obtain products and despatch to customers*'. This, along with the task of developing and improving products and processes, is likely to contain the greatest concentration of engineers. Figure 16.1 shows these tasks in diagrammatic form and how they link with the customer. Although only three of the tasks are shown with a direct interface with the customer it is important that the others are aware of and are in a position to respond to the changing needs of the customer. Remember that without the customer the organisation would not exist.

We now need to examine each of the tasks more closely and identify the need for particular functions within our company.

16.2.1 Company aims, objectives and strategy

In the same way that a new product cannot be designed without a PDS, setting the ground rules for the product, the aims and objectives of an organisation must be defined and a strategy for achieving them outlined before the organisation can stand any chance of moving towards attaining them. The direction that the company is heading towards is normally defined by the board of directors. This group of people usually consists of a director representing each of the major functions within the organisation (i.e. a sales director, an engineering director, finance director, etc.) and is led by the MD (managing director).

The decisions made at this level are generally strategic rather than detail. Typical decisions might be to invest in a new factory, to start exporting to another continent, to increase market share over the next five years at the expense of short-term profitability, to withdraw from one range of products and concentrate on another, and so on. All such decisions are made with a view to satisfying the aspirations of the owners of the company, and, of course, must take account of the needs of the customer.

16.2.2 Future products and processes

Although the final decisions regarding new products will generally be made at board level, they will be based on information and recommendations generated from certain specialist departments. There are often two areas within a company from which most of the new ideas will flow. One of these is the marketing department. This is not to be confused with sales, which is a separate function. Marketing is primarily concerned with product planning, i.e. planning the future products to be produced by the company. The work in this area will include analysis of competitors' products, contact with customers to identify reasons why particular products are selected, analysis of press reports where appropriate, indeed any activity that is likely to generate a better understanding of the market. The Ford Motor Company is an excellent example of the success of this sort of approach. They have a large product planning department which performs all the activities described. In the past they have been known to purchase a wide range of competitors' products (even including three-wheeled vehicles) which, after an analysis of their dynamic qualities, would be stripped right down to the last nut and bolt for any clues that might reveal a design or manufacturing benefit. Customer clinics would be held where mock-ups of possible future body styles would be shown to selected customers in an attempt to gain some forewarning of public reaction to a new shape. Their products are a true reflection of the company's perception of their customers' requirements. The accuracy of this perception is mirrored in their sales figures.

The second area that is likely to be a source of new ideas for future products or processes is sometimes called R&D – Research and Development. Here, work will be done on either products or manufacturing processes by those in the company whose responsibility it is for the company to maintain or establish a lead in technical expertise over its competitors. The research section of Honda will have been responsible for the initiation and development of the variable valve timing systems employed on its range of Vtec engines. The research section of Citroën will have been responsible for its Hydractive suspension developments, and the use of non-metallic materials in some of its diesel engines.

Of course, new ideas may spring from many areas of an organisation; indeed, company suggestion schemes are designed to encourage just this. However, regardless of the original source, the idea will normally be developed through the research or product planning departments before being placed before the board of directors for the authority to proceed. (Of course, board approval would only normally be required for those ideas that would require either significant investment or risk.)

16.2.3 Develop and implement new products and processes

The work in this area has already largely been introduced in Chapter 1. It should be noted that this department, often simply called *Engineering*, will deal not only with new products and processes, but will also be responsible for introducing cost savings by continual improvements to both the product and the manufacturing system. If the company produces specials in addition to standard products, this department may also contain a section dealing with quotations.

In the past, two sections would have been involved – the *Design Engineering Department*, and the *Manufacturing Engineering Department*. Although some companies still operate with two sections, many now recognise the benefits of a *Design and Manufacturing Engineering Department* in enabling an early input from those with manufacturing expertise. As with the tasks above, it is essential that those working in these areas do not lose sight of the customers' requirements.

16.2.4 Obtain customer orders

The tasks covered so far all require close attention to the needs of the customer, but this one differs in being the first that involves direct contact with the customer. Although many engineering organisations may have relatively few sales staff compared to the total employed, selling the end product is as vital a task as any other in the organisation. After all, without customers the organisation simply would not exist.

Ideally, the receipt of an order will be the signal to the manufacturing operation to start work and fulfil the requirements as soon as practicable. A company receiving many orders will need to sort the orders, obtain delivery dates, etc., so in some organisations a separate order processing section will be involved

16.2.5 Manufacture products and despatch to customers

As was mentioned earlier, this is a far more complex task than it sounds.

Production control section

The order is normally passed to this section, where the appropriate tasks will be fitted into a schedule of work currently being undertaken. The complexity of this function alone depends to a great extent on the range of products being produced. For example, a company producing single components from a plastic moulding machine, such as plastic buckets, where the only variable might be the colour, should not have much difficulty in scheduling a new order in. Imagine at the other end of the scale a company manufacturing aircraft engines; vast numbers of components, suppliers, materials and processes will be involved. Factors to be taken account of by production control will include availability of raw materials and bought-out components (some of these may be stocked, some may need to be ordered), and the availability of manufacturing resources (machines and labour). Once the work is scheduled a delivery date is generated (sometimes as a specific date but more often as a week number).

In some cases pressure may be brought to bear on the production control system to give special priority to a specific order, usually because of the importance of the customer and the expectation of further orders. Reacting to such pressure should be treated with great caution. Any significant priority given to a new order over and above those already scheduled is almost certain to have an adverse affect on the delivery dates of existing work in progress, and may well effect the efficiency of the shop-floor loading. The author recalls an experience when working for a company supplying a range of in-house designed and manufactured engineering products to a wide variety of organisations. A large order had been taken for some equipment needed by General Motors for one of their car plants in the USA. The plant had a shut-down period during which this equipment was to be installed. The managing director had given his personal assurance that there would be no problems in meeting the delivery date. However, when scheduling the work with that already in progress it was apparent that meeting the required date would not be possible. The MD instructed the author to act as project controller to progress the order through the system and ensure that the due date was met. He was to identify any necessary resources to be supplied. For several weeks the factory concentrated on this order, working extra shifts, and giving it overall priority. The delivery date was met, the customer was suitably impressed, and all concerned received hearty congratulations. The feeling of success, however, did not last long. The disruption to the previously reasonably balanced workload in the factory, and the failure to meet delivery dates for scores of other customers caused difficulties that took some six months to overcome. The moral of the story is obvious.

Stores

Any organisation that produces assembled products consisting of a number of individual components or sub-assemblies normally has one or more forms of stores. A *raw-materials store* would hold the basic materials as bought in, i.e. before any in-house work has been performed on them. For example, this store could contain metal in the form of bar, tube, sheet or coil as appropriate for the company's products. The store may also contain equipment for cutting and issuing the requisite quantity to the shop floor.

Any company having an assembly area is also likely to have a *finished components store*. Items held in this area only need to be issued to the final assembly area. It could contain machined components from the shop floor, subassemblies (of pressed and welded items or machined items), and bought-out components such as fasteners, electric motors, plastic mouldings, etc.

There may also be *intermediate stores*. A piece of bar may be issued from the raw-material stores, be turned, and then ground in-house, before being sent to a subcontractor for a process not available in-house, such as heat treatment. The stores is, in effect, a marshalling area where part completed components can wait before proceeding to the next operation.

Although these types of stores are needed by most companies, it is not because the company needs to store vast quantities of material. In fact, quite the reverse is true. Much effort is put into minimising the levels of stock held, as we shall see in the next chapter. It is, however, vitally important that the actual levels of stock held are recorded accurately, because this information will be used by the production control section to schedule new orders.

Purchasing

A function closely related to that of the stores is purchasing. Experienced buyers will order raw materials and bought-out components, and often also place orders for any sub-contract operations. Their experience regarding costs, suppliers and possible delivery times is a valuable asset to a number of areas within the organisation.

Component manufacture

The type of work carried out under this heading is highly dependent on the type of product being manufactured. It could include a press shop, a machine shop, foundry, paint shop, moulding area, welding area, polishing area, composite layup area, autoclave area, test area, or something quite different. An engineering company is likely to include a combination of some (or possibly all) of these areas. The layout and organisation in each company will tend to reflect a number of factors, including product range (type and variety), production volumes, management style, previous history, and the function of the particular area. As we will see later, the chosen layout can either help or hinder the control of the manufacturing system. At this stage, if we consider a simple component that requires a pressing and a casting to be welded together, the weld polished and then a valve seat machined, one possible system for component manufacture might involve three manufacturing areas, as shown in Figure 16.2.

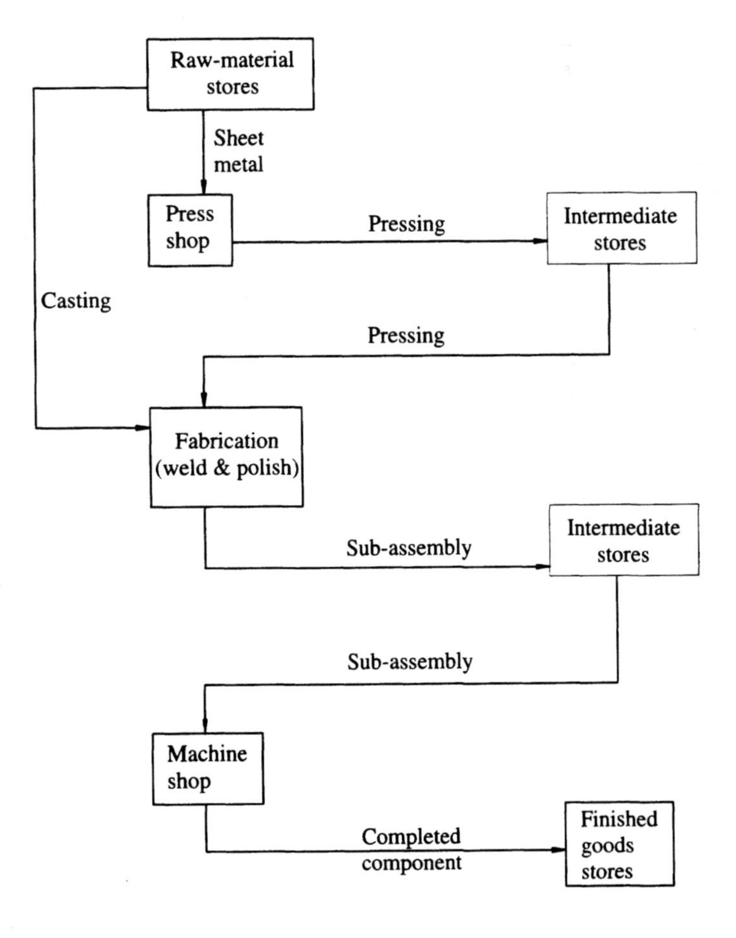

Figure 16.2 *Component manufacture*

Later we will consider possible improvements on this, but for the moment, although we will refer to component manufacture as a single function, remember that in reality it is a little more complicated.

Final assembly

Once the individual components and sub-assemblies have been produced they then have to be assembled to generate the final product. We could have considered this as the final stage of 'component' manufacture. The main reason for keeping it separate is that it will exist as a function in all the types of company that we are considering, even those organisations that will have made a strategic decision to sub-contract their component manufacture. A machine shop, for example, requires a considerable initial investment to set up, and a great deal of expertise and finance to keep it running efficiently. It may make sense for a company whose products contain only a few machined components to let a specialist produce these. This logic could be applied to the whole range of components required for the final product. The company will, of course, still need finally to assemble the components. (An organisation that buys in complete products may, of course, be a completely valid and successful company, but it is outside the scope of those that we are considering.)

Depending on the product there may well be included in the assembly procedure some means of testing that the final product functions correctly.

Packing and despatch

The final stage of the manufacturing function is to pack and despatch the final product to the customer.

16.2.6 Recover payment from the customer

Once the goods have been despatched, final payment needs to be recovered from the customer. Never forget that the survival of the company depends not only on the customer being satisfied with the goods, but also the company recovering sufficient recompense for the work undertaken. In the same way the company has an obligation to ensure that its suppliers also are paid. The job can only be said to be complete when the debts are settled. A section normally controlled by the *Finance Department*, but with close liaison with the sales function, would handle this task. It often goes under the title of *Order Processing and Invoicing*.

16.2.7 Service the system

The functions described so far are only the bare bones of a manufacturing company. There are a number of subsidiary functions that are needed to 'oil the wheels' of the system.

Personnel

Any organisation that employs people has a number of legal and moral obligations relating to each individual's employment. All but the smallest of companies will have a specialist individual or section to deal with these.

Finance

Similarly, any organisation trading as a limited company or PLC has a legal obligation to keep records of its financial transactions and pay any relevant taxes. This is the responsibility of the finance section. Although it should be made clear to all employees that their true function is to perform their particular skill in such a way that it helps maintain the long-term profitability of the company, it may be difficult for the manufacturing engineer to keep counting costs as he attempts to discover why the most recent batch of widgets has been machined undersize. The finance section will keep score of the production output and costs in order that the cash flowing within the organisation remains under control.

Legal

In these days of increasing legislation, companies need to be well advised about a wide variety of legal aspects. These can include aspects ranging from contracts (with customers or suppliers), to patent or copyright infringement.

Works engineers

Tasks such as the maintenance and repair of production machines and air-conditioning equipment, the revising of shop-floor or office layouts, the addition of guard rails, and sundry other jobs, would normally be covered by the works engineer. Depending on the size of the company this could be a designated individual, or a complete department.

Quality control

This is a vital function that can be dealt with in a number of ways in different organisations. See Chapter 11 for process control techniques.

Figure 16.1 showed in diagrammatic form the primary tasks; this has been expanded in Figure 16.3 to show the elements within the system.

Now that we have introduced the primary tasks within the system, along with those sections or departments that should deal with them, we need to consider the information that each needs to perform the various tasks.

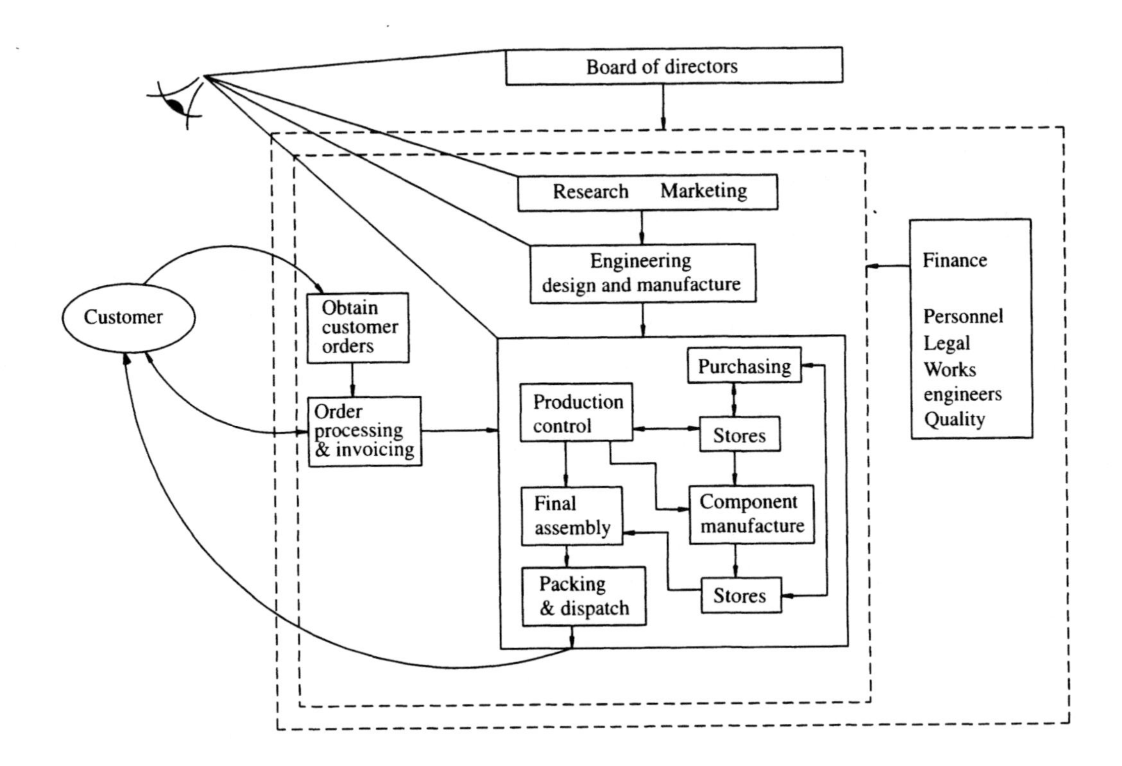

Figure 16.3 *Primary elements*

16.3 Information flows

16.3.1 Types of order

The logical way to examine the flow of information is to start with the customer's input. However, before we start, we need to differentiate between two broad groups of products, as they will each be dealt with in a slightly different way. We will call them standards, or specials. The standard product will be a product sold in reasonable volumes, each one produced in the same way, and sold for the same price (ignoring any special discounts). It may have a number of options that the customer can select from a predefined list. The special, or bespoke product, on the other hand is specially produced to meet the particular requirements of a specific customer. In this case the customer can request any options, the only limitations being technical ones and the size of the customer's wallet!

Some companies produce only standard products, some only specials, and some a mix of the two. 'Our' company will produce both standard products and specials. This is quite common, particularly among companies producing engineering components, where the specials tend to involve modifications to the standard product. For example, companies producing hydraulic dampers for motor vehicles will have a standard range to fit vehicles currently in production, but are normally willing to produce specials with different stroke lengths and different valving to meet a specific customer's need. There will, of course, be a cost penalty for this. The reasons will be apparent when we examine the information flow.

16.3.2 Special orders

Now we will discuss the implications of processing an order for a special (see Figure 16.4). Initially, a customer enquiry will be received by the sales staff, who will make a preliminary decision as to the practicalities of the enquiry. If it has potential, a quotation needs to be generated; an enquiry number will be raised and the documents passed to *Engineering* for a quotation. Here, the engineering content will be examined and an estimate made for the design and manufacturing work involved. This estimate will normally include both the time for the work to be done and a completion date for the engineering input. The documents may also need to be sent to *Purchasing* to cover any cost and delivery implications of bought items. Finally, they will be passed through *Production Control* to determine an approximate delivery date. The quotation package is then returned to *Sales*, where a selling price is calculated and a formal quotation sent to the customer. This process will vary between different companies and products. For example, if the enquiry was simply for a standard pump fitted with a heavy-duty motor, the paperwork would be much more straightforward than if it was for a high-pressure pump built to meet specific safety codes for use in a nuclear power station.

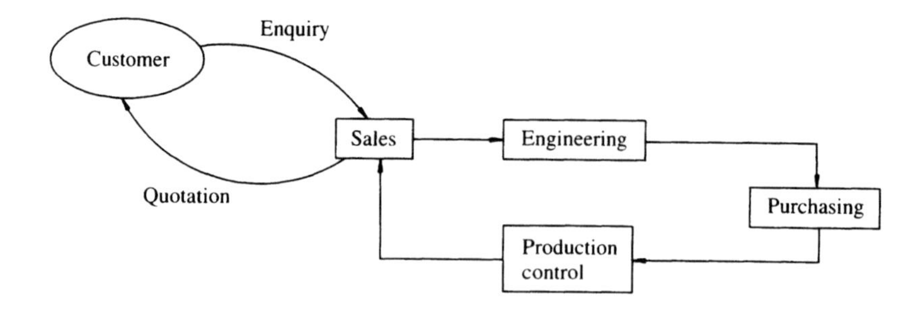

Figure 16.4 *Special quotation*

If the customer is happy with the quotation he/she will then submit an order to *Sales*, where the order processing section will raise an order number. We will call this a works order number simply to differentiate it from an order for a standard product. The estimate, together with the works order number, will then be circulated via the same route as before, this time simply to confirm or amend the delivery dates. Generally, quotations will be dealt with on a regular basis, but possibly no more than some 10 per cent will be converted into firm orders. The delivery dates on the quotations have to make some assumptions regarding the workload, and need to be confirmed for the firm order. Another benefit arising from this confirmation loop is that effectively it gives advance warning to some of the departments involved. For example, *Purchasing* will be aware that a long-lead item needs to be ordered by a particular date. The precise details of this item will need to be specified by *Engineering* who will no doubt be chased for the details by *purchasing* if the details are at all slow in appearing. The *Sales Department* will then confirm the order, giving the new delivery date to the customer (see Figure 16.5).

The *Engineering Department* will now start work on the design aspects of the 'special'. In fact, the workload within *Engineering* will need to be managed, so the work may not start immediately, if work on other specials already in the system needs to be completed first. It may, however, be necessary to do some preliminary work on identifying any non-stocked long-lead items.

The engineers concerned will attempt to minimise costs by utilising as many standard components as possible. By 'standard' in this context we include components already in volume production within our manufacturing plant. Often the older engineer who has spent many years with the company will remember a particular component, possibly one not even in current production, but which with minimum modification will suit the required task. He may even be able to locate some examples gathering dust in a corner of the stores. Whilst this may solve an immediate need, a better system is needed to locate such components.

The *part number* is invariably used to identify and locate components, so its construction is important. By selective use of code numbers, a numbering system can be used to categorise components, identifying features such as material, size range, shaft or prismatic, threaded or non-threaded, etc. In some cases the number can indicate which other products the component is used on. Often a standard, well-proven system such as that devised by Brisch will be used. Note that the part number for a component is often derived from the drawing that originally defined it. The drawing number may also contain other information, such as the drawing size, and the latest revision number.

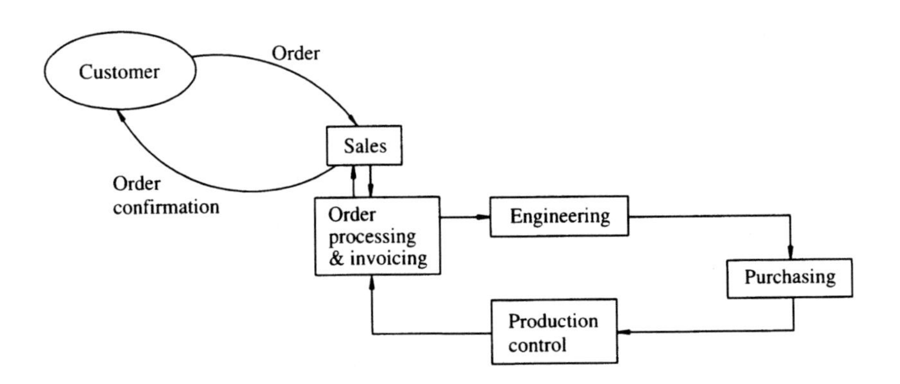

Figure 16.5 *Special order confirmation*

The engineers will also need to check the availability of components, by liaising with both *Purchasing* and *Stores*. They may arrange for Purchasing to place an advance order for any long-leadtime items prior to completion of the engineering work.

The output from this area will be a set of manufacturing drawings and planning or routing sheets. The planning sheets show the sequence of manufacturing operations that are required to produce the final product. They will identify, for each stage, the material, the operation required, the type of machine to be used, any special jigs or fixtures, appropriate tooling, set-up time and time allowed to complete the operation. These sheets can be used to calculate the cost of the product. Once the routing sheets have been generated, the planners would send the stores an advance list of the materials required. This enables those items already in stock to be reserved for the particular works order. A list of those items not in stock will be passed to *Purchasing* so that they can be obtained. The planning sheets, together with the drawings and parts list, form the package of instructions that the manufacturing system will work to.

This package will be received by *Production Control*, who will then fit this order into the manufacturing work schedule. This schedule, in effect, takes account of all the components required and lists the tasks to be done in each sub-area. Each of these areas will have its own foreman, who would normally allocate the specific tasks to those working in the area. These areas will depend on the individual organisation; in Figure 16.2, three are shown: the press shop, fabrication, and machine shop. As this example shows, the input and output to each of these areas is one or other of the stores.

It is vital that accurate records are kept as items are booked in and out of the stores. If the *Stores* is holding more than the records show, this means the company has either bought or produced components and is not aware of their

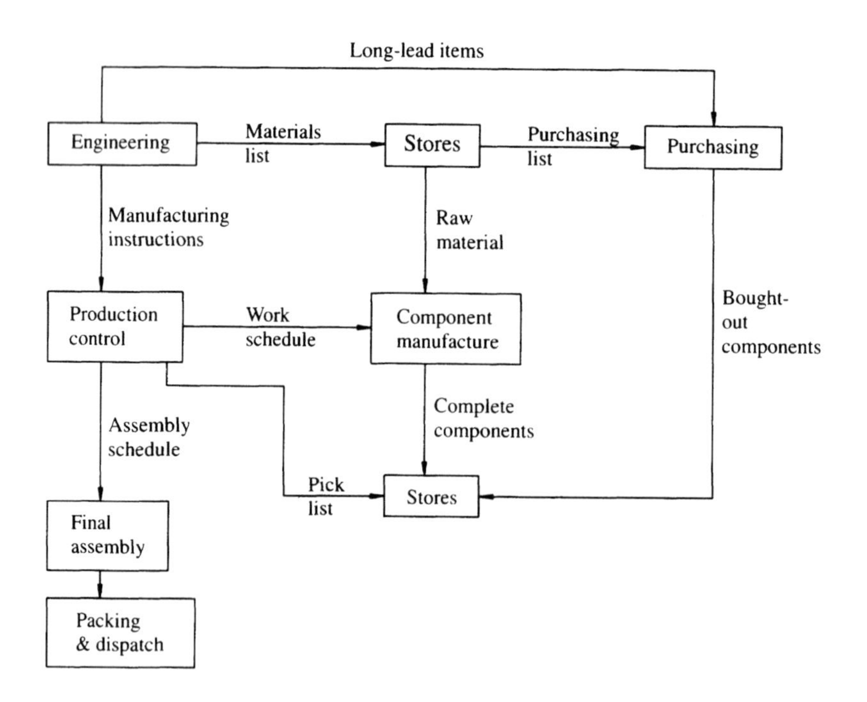

Figure 16.6 *Special component manufacture*

existence. The money spent in financing this stock effectively has been wasted. On the other hand, if the records indicate higher quantities than are actually present, it will only be a matter of time before a shop-floor operator is unable to complete his/her allocated task, because the material or component is not available to be worked on. Remember, this task was allocated as part of an overall work schedule devised to meet the delivery dates, and to maximise the plant efficiency. The more time spent waiting for non-existent items and rescheduling work, the less efficient a manufacturing plant will be. Again, there will be a financial penalty.

In some cases there will be a need for some manufacturing operations for which the facility does not exist on site. Operations such as heat treatment or a specific form of plating are often performed by specialists on a sub-contract basis. In the interests of simplicity we will treat this as simply another component manufacturing operation, which, of course, it is. The reader should, however, appreciate that such operations will involve a number of aspects outside the direct control of the manufacturing system, such as transport, quality, lead time, and cost.

Eventually (i.e. by the due date) the finished goods store should contain all the components needed for final assembly. A common procedure would be for production control to issue the finished goods store with a picking list so that a complete set (or sets) of items would be collected for delivery to the final assembly area. This would be the normal procedure for standard products, and specials that do not differ greatly from the standard item. Specials that do not fall into this category may well be handled separately. Figure 16.6 summarises the information flows for a special product.

16.3.3 Standard product order

A standard product has to progress effectively through the same stages, but without repeating all the stages each time. The engineering input is still needed, but once it has been completed it does not need to be repeated for each new order received. The information flows for an existing

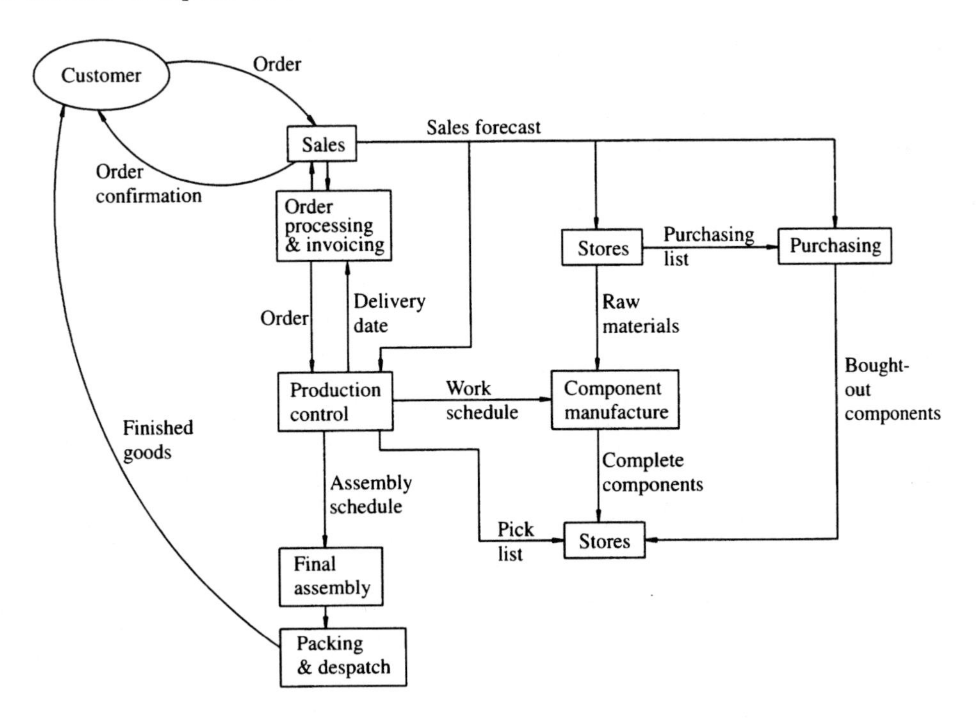

Figure 16.7 *Standard component order and manufacture*

standard product are summarised in Figure 16.7. Note that the order confirmation can be sent to the customer, simply by consulting *Production Control*. A company producing a very low volume of products with a wide range of standard options may well only manufacture when specific orders have been obtained. If the product range is produced in somewhat higher volumes, the system will often work to a sales forecast, producing some products for stock. Indeed, this may be a market requirement, in that customers require the product to be available ex-stock.

There are, however, a number of dangers in producing too much to be held in stock. First, there is the cost: the stock must be financed. Common sense dictates that the time delay between paying for stock and the receipt of sales revenue should be minimised – over-high stocks will increase the delay. Secondly, there may be a change in customer requirements. Holding vast numbers of blue washing-up bowls in stock will be of little use if the main customer demand has changed to red washing-up bowls! Thirdly, high stock levels will delay the introduction of engineering improvements to the product or its method of manufacture. Obviously the improvements cannot normally be introduced to stock that has already been completed. Finally, over-high stocks of an existing product at the end of its product life-cycle will only delay the introduction of the potentially more profitable and attractive replacement product.

So far we have only considered how the system responds to customer enquiries and orders. Previous chapters have made much of the importance and detail of the work of engineers involved in the design and manufacturing areas. We now need to see how this work interfaces with a manufacturing facility that is already up and running.

Any impression that the reader may have that once the design and manufacturing facilities for a new product have been set up, then the engineers involved are able to relax, is quite false. Apart from monitoring potential problems with the new system there will always be pressures to improve both the product and its manufacturing process. As a result, throughout a product's life-cycle there is likely to be a continuous flow of *engineering changes*. The sources of the modifications are many and varied; the section on control systems will discuss how they are handled.

The changes will normally be introduced for one of the following reasons:

- To enhance the product
- To overcome a problem with the product (e.g. over-high warranty claims)
- To reduce cost by modifying the product
- To reduce cost by modifying the manufacturing process
- To accommodate a change of component supplier
- To accommodate a significant change in production volume.

In all cases, the manufacturing system will need to be kept fully informed of the changes. (In many cases, the shop floor will be the source of manufacturing improvements and will therefore automatically be involved in the changes.) *Production Control* will need to know how any production scheduling will be affected. *Stores* and *Purchasing* also need to be involved, as the current stockholding and material order commitment situation may influence the timing of the change.

If the change involves product improvements that are apparent as benefits to the customer, *Sales* need to be kept fully informed and possibly new literature produced. In some cases, product changes may affect component interchangeability between the new and the old. Those involved with selling spare parts need to be briefed fully. Such changes should be only be introduced if the benefits outweigh the cost implications of possible dual stocking of spares, and potential problems with customers fitting the wrong spare at a later date.

At this stage the reader may feel that the more one thinks about the information flow and interchange the more complicated it becomes, even with our simple example. Although this is true, each piece of information is straightforward and has a definite purpose. In order for the system to be workable, a means for controlling the information, and hence the system, is needed.

16.4 Control systems

It should be obvious to the reader that for an organisation to be maintained under close control, much information has not only to be available to a variety of different areas, but also needs to be updated continually for maximum value. An obvious tool to help with this task is the computer. However, before we examine the benefits computers have to offer, it is important to understand the basis of the control system. We will first discuss a number of individual areas.

16.4.1 Engineering changes

As we stated earlier, a change can arise from any number of sources. Indeed, many organisations have suggestion schemes that actively encourage any employee to be vigilant for potential improvements and put forward his or her ideas. The *Engineering Department* will normally receive an *engineering change request* (ECR) which will identify the task. This needs to be considered and categorised within a range from very urgent to impractical. A very urgent change (e.g., relating to safety or a severe technical problem) would be implemented as soon as possible. In some cases the change may be so urgent that it will be implemented before the manufacturing instructions are updated. In such a case, a temporary *variance approval* would be issued by *Engineering*. This is a form describing the change and authorising a variance from the drawings issued to the shop floor. It would normally be limited to cover a specific number of components. Figure 16.8 shows the engineering change information flow.

Most changes will need to be prioritised, also taking account of a number of other factors:

Existing stocks:
: If these are high and sales are slow for this product the change will have a low priority.

Product life:
: How long is the product likely to remain in production? There is no point in spending money on a product at the end of its life-cycle.

Potential savings:
: The greater the saving, the more attractive the change will appear.

Cost:
: All changes will have an implementation cost. This needs to be recovered before any savings are generated. If the cost is very high (e.g. purchase of special-purpose equipment),

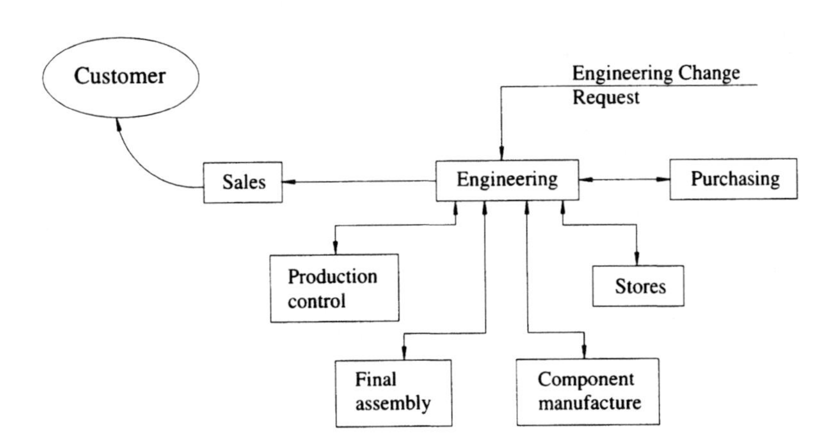

Figure 16.8 *Engineering change information flow*

the effect on the overall cash flow of the business may need to be considered.

Available resources:
Does Engineering have personnel available?

Although these criteria may seem fairly obvious, the author has experience of one organisation where the communication was so poor that some engineers were working on modifications to products that had ceased production!

The change may involve modifications to a variety of aspects, including materials, component features, dimensions, tolerances, surface finish, shop-floor layout, jigs, or fixtures. These alterations will result in modifications to the drawings, parts lists and planning sheets. The component standard cost (see Chapter 17) will also need to be amended. It is important to keep a record of the changes, partly to be able to investigate a fault with a component at some time in the future, but also to ensure that components in production are being produced to the latest drawings. As we saw earlier, careful construction of the drawing number can help in locating potentially common components, as well as giving a clue as to the physical size of the drawing. The issue number of the drawing must also be indicated clearly, possibly as a suffix to the drawing number. The drawing should also record each of the changes. Sometimes the changes are too extensive to record directly on the drawing. In this case, the drawing should record the *engineering change number*, the date of the change, and possibly a drawing grid reference for the location of the modification on the drawing sheet. *Engineering* should then have a book recording against each *engineering change number* details of the actual modification.

16.4.2 Efficiency measurement

Some measure of performance is essential if a device or system is to be controlled to maximise that performance. This is as true of tuning a car engine by measuring its power output or its exhaust emissions as it is of running a manufacturing organisation. There are three basic inputs to the system: people, machines and materials. Some of this ground will be covered in the next chapter, but it is appropriate that we introduce one or two measurements here.

Labour:

This covers all the costs associated with employing people within the organisation. Some of the employees will be working directly on fulfilling specific customer orders. The machine operator, the design engineer working on a special product, and the assembly operator will all be classified as *direct workers*. In other words, their time can be attributed directly to a customer order. On the other hand, the office cleaner, the engineer preparing a quotation for a customer, and the managing director will all be classified as *indirect workers*. Their time cannot be attributed directly to a customer order.

Measurement of the input by direct labour is simple, as each individual records his or her time against an appropriate code number. The code number may be a job number, or a code for material shortage, machine breakdown, waiting to be allocated work etc. As there should be a code number for all eventualities, all the employee's time should be booked to a number.

The efficiency of indirect labour cannot normally be monitored in such a precise fashion. It is generally left to department and section managers to ensure the performance of their areas are up to scratch.

Machines:

Machines utilisation is another useful efficiency measure. A machine tool such as a lathe is only adding value to components (i.e. earning its keep) when it is cutting metal. In a typical well-organised machine shop this may account for between 30 and 40% of the machine's availability. The rest of the time will be spent loading and unloading the component, setting the machine up for the job, changing tooling, and gauging or inspecting, apart from any delays due to scheduling problems, material shortages, etc.

One way of increasing a machine's utilisation is to run large batch sizes and ensure the machine is supplied with ample stocks of raw materials. Although this will increase the percentage of time spent metal cutting, as we shall see in the next chapter, this may well reduce the overall efficiency of the manufacturing operation.

Materials:

Materials represent a significant investment in most organisations involving manufacture of products. As with labour and machines, careful control of this investment can have a dramatic effect on the organisation's efficiency (i.e. profitability). Control (and therefore measurement) is needed over the quantity stocked, the variety used, and the means of ensuring the right material is in the right place at the right time. The next sections will introduce some of the controls, but these will be discussed further in the next chapter.

16.4.3 Stock control

As we have previously introduced the importance of controlling the levels of stockholding, we now need to discuss some of the ways in which the control can be maintained. Physical security is important, not just to protect the Stores from a burglar, but also from the over-enthusiastic foreman who, in his rush to keep his section operating efficiently, may decide to help himself to another casting to replace one that has been over-bored in error by one of his machine operators. Whilst the enthusiasm of the foreman is to be applauded, his action will mean that the Stores records will be incorrect, and no action will be taken to replace the lost casting. If this is allowed to continue, at some point the Stores will find itself unable to issue a full set of castings to meet an order. Inevitably, this order will be an important one that a future long-term contract with a customer will depend on! This kind of problem often leads to complaints about maintaining too low a level of stock, and can sometimes result in higher stockholding. This is totally the wrong action: these extra stocks have to be financed, and this represents an extremely poor use of company resources. The answer is to build a secure fence around the Stores, and only allow Stores staff to issue items. They then have the responsibility of ensuring the validity of the stock records.

Once the security is established, some methods of determining acceptable levels of stock is needed. The level of sophistication depends to a large degree on the type of component. At one end of the scale there will be a high-value item used on a product with low sales volumes (e.g. a specially wound electric motor). The customers for this product would probably not expect it to be readily available off the shelf. In this case, the normal level of stock would be zero, the motor only being ordered in response to a customer order.

At the other end of the scale would be commonly used fasteners, required by a whole range of products, and having a relatively low value. Overstocking these components would incur a minimum financial penalty that would be far outweighed by the inconvenience (and potential cost) associated with unplanned shortages. If the use of these is fairly constant, regular shipments from a supplier may be arranged, with a physical check to ensure stocking does not fall below a minimum. This check may simply be to ensure that there is always a spare bin of the items available. Such low-value items are often issued direct to the shop floor; the cost of booking out of stores and allocating to a particular job would probably cost more than the benefits such tight control could offer.

Between these value and volume limits lie those components for which individual decisions need to be made regarding suitable stock levels. Factors that need to be considered are:

Usage:

This is the expected demand for the item. If the demand is regular it may be possible for purchasing to place a large order for the items, thereby securing the best discount. As this would conflict with the objective of minimising the investment in stock, arrangements for regular deliveries are often made, paid for on a monthly basis, but still retaining the quantity discount.

Cost:
The primary reason for wishing to minimise stock levels is the cost of financing goods that are not, while simply stocked, contributing to the company's profitability. Other things being equal, the higher the component cost, the more important it is to keep the stocks low.

Leadtime:
This is the time taken from placement of order to receipt of goods. This leadtime may be greater than the maximum that customers will accept for products. An example might be a coil of stainless steel (to a specification normally only produced for our usage) that would need to be ordered direct from the steel mill. The leadtime for such an item may be several months.

Order quantity:
In some cases there will be a practical minimum order quantity. The specially rolled coil of steel, for example, would need to exceed a minimum weight for the mill to be interested in supplying the order. A batch of investment castings may also have a minimum economic batch size. In this case it may be possible to keep in-house stocks low by arranging for regular deliveries from the foundry. (The true effect of this will be to increase the stockholding at the foundry; if it does not, the supply may be at risk of becoming less than regular.)

A company will probably make individual decisions for the very-high-value items, and for the remainder use a simple formula.

Always remember the functional purpose of storing materials and components. It is simply to ensure the manufacturing and assembly operations can proceed without delays caused by stock shortages. This purpose is not achieved without cost. Keeping the stocks to an absolute minimum is the best way of value-engineering this function. An analogy might be found in today's Formula 1 racing cars, where the car's weight is a critical factor in determining its competitiveness. Saving 1kg on a top car is worth about 0.1 seconds off the lap time on a typical circuit. The fuel carried can be compared with the stock levels in a company. Its regular supply is essential for the engine to produce power. Over-high stocks of fuel will simply render the car non-competitive, whereas a shortage will just as surely prevent the car from winning. (Also see Chapter 17, section 17.4.)

16.4.4 Material stock movements

Earlier in the chapter, when discussing the primary tasks of an organisation, we included component manufacture as one of the tasks in the manufacturing area. Figure 16.2 summarised the process for a simple component. This summary includes some seven material movements, from the raw-materials store to the finished-goods store. Each of these movements has a cost in terms of both time and money. It must cost more to move a component than simply to store it (and we have already concluded that storage costs must be minimised), so it follows in our search for efficiency that the movements should also be minimised. The number of movements will be determined by a variety of factors:

Original design:
This will define tolerance and surface finish requirements, which will in turn call on the planning sheets for particular operations. For example, increasing the surface finish quality may add a grinding operation. Reducing the requirement slightly to that achievable on the previous machining operation could save the grinding operation, together with the material movements.

Factory layout:
Many factories group like machines or operations together. If all the lathes are together it is often possible for one operator to control two (or more) adjacent machines. Possibly one machine setter will be able to cater for a number of like machines. The effect of a breakdown on one machine out of possibly a dozen may be minimised by careful rescheduling by the foreman. Welding

can be a dirty operation, with shielding required to prevent eye damage, so it could make sense to group all types of welding together. Although these are all valid points, a disadvantage can be an over-high level of component movements.

If components can be categorised into groups or families requiring similar manufacturing processes it should be possible to create manufacturing cells, within which the components could be completed. Within the cell, components can move directly from one machine to the next, eliminating the need for some of the intermediate stores. This is often called *group technology*. The advent of the computer is making this form of working more popular, as we shall see later.

Material form:

Some designs can be produced via different, but equally valid, manufacturing processes. One company producing centrifugal pumps fabricated the main body component from four items: a pressing and three castings. The operations involved pressing, machining, welding and polishing. A total of 23 operations were needed and the total distance travelled by the various items amounted to more than a mile within the factory! The time taken to process this item was some five months. As an alternative, an investment casting was introduced. This required only four machining operations, with a consequent dramatic reduction in the leadtime. The point here is not that one material format is better than another, only that it may have a significant effect on the number of operations and movements required.

16.5 Computer aided

16.5.1 Background

Here, let us review briefly the points covered in this chapter. We have introduced the main overall tasks that need to be performed in an organisation designing, producing and selling products. This led to the need for individual sections or departments with their own responsibilities. Consequently, a complex system of information flows is required between the various departments. Finally, for the system to operate as a whole, performance measurements need to be taken, analysed, and appropriate action taken.

Prior to the advent of the microchip, many innovative techniques were introduced to control systems. Henry Ford, for example, by his offer of 'any colour – so long as it's black' on the Model T Ford managed to simplify the manufacturing system and thus increase his level of control. His moving assembly lines, which brought the vehicle to the Stores rather than the components to the vehicle, simplified stock movements, and introduced a visible means of control. For instance, if the stock of wheels was running low it would be immediately obvious.

However, the mass of information-processing required put an upper limit on the effectiveness of such innovations. Even in the over-simplified model of an organisation that we are using it should be apparent that a great deal of effort would be needed to gather, let alone analyse, the performance measurements. Even ignoring the effort required, the time taken to do the job manually would be such that, by the time a decision was made, the situation would probably have changed, thus devaluing the process considerably. The advent of the computer has had a dramatic effect, particularly on the time taken to analyse such data.

In fact, the influence of the microchip has been so great that today it would be almost impossible for a manufacturing operation of any size to survive among the competition without some help from the computer. In other words, if one of your competitors is effective in employing the power of the computer to help run his or her business, you will need to do likewise if your business is to survive.

In the 1960s a number of mainly large organisations could see the benefits, in terms of their 'number crunching' ability, of computers for handling simple but repetitive tasks such as wages. At this time the only computers were large, very expensive and required staff with specialised knowledge to operate them. This remained the case during the 1970s, when a number of organisations also used the computers for technical work such as stress analysis. During this period the cost of computing power and memory reduced each year at a seemingly ever increasing rate (and has continued to do so ever since) to the point that the early 1980s saw the introduction of low-cost desk-top computers in companies, schools and homes. By the 1990s the power and memory of a fairly standard desk-top machine exceeded the performance of the large mainframes of 20 years earlier!

We will examine how computer systems have influenced two main areas: product development and manufacturing control.

16.5.2 Product development

At various points in the text we have covered the process of product development, from the initial specification (PDS), through evolving a concept, evaluating the concept with an engineering model, producing the detail design, producing and testing prototypes, and creating a manufacturing system capable of producing the products repeatedly.

During this process consideration must be given to functionality, cost and ease of manufacture. A steady flow of information is needed regarding manufacturing capability, material costs, availability of material form, common raw material, etc. Constant liaison is required between all areas in the system – designers, buyers, production engineers, manufacturing management, and so on.

Now consider those tasks that a computer excels at. It can:

- perform calculations at great speed
- store and retrieve information very quickly
- display information on the screen
- display information on hard copy.

How can this be used to help?

Early developments

An engineering drawing is an unambiguous representation of an artefact on a 2D surface. As it is unambiguous, the start and finish points for each line can be defined by coordinates. The combination of the realisation of this with the development of electronic processing power, and the emergence of suitable display hardware, led to 'Sketchpad' (in effect an early drawing system, i.e. CAD – Computer Aided Drawing) produced by Ivan Sutherland at MIT in 1962. At around the same time, a team at General Motors Research Laboratories was developing a system which not only displayed shapes on a screen, but also linked this information to NC controlled machines. This led in 1964 to the construction of the first CADCAM (Computer Aided Design and Computer Aided Manufacture) system called DAC1 (Design Augmented by Computer). So even at this early stage of the evolution of computer-aided systems it was recognised that the design and manufacture functions could be linked.

The development of the early systems tended to be directed towards the requirements of particular applications in those large capital-intensive major industries such as aircraft, motor, space, and defence, who could afford the high costs involved. A typical need would be for a system that would generate complex shapes and surfaces able to satisfy the requirements of aerodynamics, structural strength and visual appeal, which would often be in conflict with each other. The size of theses industries and the competitive environment in which they operated meant that they were prepared to invest in what were then highly expensive computer facilities. In many cases the software produced, whilst appropriate for its designed task, was severely limited in terms of flexibility and adaptability to other tasks.

The potential benefits available from an integrated design and manufacturing system were readily recognised by the increasing number of computer-literate companies. However, as the following example illustrates, blazing a trail with a new technology has its fair share of potential pitfalls.

One such computer-literate organisation was concerned with supplying equipment for the various process industries. In the 1970s work had been done on a computer-aided system for generating pipework layouts, so the potential benefits of such systems were well known to this company. Around 1980 another type of heat exchanger was to be added to the range: a spiral heat exchanger. Each one was a bespoke design with a limited number of variables, namely:

- *material*
- *sheet thickness*
- *sheet width*
- *sheet spacing*
- *stud spacing*
- *stud size*
- *number of turns*
- *position and type of connections.*

The software was written by a number of graduates. It could produce

- *A preliminary design for costing/quotation*
- *A fully detailed design*
- *Complete set of working drawings*
- *Routing sheets*
- *Purchasing information*
- *NC program for studding machine (The studding machine was a special-purpose piece of equipment that projection welded studs onto a metal sheet in a pattern specific to a particular design. The machine was computer controlled and responded to the NC program defining the required pattern.)*

After some initial teething problems the system was capable of generating all of the above. Although this sounds good, it was in effect a Computer Aided Disaster. Why?

The design and manufacture of these heat exchangers had been undertaken by the company some 20 years earlier, then sold to a Japanese company, from whom they bought back the process in 1980. Despite their technical reputation the Japanese had not progressed the design very much, but this was not appreciated by those who repurchased the system. The CADCAM system was set up right from the start incorporating most of the Japanese methods of design and manufacture.

When put into practice in the UK, production times were too long and costs were too high. The product was re-engineered, involving changes both to the design and manufacture methods. The changes gradually had the effect of making the product a profit earner, but they could not easily be incorporated into the CADCAM system. Alterations could only be made by computer programmers. As these took time, the CAD drawings had to be passed to a manual drawing office for modification, which at times seemed to take as long as starting the drawing from scratch. Errors in the costings caused by the changes increased the cost of introduction of this product to the extent that for a period its future was in doubt.

The first problem was that this early type of CAD was too inflexible; it could not be modified easily.

The second problem was that the first problem was not at the time appreciated: that sort of CADCAM system should not have been introduced until the product had been developed fully.

This type of CADCAM could really be considered to be semi-dedicated electronic tooling.

As with any new technology, those who blaze an advanced trail tend to discover the limitations of at least the early versions of the technology. The complexity of the software involved resulted in progress being made, often independently, in individual areas. We will start by looking at some of these elements.

CAD (Computer Aided Design)

Many organisations, particularly during the 1980s, introduced CAD systems, often with the simple initial aim of improving productivity in the drawing office. Over-optimistic productivity gains of up to 4 or 5 times were advertised. In fact, producing a drawing from scratch on CAD could easily take as long as an experienced draughtsman would take with a drawing board, although adding the dimensions should be quicker on the CAD system. However, this misses the point of CAD. Most drawings contain repeated features, from either another drawing or another part of the existing drawing. Such features might be gear teeth, an assembly containing standard parts such as electric motors, or commonly occurring items such as nuts and bolts. The CAD system can make use of a vast array of highly sophisticated templates. In addition, the drawing layout can easily be standardised, the standard of linework is always constant, microfilming is easier, and a more high-tech image is passed to customers.

These early CAD systems were initially only 2D systems, but as the price of computing power fell the developments included 3D draughting systems, surface modelling systems and eventually programs that could generate solid models. The 3D draughting system could generate vectors in three dimensions and display them as lines or curves on the screen, thus producing an image of the final product looking much as a model would appear generated from pieces of wire (*wire frame model*). The number of lines on the screen could make it difficult to visualise a complex object with this system. A *surface modelling* system enabled surfaces to be created between appropriate vectors. The surfaces could help visualisation, either by being coloured, or by enabling the computer to remove the hidden lines from the display. The next stage in development was the introduction of the *solid modeller* where the computer held information not simply about the vertices of components, but also about which coordinates in space were thin air and which were not. It should be apparent that the quantity of information the computer must hold to generate a solid model is vastly greater than that required for a simple 2D representation.

The D of the early CAD systems really only referred to draughting, but for the more sophisticated systems the D could truly represent design. The level of *visualisation* on today's systems is excellent, rivalling that of true photographs with the better systems. Companies producing consumer products where the aesthetic appeal of the end product is vital for its success will often use a CAD system for this reason alone. Shoes and domestic lawn mowers are often first 'seen' on a CAD system.

The design aspects are not limited to visualisation. FEA (*Finite Element Analysis*) is a technique for analysing the effects of applying loads to components, which, because of the high level of number-crunching required, is an eminently suitable task for the computer. Early systems required the engineer to subdivide the component into a mesh of suitable elements. Data defining the mesh had to be entered into the computer, with details of the loads, so that the computer could perform its calculations. The output tended to be in the format of a vast table of numbers that the engineer would have to interpret into meaningful information. The current systems are usually linked with a drawing or modelling system to make the mesh-generation and interpretation tasks easier. Some systems will create the mesh automatically, and produce an output showing the deflections and areas of different stress levels. However, it must be recognised that the validity of the output is totally dependent on the validity of the model. An automatically generated mesh may need to be modified to avoid errors in the model. Computers have not displaced the skill of the analyst, at least not yet!

When creating assemblies of components that move relative to each other, the designer has to ensure that adequate clearances are allowed to avoid collisions. In some complex three-dimensional systems these can often be difficult to visualise. In the past it was not unknown for an engineer to create a scale model of the system or linkage to confirm the design. Remember, the cheapest time to catch an error is at the drawing stage. Some of the more sophisticated systems are able to offer a *Kinematics* facility where the computer is able to simulate movement of the various components and enable any collisions to be identified. Many of the solid modellers can also be used to identify dimension problems when components are assembled. The current state-of-the art systems will identify collisions arising from errors in the dimensions, but do not yet have the facility to identify easily errors relating from poor tolerancing.

CAM (Computer Aided Manufacture)

CAM is a very wide heading that can include almost any piece of software designed to help the manufacturing process. Here we will for the moment restrict ourselves to the control of machine tools.

Machine tool control: Over the years, various attempts at automating the tool-path-movement controls on machine tools involved a variety of principles, including cams, template copying, and numerically controlled systems (via a plugboard). By the mid-1970s small computers were being added to various types of machine tools (particularly lathes and milling machines) to control features such as tool-path movement, cutting speed, and feed rates. Such CNC (*Computer Numerically Controlled*) machines offered improvements in terms of both rate of production and consistent quality, and reduced changeover time from one job to the next. Although this tended to de-skill the machinist's job, it meant an NC (or CNC) programmer was required to generate the program for the machine. The program when written was used to generate a punched tape which would then be taken to the machine tool, where a tape reader would load the program into the machine's computer (controller). Some companies developed systems where a larger computer would be able to control a number of machine tools in real time. This system, called DNC (*Direct Numerical Control*), had the advantage of eliminating the need to use paper tape to load the individual machine with the appropriate data. However, this system was quickly overtaken by the increasing power of the small computers fitted to the machines. By way of simplifying the information transfer to the individual machines, these controllers can be networked, so that the tool-path programs can be directly down-loaded to the machine, thus avoiding the need for paper tape. This is the current form of DNC (*Distributed Numerical Control*).

Over the years, the computing power of the machine-tool controller increased, and graphics displays were added, which enabled the machine operator to program the machine. Note that there are different schools of thought as to what is expected of a machine operator. The operator who is able to program his/her own machine would be of great benefit in a prototype area, or at least an environment involving relatively short production runs. However, a high-volume production area might wish to employ a less skilled (and hence cheaper) machine-minder to load and unload the machines, whilst using those with the programming skills to concentrate on their areas of expertise. The power of the machine-tool controllers has increased, so that today it is quite common to be able to create the component geometry, add the machine-tool-path movements and even check for possible collisions. The software can run in real time and thus give the production time for the component, before any metal is cut.

The developments in the machine-tool controllers were largely pursued by the machine-tool manufacturers, with very little liaison with the CAD creators, as the controllers were generally perceived as part of the ever-increasing sophistication of the machine tool, rather than as part of an overall integrated system. The CAD creators recognised that once the geometry of the part was computerised this could be used to generate the tool-path movements (after all, this had been demonstrated in 1964). Many systems therefore added a further machining element that took the component geometry and enabled tool-path movements to be created. They also recognised that other manufacturing operations, such as sheet metal work, could benefit from the CAD system calculating the developed shape, and even providing the commands for a CNC machine. However, in many cases, these facilities were simply added in the form of an additional module.

Integration

One of the limitations of the early CAD systems was their lack of integration between the various elements: 2D geometry, surfacing, solid modelling, FEA, kinematics and machining almost invariably appeared to the user as separate pieces of software that required some dexterity on the part of either the software or the user to transfer the data from one to the other. In many cases the transfer of information was only possible in one direction.

By the early 1990s the leading software companies were offering products that greatly smoothed the transition from one module to the next. These systems frequently use the solid model as their information base. Any 2D drawings are created from the solid. The generation of three different views – front, plan and side elevation – can be automatic. The views are said to exhibit associativity, i.e. when one is altered the others are updated automatically to remain compatible. The system can often be set so that such modification will also update the solid model. In fact, one company offered what it termed bi-directional associativity, where the component geometry could be moved between all the modules and modified within any module, the modifications being reflected automatically in the other modules. This is a very powerful facility that needs to be handled with care and is therefore equipped with various levels of protection to prevent unauthorised amendments.

This sophisticated integration falters somewhat when the information is transferred to the machine-tool controller. Generally, a *post processor* is used to convert the CAD-generated tool-path movements into a format understood by the machine controller. Unfortunately, a different post processor is needed for each machine tool and controller combination. Generic post processors are available to help with this problem. They offer a simple menu system to create the program automatically. However, their disadvantage is that they may be unable to take full advantage of the machine tool's capabilities. This is because the latest generation of machine tools uses very sophisticated combinations of operations that have resulted from a mix of mechanical and software developments. The generic post processors tend to cover only standard common operations.

The problems of integration were further complicated by the fact that some early users of 2D CAD had focused initially on a local need to improve efficiency in the drawing office. One significant difference between a CAD drawing and a manual drawing is that the dimensions in the CAD system are represented by the geometry. Conventional dimensions will be added to the CAD drawing and will, of course, be used by those reading the drawing. However, if the drawing geometry is used to generate tool-path movements within the CAD system, the sizes taken will be those from the geometry and not the dimensions. The problem arises when for expediency the draughtsman, when modifying a CAD drawing, has simply altered the dimensions, leaving the geometry unchanged. This perfectly acceptable technique for manual drawings was satisfactory while the CAD system was only used to produce drawings. Subsequent later uprating to a more sophisticated system can therefore be overly complicated unless possible future upgrades are considered at an early stage.

16.5.3 Manufacturing control

Stand alone systems

Much of the early development of software systems was undertaken by those large organisations able to afford the then very expensive mainframe computer systems and the teams of computer programmers that were necessary for development and maintenance of the software. In general, most organisations looked on the computer system as a means of solving particular problems in the existing organisation. As the price of computing power reduced, more organisations became convinced of its potential benefits, each being eager to implement systems to solve particular problems with minimum delay. A number of computer programmers recognised the need for standard software that could be tailored to meet a specific company's needs, so the software house was born. While it is true to say that in the early days of computing a potential user would start by contacting the hardware supplier, who would recommend a suitable combination of 'black boxes', the software then being written around the hardware capabilities. Today, most potential users would start with a problem, select the most appropriate software, and then choose a suitable hardware platform. We will now look at some of these individual problems that the computer can help to solve. As there was very little integration initially between the different packages, they are termed 'stand alone' systems.

Time and attendance

As we mentioned earlier (section 16.4.2), one of the elements that needs to be controlled is direct labour. Those in control need to know any problem areas so that corrective action may be taken. Prior to the advent of computers the operator would complete a card recording how his or her time had been spent, i.e. working, waiting for material, waiting for a machine to be repaired, etc. The cards at the end of a shift would be collated by a clerk and the results submitted to the foreman. The section manager would probably see a weekly summary defining the efficiency levels achieved. Many of the shop-floor workers would clock in (i.e. have their card stamped) at the start and end of each shift. These cards would be collected and the stamped times recorded by a clerk. If they were paid on an hourly basis, this information would be needed by the wages department to make up the payslips. It soon became apparent that this procedure could be automated by getting them to 'clock in' with a particular code number, not simply at the start and finish of the shift, but also at the start and finish of individual tasks. The same information could be used to create both the payslips and the efficiency reports.

Simulation and scheduling

It is obviously important for work to be scheduled in such a manner as to use the available resources in the most efficient way. Devising a suitable schedule for possibly a wide range of differing components would be likely to prove a time-consuming operation. The foreman of a particular area – say, a machine shop with 15 operators looking after 20 machines – will have at the start of his or her shift a target set of work to complete. It is not uncommon for the foreman to be left to assign the work to particular individuals under his/her control; his knowledge of the components, machines and people concerned will normally mean that he or she is the best person to make the assignments. Some organisations use a simulation package for the foreman to investigate some 'what if' situations. In this case, the simulation package will previously have been set up to simulate his or her work area, possibly in terms of layout, material movements, and product processes. The data for the current situation will be entered by the foreman, and the program will simulate, say, one shift's work, giving a range of statistics such as times of completion of each batch, machine utilisation, etc. This type of program is particularly helpful in simulating the effect of a machine breakdown by predicting the consequences of alternative courses of action. Such a package is used in this way by one of the UK car manufacturers.

A possibly more common use for such a package is at the planning stage for a shop-floor layout where the effects of different levels of automation, numbers of machines, and staffing levels can all be analysed. As everyday examples, such systems are used to plan the number of checkouts needed in large supermarkets, or the layout of the passenger-handling areas of airports.

CAPP (Computer Aided Process Planning)

We have already come across the planning sheet as one of the elements in the manufacturing package (drawings and parts lists are the other two) sent from *engineering* into the manufacturing system. These sheets detailing the operations required and the time taken for each are normally completed by planning engineers. However, if Group Technology is employed there is scope for automating this process.

Group Technology involves examining the components manufactured and wherever possible grouping them together in families of like components. This should, of course, be done at the design stage (remember the importance of the part number in identifying component types). A whole series of benefits flow from such a system, including better use of common components, potential for limiting the range of raw material stocked, more efficient use of CADCAM via parametrics, and the possibility of using CAPP. In addition to these, the chance of setting up manufacturing cells (see section 16.4.4) can offer dramatic benefits in terms of simplified work control, reduced component movements and reduced leadtimes, and thus reduced levels of WIP. Within CAD a *parametric* is a program that will allow an engineer quickly to define a component that is a member of a family. For example, a parametric defining a family of shafts would ask the user a series of questions, such as: diameter? length? threaded end? length of thread? etc. The freedom of choice given for the answers will be depend on how the parametric was written. On completing the answers, the CAD system will create a solid model drawing of the component. It could also quite easily automatically generate the tool-path movements for the turning machine. In the same way, it could also generate the planning sheet. Remember, it is not actually intelligent enough to create the planning sheet (or tool-path movements) from scratch; it will simply use the parameters that were supplied by the engineer to produce the planning sheet.

Normally, once planning sheets have been generated by CAPP they would be checked and edited as appropriate before being issued straight to the shop-floor. This check would only be eliminated after a suitable period of validation of the system.

MRP (Materials Requirements Planning)

In section 16.4.3 we identified the importance of controlling stock, particularly in avoiding excessive levels without creating shortages. MRP offers a form of control for all types of stock or inventory, including part-completed work, WIP (*work in progress*). The input to an MRP system is the *master schedule*, which is a prediction of future demand for all items that will be required as an output from the manufacturing system. The MRP program will have details of the breakdown of these items into their individual components (i.e. the *bill of materials*), and will calculate the numbers of each component required, complete with its due date. Prior to the introduction of such a system, stock control tended to rely on predictions of annual usage, which in many cases were based on the previous year's usage. Control of inventory in this way is a little like steering a ship by referring to its wake! MRP actively encourages (indeed requires) a sales forecast. The program is run at regular intervals, possibly weekly or fortnightly, when amendments to the schedule can be made.

It did not take long for those involved with such programs to realise that if the *bill of material* could be used to break down the final products into their component parts, then in a similar way the planning sheets could also be used to define the usage requirements of labour and machines. In fact, with sufficient information it would be able to define all the manufacturing resources. The next stage of computing assistance was called MRP (*Manufacturing Resource Planning*). These systems cover the scheduling of production orders (from manufacture of the individual components to despatch), and the release of purchase orders. It can also be used to monitor the status of stock, WIP and various performance indicators such as machine down time and levels of scrap. These systems have the inherent limitations of any planning system in that they are run in a batch mode. That is, the program will be run on possibly a weekly basis and will produce a plan for the next week. It will not take account of – and replan to accommodate – changes that occur during that week. Later versions, called MRPII added the facility of feedback where some such changes could be accommodated.

There is a danger that some may see such computer aids as the panacea to solve all problems. This is not the case: all these aids can do is to offer advice, much in the way that traffic lights control traffic. They work really well provided that all who use them stop when the light is red and go when it is green. If all operations within the manufacturing system were performed as the planning scheme expected then the results would be as predicted. There are many well-documented cases of so-called failures of MRP systems where the failures are not due to any direct fault in the software, but lie in the implementation. Adequate training is vital for all who will use the system, so that they appreciate that any errors fed into the system will simply degrade the quality of the output. The poor output will not be the result of a computer error, just a reflection of the human error.

A manufacturing system is an immensely complicated operation that is forever changing in detail. Even with today's sophisticated software coupled with some very powerful computers, the perfect system for controlling resources is still a dream for the future.

16.5.4 Computer Integrated Manufacture (CIM)

The next logical step for the computer in advancing its help for our organisation is in linking all the different functions together. We have already seen in the *engineering* area that it is possible to use the geometry created at the design stage to control the tool-path movements on an appropriate machine tool, so it should be possible to extend this philosophy further. The assembly drawing will have a parts list generated at the drawing stage. The MRP system will require a breakdown of parts for each assembly. Rather than key in this information a second time, a truly integrated system will allow this information to be passed from the CAD directly to the MRP system. We are now creating a *common database*, i.e. a single source of information that all functions in the system will work from. This has immense potential advantages. Errors caused by copying information manually from one system to another will be eliminated, time wasted through this copying process will be saved, problems arising from different issues of information in circulation at the same time will be eliminated, and so on. The overall purpose of a CIM is to convert product ideas into goods that can be supplied to the customer in the shortest possible time with minimum cost.

The type of organisation likely to embrace the concepts of CIM is almost certain to be computer literate, and will therefore have installed a number of existing systems. The immediate practical problem with CIM is that the systems need to be interfaced, to allow the required flow of information. As they will have been installed at different times, communication between them will be either costly or impossible. In an attempt to deal with this problem, General Motors, as one of the world's largest users of computers in manufacturing, initiated a specification defining a network interface for all such equipment. This was called MAP (*Manufacturing Automation Protocol*). MAP uses a seven-layer reference model to define the protocols used to pass the data to and from the network connecting the computers. Level 1 defines the physical connection to a coaxial cable, while level 7 covers the communication with the application program. The intermediate levels cover the protocols for restructuring the data to a format recognised by the network, adding the address, verifying that the information has been transferred, routing the data between networks, and routing the data within a local network.

The principles behind CIM were heavily promoted during the early to mid-1980s, but by the mid-1990s very few factories have truly embraced the concept. This should not be seen as an adverse judgement on CIM; simply that the advances in computing capabilities have been so great that expectations have in some cases run a little in front of reality.

16.6 Summary

The reader should at this point have a global understanding of the main tasks that need to be performed in an organisation designing and manufacturing products. For simplicity we have concentrated on the primary tasks, but the reader should also appreciate that there are other important elements: technical publications, and servicing, to name but two.

Even an organisation of a modest size will require a fairly comprehensive system for distributing information. The reader should appreciate that there is a need both for the information and for some means of controlling it if the organisation is to be in any way efficient. The controls are geared to a single central aim: that of enabling the organisation to generate products in the most sustainable cost-effective manner.

The advent of the computer has raised the stakes in terms of the potential control that can be maintained over an organisation. Although there are few organisations using the full benefits of computerisation, it is very unlikely that any company designing and manufacturing products can currently do so without the aid of computers in some form. The reader should recognise that this chapter has introduced a number of areas where the computer can help, but also that it is only an aid. People are still involved in the system and are likely to be needed for the foreseeable future.

16.7 Questions

Now use the following questions to check your understanding of the chapter.

1. List the primary tasks for an organisation that is developing, producing and selling products. (16.2)
2. What factors should be taken into account when prioritising engineering changes? (16.4.1)
3. If CAD is considered simply as a replacement for manual draughting, what advantages could such a system offer? Do some of the advantages depend on the types of drawing? (16.5.2)
4. What is the first input to an MRP system and what does the system attempt to control? (16.5.3)
5. What do you understand by the term 'bi-directional associativity' in relation to CAD? What are some of the advantages and dangers of such a facility? (16.5.4)
6. Name four factors that need to be considered in controlling levels of stock. (16.4.3)
7. Engineering changes can be introduced for a variety of reasons. Name six. (16.3.3)
8. Sketch a flow diagram showing typical flows of information when a quotation for a special order is accepted by a customer. (16.3.2)
9. What do understand by the term 'common database'? Describe some of the benefits that might arise from having such a database. (16.5.4)

17 The Business Context (Organisation and Costing)

17.1 Introduction

In Chapter 16 we looked at the primary tasks and the resulting information flows that needed to be recognised for the efficient running of an organisation designing and manufacturing products. There are, however, a number of further areas that need to be introduced which are essential to the continuing survival of our organisation.

17.2 Basics of costing

The long-term survival of any commercially based organisation is dependent on its ability to generate sufficient funds to cover not only the day-to-day needs but also to provide for the future – new products, plant, equipment, training – and, finally, to satisfy the aspirations of the shareholders. Obviously these funds have to be generated on a regular and sustainable basis. In order to achieve this happy state, the organisation must cover its costs and produce an appropriate level of profit. Tax laws will have specific rules that define closely terms such as 'profit', 'income' and 'outgoings'. These definitions may vary slightly between different countries and from year to year, but the essence of the basic definitions that we shall use should be universal.

17.2.1 Profit

This we will define as the difference between the income and the outgoings.

17.2.2 Income

Income is the money coming into the company. In practice this can arise from a number of sources: sales of capital assets (e.g. unwanted company buildings), from the stock market via a share issue, or even via government grants, to name but three. However, remembering the need for this income to be sustainable in the long term, for 'our' company we will assume that this will be generated from the revenue arising from the sales of our products widgets. This will obviously depend on the selling price and the number sold. The selling price or NSV (*net sales value*) is the actual price received by 'our' company; it is not the same as the list price. Many manufacturing companies sell their products through third parties such as retail outlets. The list price in the sales brochure for a new car may be the price that the car dealer will quote you, but (ignoring taxes) this certainly does not represent the income to the car manufacturer. He will have sold the car to the dealer at a discount, so that he, the dealer, may remain in business to sell even more cars. This discount may not be the same for all dealers; the manufacturer may operate an incentive scheme whereby the more cars sold, the greater the discount. Even if our company only sells one version of the world-renowned widget, it is likely to be sold for different amounts to different customers. The NSV therefore represents the average actual sales revenue from each widget. Hence

$$\textit{Income} = \textit{Sales Volume} \times \textit{Net Sales Value}$$

For most products there is some relationship between sales volume and net sales value – increasing the selling price will tend to reduce sales, and vice versa. The aim is to maximise their product (sales volume and net sales value), the income, on a sustainable basis. Note that the selling price will be determined by the strength of the product in the market place, i.e. its value as seen by the customer – the *perceived value*. It is almost always independent of the manufacturing cost. (In the past, some government contracts were issued on a cost plus basis, whereby the customer paid the producer's costs plus an element for profit. These are now the exception rather than the rule, the supplier having to quote a selling price at the start of the contract.)

Now, how can we as engineers involved in the design and manufacturing processes improve the income? Generally, the most we can do is to ensure the perceived value is as high as possible, by making sure at the PDS stage that all the necessary functional, aesthetic, reliability and any other criteria are included, and are fully met at the subsequent stages in the product introduction process. Note that this must, of course, be done whilst still meeting the cost constraints of the specification.

17.2.3 Outgoings

These represent all the costs incurred in running the business. This will include the wages bill, all material costs, buildings and their associated charges (rent, rates, heating, lighting, maintenance, etc.), plant and machinery purchases, consumable tools, any taxes due; in fact, the total of all of the costs. In simple terms, over a given time period the outgoings will represent the cost of producing those products that were manufactured over that period. So, if a company producing a single range of products produces 200 000 widgets over, say, a twelve-month period, and incurs outgoings during this period of some £1 million, the total cost of each widget will have been £5. Now, as engineers making decisions that will have a direct effect on the outgoings (such as the design of the widget, the type and quantity of materials used, the raw-material format, the manufacturing processes, the types of special tools, jigs and fixtures, the factory layout, and so on) we should be in a strong position to influence the manufacturing cost of each widget. To do this effectively we need to know how each feature of the design, and each element of the manufacturing process, contributes to the overall cost of £5. An understanding of how the costs arise and a form of costing system are needed.

17.2.4 Fixed and variable costs

The outgoings can be thought of as the sum of *fixed costs* and *variable costs*. In simple terms, variable costs are those that vary with the production rate, whereas fixed costs are those that do not. For example, the rates, rent and heating costs for a building could be termed fixed costs, but the costs associated with material purchases, and direct labour costs could be considered to be variable costs. (Note that, in practice, the demarcation lines between fixed and variable costs can vary according to circumstances.) For the business to be run efficiently there must be continual pressure to reduce all types of costs. The reason for dividing the costs into these two categories is simply to focus the mind on those costs that can be influenced by engineering decisions.

As an example, consider a factory with a surplus of orders and operating at full capacity. It would make sense to purchase a power tool, so that the direct labour content on assembling several components is reduced, thus allowing the operator to build more assemblies during the shift. Provided the cost of the power tool was less than that saved on the direct labour cost (over a defined period) this would obviously be a sensible decision. In this case the direct labour cost is a variable cost.

Now consider a situation where, in the boom times, in order to attract direct labour the company puts the total workforce on a monthly paid staff basis, and guarantees to pay its staff full pay for eight weeks in the event of any layoffs caused by shortage of work. Sadly, the world falls into a recession, orders fall off, and the factory is no longer working at full capacity. Pressure is on the engineers to reduce the outgoings. Reducing the direct labour content by providing power tools no longer looks a sensible option. Why? Because, at least in the short term, direct labour is no longer a variable cost; the benevolent company agreement made in the good times has converted it effectively into a fixed cost. In these circumstances, perhaps the engineering effort would be better directed towards either improving the product (and thereby generating more sales volume), or possibly reducing the material content, which is still a variable cost.

The reader should not, however, gain the impression that the engineer is totally helpless against the fixed costs, because in the long term all costs become variable and can therefore be reduced. As an extreme example, a company could be taken over by a competitor, its staff made redundant, buildings, plant, etc. sold and the manufacture of the products transferred to the predator company's facilities to reduce its over-capacity problems. Thankfully such examples are not commonplace.

At this point we should remind ourselves of the concept of *direct* and *indirect* costs. A cost that can be directly attributed to a specific product is called a direct cost. Direct costs would, for example, include the material costs and the labour costs of the production and assembly workers. The costs associated with employing the cleaner and the managing director would be termed indirect costs, as they could not be attributable directly to a specific product. The costs of running the design office would normally also be termed indirect costs. However, if the company was producing bespoke products (i.e. specifically designed for a particular customer's needs) and the designer's time spent on that particular project was recorded, this would come into the direct cost category.

The division into direct and indirect simply identifies those costs that can be attributed directly to a product. This is different from the division between fixed and variable costs, which identifies those costs that can be altered in the short term.

Remember, the point of dividing costs into fixed, variable, direct or indirect categories is simply to highlight those costs that can be influenced by engineering decisions. Engineers within a design and manufacturing environment will undertake a variety of tasks, all of which will be influenced by cost to some degree. In the next section we will look at three such tasks and examine the effects of costs.

a. Detail changes to an existing product or process
b. Introducing a new process or piece of plant
c. Introducing a new product.

17.3 Applications

17.3.1 Changes to an existing product

Changes may be made for a variety of reasons (in response to competition, to overcome a manufacturing difficulty, to compensate for change of supplier, to save cost, etc.), but without doubt, in all cases the engineers will be under pressure to minimise the cost of the end product. To allow this to be done there needs to be a system that relates costs to the component being manufactured, its features, and the manufacturing processes involved. Standard costing is one such system.

Standard costing

This system divides the component costs into three elements: *materials*, *labour* and *overheads*. It assumes that each component contains a certain amount of material, and requires a known amount of labour time for its manufacture. To operate such a system, a means of establishing a cost rate for both materials and labour time must first be established. Different organisations will vary in the detail of their cost rates, but would be broadly similar to the following.

Materials costs: Take a component requiring an electric motor. If the number required is high enough, these may be ordered on an annual basis, and delivered in batches weekly, directly to the assembly line. The price may be renegotiated each time a new order is placed. In this case, assuming the standard cost rate is updated annually it would seem quite reasonable to set the standard cost for the motor to be the cost price for the forthcoming year. During the year the standard cost and the actual cost would therefore be the same.

Now consider a different component, requiring a pressing made from a type of stainless steel not readily available from the stockist. The material has to be ordered as a coil direct from the mill and has a variable delivery time of between three and six months. The component is strategically important to the business, so a decision has been made to stock it. One coil will provide enough material to satisfy production requirements for eighteen months. In this case, determining a standard cost is not so easy. The coil will be stored (i.e. it will use valuable floor space); it will be held in the raw-material stores, an area that needs to be manned so that material can be cut and issued to particular jobs, and appropriate records kept. The coil will be paid for well in advance of the income generated by many of those components that will be manufactured from it. In this instance, to be realistic, the standard cost should take some account of the original cost, the storage costs, the financing costs, and possibly any price inflation of the raw material. (This last element was added to avoid too sharp a price change when the next coil was ordered.)

These two components are perhaps at opposite extremes, but they do illustrate that the material cost may well not be the same as the actual price paid. In practice, most organisations would find it too time-consuming to have a separate set of rules for calculating the standard cost of each type of material. One simple rule adopted by some companies is to add a fixed percentage to the purchase price for each item stocked.

So far we have only discussed the costs of the material. Obviously the quantity required by each component must be known. Of course, this is precisely known in the electric motor example above, but for raw materials cutting allowances must be given. The standard cost would therefore normally include the maximum acceptable wastage.

Labour costs: The planning sheets will normally define the standard time allowed for each operation in the manufacture of the component. This is the time that one could reasonably expect a worker to take when maintaining a work rate that could be sustained throughout a shift. It will include allowances for relaxation, which may vary according to the demands of the task. Many companies will have comprehensive sets of rules as to the allowances appropriate for different tasks, often created in consultation with the relevant trade union.

The times used would usually be generated by the *manufacturing engineers* and will normally consist of two elements, a set-up time and a run time. The set-up time will include those tasks necessary to change from producing one component to another. This could involve changing work-holding devices, cutting tools, and the NC program on a turning machine, or for a hand operation collecting work sheet from the charge hand, obtaining material from the stores, and collecting appropriate drills from the tool stores. These are the jobs that need to be carried out, normally only once, at the start of each batch of components.

The run time, on the other hand, is the time taken to perform the operation on each component. For an item on a turning machine this would include the loading time, the time the machine takes to complete the operation and the time to unload the component. The total time for each component could be said to be the run time plus a proportion of the set-up time. That is,

Operation time per component
= (set-up time)/(batch size) + run time

From this it is apparent that as the batch size increases the set-up time becomes less significant. This is the reason why for many years manufacturing engineers concentrated their efforts on reducing the run time, largely ignoring the time taken to set the job up. To ensure that the set-up time per component is kept within acceptable bounds, many organisations make sure that small, uneconomic batch sizes are avoided by calculating an EBQ (*economic batch quantity*). This is discussed further in section 17.4 of this chapter.

The run times for automatic machines are easily calculated, but for manual operations the manufacturing engineer will rely on previous experience and practice within the organisation. To help the engineer, tables of standard times for defined operations are available and can be used to build up an estimate for the whole operation. These times are called *synthetics*. They are of particular use in estimating the times for operations or techniques new to the organisation. Many large companies employ specialists such as *Industrial Engineers* (sometimes called *Work Study* or *Time Study Engineers*) who analyse all aspects of the workplace environment with a view to reducing the time taken to perform the various tasks. Their work may involve examining the flow of material from one work centre to another and introducing modifications that will simplify the flow. It may cover a detailed study of an assembly operation, where the number of hand movements, the location of components, and the use of assembly aids are all closely examined. Times will be allocated to each element of the assembly operation, and modifications introduced to reduce these times.

Overhead costs: Although we have accounted for the direct costs associated with our component, there are many other costs within the organisation which we have so far ignored. These costs, called overheads, will include rent, taxes, heating, lighting, costs of indirect labour, etc. If we were not producing any products, the company would close and these costs would cease to exist, so it follows that they only arise because we are producing products. The problem is how to allocate them to individual products.

The traditional way to allocate overheads is to identify those costs which are associated with specific processes or machines on the shop floor. For example, it may be possible to identify for a particular machine the costs associated with depreciation, power consumption and servicing. These can be considered as directly attributable costs. The remaining costs are summed and divided by the floor area of the manufacturing section to derive a cost per unit area. Then by sharing the total floor area amongst the machines or work-centres, and adding the directly attributable costs to the appropriate floor charge a figure for the annual overhead cost for each machine or work centre in the factory can be calculated. Note that this is not the actual cost for running that area of the factory, simply a means of attempting to spread the overheads in an equitable manner. Although this principle is frequently used, the detail and the sophistication of the allocation method can vary widely between different organisations.

Identifying the annual charge for each area is only the first step. The accountant will then, from the sales forecast and the planning sheets, estimate the number of direct labour hours likely to be required for each of the work centres. Simply dividing the annual overhead allocation for a work centre by the estimated hours for that work centre will yield an hourly rate for overhead allocation.

We now have the basis for the standard cost of our product. It will be the material cost plus, for each operation, the direct hours × (direct labour rate + overhead rate). Note that some operations may need to performed on a sub-contract basis. For example, if there is no on-site facility for heat treatment, this will need to be subcontracted. As far as costing is concerned, the associated charges will be categorised as material costs.

Standard costing has a number of positive benefits. It was developed originally to provide a yardstick by which day-to-day performance could be monitored. The basic standard cost would be determined on an annual basis and would remain fixed until the next annual review. Obviously, throughout the year some of the costs will change – a price rise from a supplier, for example. The difference between the actual cost and the standard cost is termed a *cost variance*. This variance could be positive or negative and could apply to any of the elements of the cost. If, for example, a particular work centre was consistently showing a negative cost variance, this would highlight an area with a problem needing attention.

The system is liked by accountants, largely because it provides a cost for individual components which appears correct, because at the end of any period the total of the standard costs, taking account of any cost variances, for all the components produced will equal costs incurred in running the company. The books balance.

Remember that standard costing was developed largely to provide a means of monitoring day-to-day performance by cost accountants; it was not designed to help the engineer in making decisions. Despite this, it is often the only system available to the engineer, and is even used to judge the success of some engineering decisions.

Standard costing works reasonably well if:

(a) The organisation is working at full capacity, and
(b) There is a reasonably high labour content in the product.

Standard costing can lead to poor decisions if:

(a) The organisation has significant spare capacity, and
(b) There is a low labour content: for example, if the manufacturing process is highly automated.

This is best demonstrated with an example.[1] Year 1 is a boom year, the company is working two shifts and has a full order book. A new machine has been installed, and the accountant works out the overhead rate as follows:

Annual overhead allocation £60 000
Anticipated annual usage 3000 hours
Overhead rate (60 000/3000) = £20 per hour

Sales are good throughout most of the year and there are no problems with recovering this level of overhead on the machine. Towards the end of the year, sales are hit by the combination of the arrival of a competitor's lower cost product, and a general economic downturn. The sales forecast for the next year is reduced. At year end the accountants perform their annual adjustment of the overhead rates as follows:

Annual overhead allocation £60 000
Anticipated annual usage 2000 hours
Overhead rate (60 000/2000) = £30 per hour

Note that for the new machine, although the total overhead allocation is unchanged, the reduced sales volume has increased the overhead rate by 50%. The pressure is now on the product engineers to find a way of reducing the manufacturing cost. The managing director has asked for a method that requires minimum investment and that can be introduced almost immediately! It does not take long for one of the engineers to propose a solution that meets these criteria. He has found a subcontractor with an equivalent machine who is offering to complete the job for £25 per hour. He calculates the annual saving as follows:

Cost for in-house machining
(2000 × 30) = £60 000
Cost for subcontract machining
(2000 × 25) = £50 000
Annual saving £10 000

He is quite correct in calculating that the standard cost of this product over the year will show a total saving of £10 000. Should he be congratulated, or chastised?

In practice, the overhead costs of the machine are costs that will exist whether or not the machine is used. So if the operation is subcontracted out, the costs that the company will see are:

In-house machine overheads	£60 000
Subcontract costs	£50 000
Total costs	£110 000

Rather than the saving of £10 000 indicated by the standard cost, the actual cost on the company has increased by £50 000!

The reader should not conclude from this example that subcontracting is always a bad idea. It offers an excellent means of providing capacity for additional work, when that capacity is not available in-house. Nor should the reader conclude that standard costing is of little use; as a tool for providing cost information for engineers, it should be used with caution, particularly with the overhead element.

An alternative system used by some organisations is the concept of *prime cost*. This only includes the materials and direct labour elements of the standard cost, both of which are likely to be affected by engineering changes made to the product. The danger in using this cost is that it does not reflect the full cost of manufacture. It should only be used as an in-house guide to the relative merits of alternative engineering decisions.

1. *Note*: For clarity we will ignore the labour and material costs in this example.

17.3.2 Introducing a new process or piece of plant

When a change is made to the detail of an existing product, probably the main economic concern to the engineer is the effect of the change on the manufacturing cost of the product. In some cases the change may require a modification to the manufacturing process that will involve some form of capital investment. This could range from the provision of a simple jig or fixture to the purchase of new production plant or machinery. In these cases the economic aspects must look somewhat further than simply the effect on the product manufacturing cost.

Let us take an example where an engineer is proposing that an old machine be replaced by a new piece of plant, because the higher production rate of the new machine will result in a reduction of the unit costs for the operation that the machine performs. Undoubtedly, the new machine will require the company to make a capital investment. A case needs to be made by the engineer, initially to convince himself of the need, then to convince those who hold the purse strings. The case should examine both the costs and the benefits.

The benefits are almost certain to be more complex than a simple unit cost reduction. They may include improved reliability, both in terms of machine availability through fewer breakdowns, and more consistent quality of output resulting in fewer out-of-tolerance rejections. The new machine might also be able to combine operations previously performed on two machines, with a possible effect on manning levels and manufacturing leadtimes. These cost savings can be put with the product volumes to give an estimate for a total annual saving.

The benefits then need to be compared with the costs associated with installing the new machine. The most significant cost will be the purchase price of the machine, but any associated costs should also be included. There may be extra training required for operators or maintenance personnel. If the equipment is a turning centre or a machining centre, has the cost of tooling been included? Will production be disrupted during installation of the machine? Obtaining these costs

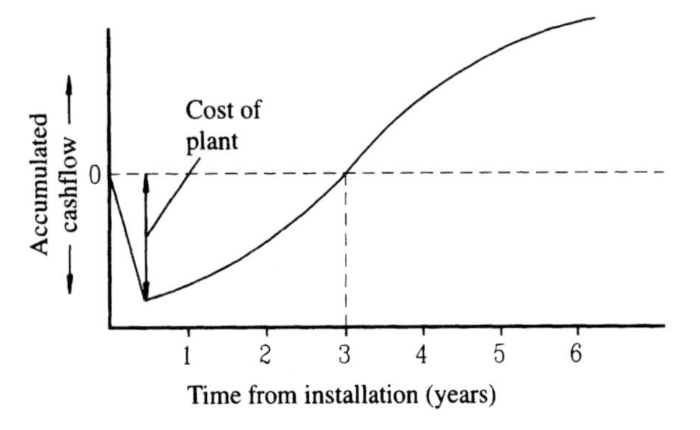

Figure 17.1 *Typical cashflow*

should be straightforward, and provided they are significantly less than the annual savings, the case for the investment is easily proved.

In almost all cases, the savings will be generated over a period of time after the installation of the new plant. This means that the company has to make the investment in advance of generating the savings. Figure 17.1 shows the typical cashflow for a new piece of plant. The initial flow is negative, as the plant is paid for, and turns positive as the benefits from the plant come on stream. In this instance, it indicates that the investment will be repaid in 3 years. The proposal is said to have a *payback* period of 3 years. Alternative proposals could be compared on the basis of payback period, the shortest period indicating the 'best' investment. This is, however, a very crude indicator, as it ignores the effect that time has on the value of money.

To combat the ever-decreasing value of money, interest is normally paid on money saved. The sum accumulated (if the interest is compounded) is given by:

$$S = P\left(1 + \frac{R}{100}\right)^n$$

where S is the total sum accumulated after n years having started with a principle of P accruing interest at an annual rate of $R\%$.

If the interest is simply maintaining the purchasing power of the money, S would represent the sum that in n years' time would have the same purchasing power as P would have today. This could be written as:

$$\text{Future worth} = \text{Present worth}\left(1 + \frac{R}{100}\right)^n$$

The concept of *present worth* is used to adjust the money values to a common base on the cashflow diagram.

This is best illustrated by an example. Assume that a machine is being installed at a total cost of £20 000. Once on-stream, the machine is expected to save the company £5000 during each year that it is operating. In reality, the savings may change over the years, but in the interests of simplicity we will assume they remain constant. For the same reason we will assume that the £20 000 for the machine is paid in total at the start of its first year of operation, and that the savings are collected at the end of each year.

At the start of operation there is a cash outflow of £20 000, and after one year's production savings of £5000 have been created. If we ignore the decreasing value of money, the net outflow after one year has been £15 000.

Now assume that in order to maintain the same purchasing value as the original £20 000 had at the start of the year, interest has to be added. In this example we will use an annual interest rate of 10%. The £5000 received at the end of the year is in effect the equivalent of only £4545 (i.e. £5000/1.1) at the start of the year. In present-worth

Table 17.1

Years from start	*Future worth (£)*	*Present worth (£)*			
		2.5%	*5%*	*10%*	*15%*
0	5000	5000	5000	5000	5000
1	5000	4878	4762	4545	4348
2	5000	4759	4535	4132	3781
3	5000	4643	4319	3757	3288
4	5000	4530	4114	3415	2859
5	5000	4419	3918	3105	2486
6	5000	4311	3731	2822	2162
7	5000	4206	3553	2566	1880
8	5000	4104	3384	2333	1635
9	5000	4004	3223	2120	1421

Table 17.2

Years from start	*Cashflow (£) for different interest rates*				
	0%	*2.5%*	*5%*	*10%*	*15%*
0	−20 000	−20 000	−20000	−20 000	−20 000
1	−15 000	−15 122	−15238	−15 455	−15 652
2	−10 000	−10 363	−10703	−11 322	−11 871
3	−5 000	−5 720	−6384	−7 566	−8 584
4	0	−1 190	−2 270	−4 151	−5 725
5	5 000	3 229	1 647	−1 046	−3 239
6	10 000	7 541	5 378	1 176	−1 078
7	15 000	11 747	8 932	4 342	802
8	20 000	15 851	12 316	6 675	2 437
9	25 000	19 854	15 539	8 792	3 858
Payback (years)	4	4.27	4.58	5.37	6.73

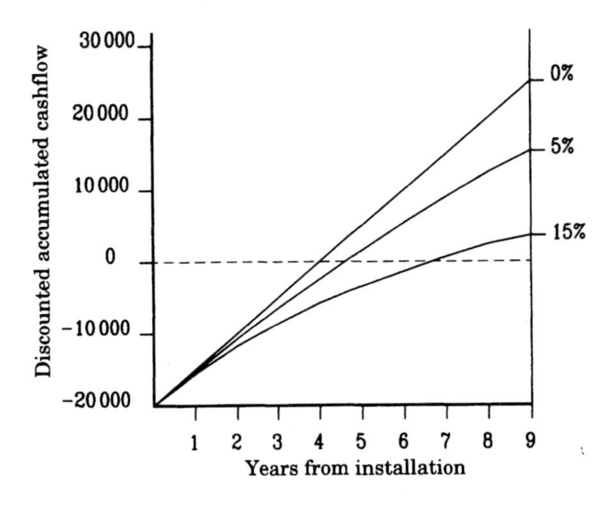

Figure 17.2 *Interest rates and discounted cash flow*

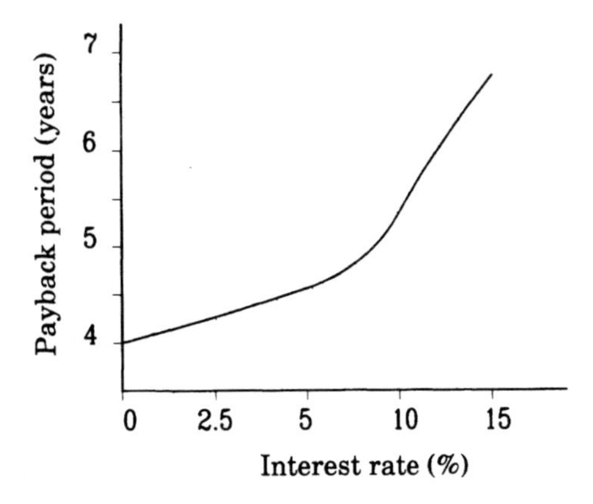

Figure 17.3 *Interest rates and payback*

terms the net cash outflow is £15 455 (i.e. £20 000 – £4545). Including the decreasing value of money will estimate the true worth of monetary benefits received in the future. Table 17.1 shows the present worth of £5000 received after varying numbers of years. The present value is affected by both the interest rate and the time period.

Now let us consider for our example the cashflow, assuming savings of £5000 are made at the end of each year. As Table 17.2 and Figure 17.2 show, the higher the interest rate (i.e. the faster money depreciates) the longer the payback period and hence the less attractive the investment looks. Figure 17.3 illustrates graphically how, for this example, the payback period is affected by the interest rate.

In our example the effect of depreciation of the value of money is dramatic. At 2.5% there is a payback period of 4.27 years, and after 7 years there has been an overall financial benefit of over £11 000. If the interest rate was 15%, it would take 6.73 years simply to cover the costs, and after 7 years the overall financial benefit would only be £802. In one case, the investment may be considered worthwhile, and in the other, a waste of time. In this type of analysis (present-worth) a decision has to be made by the organisation as to the value for the level of interest to be used. This is normally called the minimum acceptable rate of return.

Another method is to set an acceptable payback period and calculate the rates of interest required to achieve it as a means of comparing the costs of alternative engineering solutions. The calculated rate is called the *internal rate of return.* A company may, for example, have a generalised rule that a payback period of 3 years is acceptable for certain investments. The engineer, when comparing alternative engineering investments, would calculate from the predicted costs and savings the interest rate that would generate a cashflow with a 3-year payback. This interest rate is called the internal rate of return. The higher the internal rate of return, the more attractive the investment. Take an example where a choice has to be made between two machines, one costing £5000, offering annual savings of £2300, and the other costing £6000, offering savings of £2600. Assuming a payback period of 3 years in each case, the total cashflow would be:

Year	*Machine A*	*Machine B*
0	–5 000	–6 000
1	–3 051	–3 727
2	–1 400	–1 738
3	0	0

These values were calculated using the present-worth formula on a trial-and-error basis to arrive at internal rates of return values of 18.01% for machine A and 14.56% for machine B. On this basis, machine A appears to be the better solution. (Also see Appendix C.)

As a general point, although there are a number of ways of comparing the costs of different machines, these should not overrule the many other factors that must also come into the selection process. For example, in a world where technology is advancing rapidly, it would not be sensible to plan for ongoing savings to be generated by the machine if either the product or the process might become obsolete before the payback period was reached.

17.3.3 Introducing a new product

In economic terms, introducing a new product is really a combination of modifying an existing product and introducing manufacturing changes, but on a more comprehensive scale. Hence the benefits must be defined, along with any costs, and a view of the overall cashflow taken before the project can start.

The income generated will be defined by the sales volumes and the selling prices. As the value of money is depreciating continually, these figures need to be estimated on a time basis. By assuming a launch date and predicting sales volumes for, say, the next three years, an estimate of the incoming cashflow can be made. At this point, if the market information is good enough it may even be possible to indicate a level of price–volume sensitivity. Remember that the income is that derived from the net sales value, not the list price. It is important to appreciate that not all sales of a particular product will generate the same income. For example, export sales may be via an import agency who will require a certain share of the selling price. In this case overseas sales should be shown separately; indeed, the launch dates for export products are very often some months later than those for products on the home market. (This is prudent business practice in that any teething problems tend to be easier to cure on the home market.)

As the project progresses it may be necessary to update these income figures, so records must be kept of any assumptions made at each review point. If the product is totally new to the company, the assumptions regarding the source of the projected sales should be stated. Are they from other existing manufacturers? If so, they should be identified and consideration given to their likely response to the increased competition. Is a price war with a consequent income reduction a possibility? Are there any other in-house products whose sales will suffer as a result of the new product's introduction? The best possible overall picture of the potential income needs to be generated.

At the same time, the costs of introducing the new product need to be identified. Obviously, when the product is little more than a twinkle in the designer's eye, it is difficult to be precise about the level of expenditure needed. As the development process progresses the estimates for the manufacturing costs, the testing time, the tooling requirements, indeed the launch dates, gradually home in on the actual figures. In order to keep the project under control, review points must be introduced.

Timing of the review points will obviously depend on the organisation and the product being produced. However, each review point should normally be prior to the commitment of a significant level of company resources. Following an instruction from the Company Board to investigate the possibility of introducing a new widget, typical review points would be:

Marketing appraisal:

At this point, expenditure will have been limited to internal paper exercises, and possibly some customer surveys by the product planning section. Some preliminary consideration of development time and capital requirements will also have been made, but at this stage only budget figures will have been proposed by the engineering sections. If the data looks promising, authority to move to the next stage will be granted.

Technical feasibility:

The design process is now started with the generation of the PDS, followed by the development of conceptual ideas. These ideas may need proving for technical feasibility: that is, checking to see if they are practical and will work by building an engineering model and performing some basic tests. This stage will be more expensive than the marketing appraisal, but is still very modest in terms of the total cost of the project. In parallel with the tests, the estimates for progressing through the next stages will be refined, as will the estimated manufacturing cost price, and indeed the sales revenue data. Assuming the picture is satisfactory, the project will move to the next phase.

Prototype:
During this stage the company will be making a substantial commitment in resources. The detail design work will be undertaken, some tooling will be ordered, and initial prototypes built. Unlike the engineering models, these will more closely replicate the final product. They will, however, be produced without production tooling.

Field trials:
Once a level of confidence regarding the performance of the prototypes has been established, many organisations place some of the initial prototypes with selected customers, to ensure that there are no glaring faults in the product. Changes from this point on are likely to be minor. Acceptance at this point will clear the way to setting up the production process.

Pre-production:
Once the production equipment has been installed, an initial small batch is produced, mainly to iron out any production snags, before a full production run is undertaken. A green light at this stage will set the product into production.

The reader should note that these stages are similar to the stages of the product introduction process as described in Chapter 1. At each review stage, although there may be much discussion covering details of the project, the real purpose of the review is to ensure that the risks and costs of moving to the next stage are as well defined as possible.

The risks can appear under a number of categories, such as:

Sales:
Will the sales volumes be achieved? Will the target selling prices be met? Is a competitor likely to introduce a similar product? As the project proceeds towards the launch date, estimates of how the market is likely to respond to the new product become firmer, partly through increased confidence in the new product and partly through better knowledge of the competitors already in the market.

Design:
Is the design technically feasible at the required price? Are patents going to prove a problem? Will existing and future legislation be fully met? Is the product going to prove reliable? Are the design costs within budget? As with the sales risks, the answers to questions of this sort can only be answered fully after the product has been in production for some little time. However, as the project proceeds, the design knowledge continually increases to the point where at each of the review stages the risks or chances of product failure are reduced to an acceptable level before authorisation is given to move to the next stage.

Production:
Will the production costs be met? Will the introduction date be met? Will the estimated investment in tooling be sufficient? Are there any doubts about new production techniques or materials?

Service:
Are new skills required to service the product? Will service training be adequate? What investment in spare parts should be made prior to launch? What are the likely warranty costs?

As the project starts, many of the factors affecting these points are difficult to quantify, so the risk of the project not meeting its objectives and therefore being aborted is high. At each of the review points this risk reduces as work on the project proceeds. In fact, it is essential that it does reduce, for at each of the review points an ever-increasing commitment to the new product is being made by the company. At the technical feasibility stage, for example, the estimated manufacturing cost may include an error of 20%, but by the time it enters production this error is likely to be no more than, say 5%. Figure 17.4 shows how the risk of abandoning the project varies as it proceeds.

Figure 17.5 shows a typical new product appraisal summary form that might be presented at each of the review stages. At a glance it shows the estimated volumes, profitability and risk with the resource requirements section summarising the project plan in basic terms of how much and when. Note that the expenditure is divided into two categories, capital and revenue. Capital is generally the money spent on hardware, i.e. machines, jigs, fixtures, tooling, etc. Revenue tends to be all other costs, such as labour, maintenance, etc. The division is mainly for tax purposes.

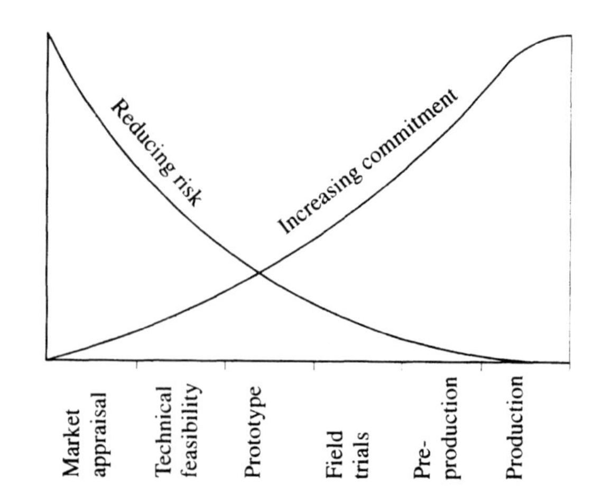

Figure 17.4 *Risk as project proceeds*

New product sales	Year 1 Home	Year 1 Export	Year 2 Home	Year 2 Export	Year 3 Home	Year 3 Export
Sales (units)						
Sales (value)						
Market share %						
Gross margin						
Overheads						
Gross profit (GP)						

Affected profits

Replaced gross profit	
Changed gross profit	

Cashflow	Pre-launch costs	

Risk analysis

Activity \ Field	Existing	Related	New
Sales/Mktg			
Design			
Production			
Service			

Investment required

Development costs	
Tooling	
Mean working capital	
Development as % Sales	

Resource requirements		Marketing appraisal	Technical feasibility	Prototype	Field trials	Pre-production	Total
Project totals	Capital/Tooling						
	Revenue						
	Est. Comp. Date						
Spent to date	Capital/Tooling						
	Revenue						
	Est. Comp. Date						

Figure 17.5 *Project review summary*

17.4 Quantity

In Chapter 14 we introduced the link between engineering decisions and the required production volumes. In this chapter, as we discussed labour costs (section 17.3.1), the need for an economic batch quantity based on the relationship between run and set times was raised. For many years engineering efforts were concentrated on the reduction of the run times in the belief that if the production volumes were sufficiently high, the set-up time was largely irrelevant. Hence the often quoted phrase of Henry Ford's that 'any colour – so long as it's black'. The aim was to keep the production run as high as possible. However, in today's world, high levels of competition have led to the need for a more sophisticated approach.

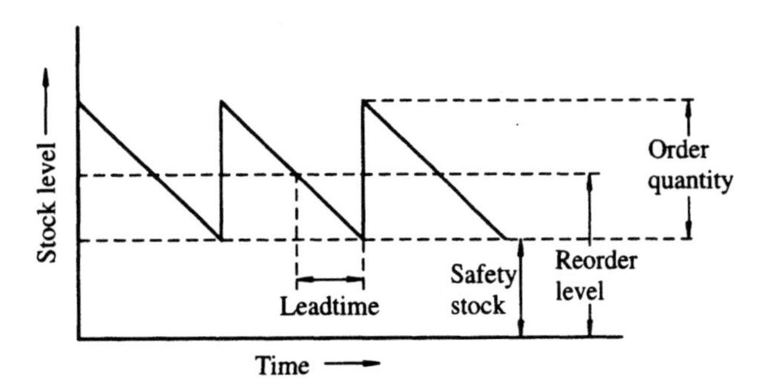

Figure 17.6 *Inventory levels*

First there is an ever-increasing expectation of choice on the part of the customer. A current car producer must offer a range of colours to attract customers. The engineers must concentrate not only on reducing the time taken to paint the car, but also the time taken to change from one colour to the next, the set-up time.

Perhaps more importantly, increasing levels of competition have led to a rigorous search for cost savings. Batch sizes have an effect on the levels of stock held. The greater the amount of stock held, the greater the proportion of the company's resources required to finance the stockholding. If some of these resources could be freed without affecting the overall output we have a means of saving cost. Figure 17.6 shows the traditional saw-tooth curve for stock levels of an individual component. The downward slope is determined by the rate of usage of the component. Once the stockholding has fallen to a predetermined minimum level, a new batch is ordered. The stockholding continues to fall during the lead time for producing the batch, and then rises to a peak once the new batch is booked into the stores. The reorder quantity is selected to ensure that the stock levels never fall below the safety level, by considering the usage rate and the leadtime.

Now consider an example where the batch size and the lead-time have been reduced to a bare minimum. The stockholding graph will be reduced to a very fine saw-tooth curve (Figure 17.7). The area under the curve represents the average levels of stock held, and hence the costs of financing that stock. From the two graphs it is apparent that reducing the batch size and the production lead times will both result in cost savings. (Note that, for an individual item, reducing the levels of its stockholding will not have a direct effect on its standard cost. However, a general reduction in inventory levels will reduce the overall overhead costs.)

We can also see from Figure 17.7 that the limiting factor in reducing inventory is the acceptable level of safety stocks. The philosophy of JIT (*just in time*) manufacturing is to reduce the safety stock levels to zero and produce the required components just as they are needed. To operate such a method, very close control of the manufacturing system is required, almost certainly employing the use of computers, as discussed in Chapter 16. Some organisations concentrating largely on in-house assembly operations employ a JIT system with their component suppliers. The suppliers are required to deliver specified stocks at specified times, so that the assembling company minimises its stockholding. This works well provided the suppliers are not simply holding a safety stock to guard against any supply problems they may have.

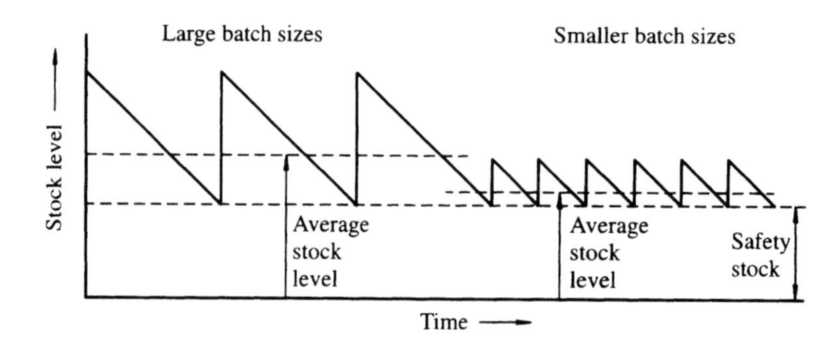

Figure 17.7 *Effects of batch size on inventory levels*

17.5 Summary

By the end of this chapter you should understand where some of the running costs of a design and manufacturing organisation arise from, and appreciate that the costs implications of engineering decisions are equally (if not more) important than the technical implications. You should appreciate the following points:

- That probably the single most important feature of any design is its ability to be produced at an acceptable cost. Although it is relatively easy to understand the source of costs in a design and manufacturing environment, it is much more difficult to apportion costs to particular products or processes. Standard costing is a method of allocating costs which has many advantages, but should be treated with caution if used to guide engineering decisions.

- If a new piece of manufacturing plant having a life span of a number of years is to be introduced, any savings generated should take account of the continuing decline in the value of money over time.

- When developing a new product, the organisation must have as good a view as is practical of the risks involved, be they technical or financial. A procedure should be used that involves review points, particularly at stages where the company has to make significant resource commitments to proceed further.

- There are inherent costs that are not reflected in conventional component costing techniques. Batch size, for example, can have a significant effect on the operating costs of a company, although this significance may not be apparent in the individual component cost.

17.6 Questions

1. A company introducing a new product will want to review progress on a regular basis. What will the critical review points be? (17.3.3)
2. What is meant by the term 'internal rate of return'? (17.3.2)
3. What is the difference between a direct and an indirect cost? (17.2.4)
4. Referring to Question 3, in which category would you place the following: production worker, cleaner, managing director, design engineer, manufacturing engineer, accountant? Would they always be in the same category? (17.2)
5. Standard costing is one of the most popular forms of costing in industry. Under which circumstances should the engineer be wary of

its shortcomings from an engineering standpoint? (17.3.1)

6. As a product proceeds down its introduction path, the risks associated with the programme being cancelled gradually decrease. Is this because once the company has committed a large proportion of its resources to a project it is unlikely to cancel the project, or is there another reason? (17.3.3)

7. A company is looking to purchase a new special purpose production machine. There are two alternatives. One machine will cost £10 000 and is predicted to save £4500 per year of operation. The other machine will cost £13 000, but is expected to generate annual savings of £6000. If a payback period of three years is specified, what is the internal rate of return in each case? Assume that each machine will be paid for on installation and should start generating the savings from that point. (17.3.2)

8. Explain the difference between net sales value and list price. (17.2.2)

9. Explain the relationship between batch size and stock levels. (17.4)

Part IV
Appendices

Appendix A Units and Definitions

A.1 Introduction

The essence of this appendix is to provide a user-friendly guide to a number of aspects of engineering science that engineers working in the design and manufacturing areas should have at their fingertips. Students should have already covered all the content at an earlier stage of their careers.

A.2 Units

A.2.1 SI system

Over the years, a number of alternative systems of units have evolved, along with the potential for confusion. The agreed system in the UK and a number of other countries is the SI system (Système International d'Unités). The SI system uses seven basic units, with two additional units from which the others are derived. These are:

Dimension	*Unit name*	*Symbol*
Length	Metre	m
Mass	Kilogramme	kg
Time	Second	s
Electric current	Ampere	A
Temperature	Kelvin	K
Luminous intensity	Candela	cd
Amount of substance	Mole	mol
Plane angle	Radian	rad
Solid angle	Steradian	sr

To enable the system to cater for a wide range of science and engineering activities, the following multipliers can be used. Where possible, preferred multipliers should be used. These are the ones that multiply in units of 10^3.

Prefix name	*Prefix symbol*	*Multiplier*
tera	T	10^{12}
giga	G	10^9
mega	M	10^6
kilo	k	10^3
hecto	h	10^2
deca	da	10
deci	d	10^{-1}
centi	c	10^{-2}
milli	m	10^{-3}
micro	μ	10^{-6}
nano	n	10^{-9}
pico	p	10^{-12}
femto	f	10^{-15}
atto	a	10^{-18}

These units are combined in various formats to represent a further sequence of dimensions with logical unit names. Some commonly used examples are:

Linear velocity: Distance per unit time, metre per second (m/s or ms^{-1})

Angular velocity: Angular change per unit time, radian per second (rad/s)

Acceleration: Change in velocity per unit time, metre per second per second (m/s^2)

Density: Mass per unit volume, kilogramme per cubic metre (kg/m^3)

Moment of inertia: Mass × length squared, kilogramme metre squared ($kg\ m^2$)

Momentum: Mass × velocity, kilogramme metre per second (kg m/s)

In some cases, the logical unit names are replaced with further derived units. These unit names can also be combined. Some commonly used examples are:

Force: The SI unit of force is the newton (N), which is defined as that force which, when applied to a body having a mass of 1 kg, gives it an acceleration of 1 m/s^2.

$$1 \text{ newton} = 1 \text{ kgms}^{-2}$$

Torque: Moment of force, newton metre (Nm)

Pressure: The SI unit of pressure is the pascal (Pa), which is defined as the pressure produced by a force of 1N applied, uniformly distributed over an area of 1m^2.

$$1 \text{ pascal} = 1 \text{ N/m}^2$$

Stress: As stress is also defined as force per unit area, the SI unit for stress is the pascal, even though it is a different physical quantity.

Energy, work: The SI unit of energy is the joule (J), which is defined as the work done when the point of application of a force of 1N is displaced through a distance of 1m.

$$1 \text{ joule} = 1 \text{ Nm}$$

Power: The SI unit of power is the watt (W), which is defined as the power that is generated by the application of 1J for 1s (i.e. power is the rate of doing work).

$$1 \text{ watt} = 1 \text{ J/s}$$

Frequency: The SI unit is the hertz (Hz), which is the number of repetitions per second

$$1 \text{ hertz} = 1/\text{s}$$

Quantity of electricity: The SI unit is the coulomb (C), which is defined as the quantity of electricity carried in 1s by a current of 1A.

$$1 \text{ coulomb} = 1\text{As}$$

Electrical potential: The SI unit of electrical potential is the volt (V), which is defined as the potential difference between two points in a wire carrying a current of 1A, when the power dissipated between these points is 1W.

$$1 \text{ volt} = 1 \text{ W/A}$$

Electrical resistance: The SI unit is the ohm (Ω), which is the electrical resistance between two points in a conductor through which a current of 1A flows when the points are subjected to a potential difference of 1V.

$$1\,\Omega = 1 \text{ V/A}$$

Electrical capacitance: The SI unit is the farad (F), which is the capacitance between two plates which, when subjected to a potential difference of 1V, become charged with 1C of electricity.

$$1 \text{ farad} = 1 \text{ C/V}$$

Inductance: The SI unit is the henry (H), which is the inductance of a closed circuit within which an electromotive force of 1V is generated when the current flowing through the circuit changes at a rate of 1A/s.

$$1 \text{ henry} = 1 \text{ Vs/A}$$

Magnetic flux: The SI unit is the weber (Wb), which is the magnetic flux which, linking a circuit of one turn, would produce in it an electromotive force of 1V if it were reduced to zero in one second.

1 weber = 1Vs

Magnetic flux density: The SI unit is the telsa (T), which is defined as 1W per square metre.

1 telsa = 1 Wb/m^2

A.2.2 Imperial system

Although the SI system has wide acceptance, engineers are almost certain to be confronted by other systems, particularly the Imperial system, at some point. A popular area for confusion lies around the differences between the terms 'force', 'mass' and 'weight'.

Mass is the amount of matter in a body. It is independent of the body's location in space. The SI primary unit is the kilogramme, and the Imperial primary unit is the pound (lb).

Force when applied to a body will impart it with an acceleration. As above in the SI system, the newton is the force which, when applied to a body of mass 1kg, will generate an acceleration of $1 m/s^2$. The unit of force in the Imperial system is the poundal (pdl), which is that force which, when applied to a mass of 1lb, will result in an acceleration of $1 ft/s^2$. Within the Imperial system the unit of force normally used is the pound-force (lbf) as defined below. (1lbf = 32.174 pdl).

Weight is a specific instance of force. It is the force acting on a body due to gravity. This will vary according to the location of the body within space. A body in space may have a weight of almost zero, but its mass will be the same as if it was on the surface of the earth.

The metric primary unit is the kilogramme-force (kgf) which is that force which, when applied to a body of mass 1kg, will produce an acceleration equal to that provided by gravity (g) i.e. $9.80665 m/s^2$.

The Imperial primary unit is the pound-force (lbf), which is that force which when applied to a body of mass 1lb will impart an acceleration of g ($32.174 ft/s^2$).

Some of the more popular Imperial units are listed in Table A.I:

Table A.1 *Imperial units*

Dimension primary SI unit	*Imperial unit*	*Conversion between IMP units*	*Nearest pref. SI unit*	*Other metric units*
Length metre (m)	thou	1 thou = 10^{-3} in	= 25.4 μm	
	inch (in)	1 in = 1000 thou	= 25.4 mm	
	foot (ft)	1 ft = 12 in	= 0.3048 m	
	yard (yd)	1 yd = 3 ft	= 0.9144 m	
	chain (ch)	1 ch = 22 yd	= 20.12 m	
	furlong (f)	1 f = 10 ch	= 201.2 m	
	mile	1 mile = 1760 yd	= 1.609 km	
Area square metre (m^2)	sq. inch (in^2)	1 in^2	= 645.2 mm^2	= 6.452 cm^2
	sq. foot (ft^2)	1 ft^2 = 144 in^2	= 9.29 10^{-2} m^2	
	sq. yard (yd^2)	1 yd^2 = 9 ft^2	= 0.8361 m^2	hectare (ha)
	acre	1 acre = 4840 yd^2	= 4047 m^2	1 ha = 10^4m^2
	sq. miles	1 mile^2 = 640 a	= 2.589 km^2	1 km^2 = 100 ha
Velocity metre per second (m/s)	foot per minute	1 ft/min	= 5.08 mm/s	
	foot per second	1 ft/s = 60 ft/min	= 0.3048 m/s	
	miles per hour	1 mph = 88 ft/min	= 1.609 kph	
Volume[1,2,3,4] cubic metre (m^3)	Dry measures			
	cu. inch (in^3)	1 in^3	= 16387 mm^3	= 16.34 cm^3
	gallon (gal)	1 gal = 277.4 in^3	= 4.546 10^{-3} m^3	= 4.546 litres
	cu. foot (ft^3)	1 ft^3 = 6.229 gals	= 2.831 10^{-2} m^3	
	cu. yards (yd^3)	1 yd^3 = 27 ft^3	= 0.7645 m^3	
	Liquid measures			
	cu. inch (in^3)	1 in^3	= 16387 mm^3	= 16.34 cm^3
	fluid ounce	1 fl oz = 1.734 in^3	= 28415 mm^3	= 28.33 cm^3
	gills	1 gill = 5 fl oz	= 2.84 10^{-3} m^3	= 141.65 cm^3
	pint (pt)	1 pint = 4 gills	= 1.137 10^{-3} m^3	= 1.137 litres
	quart (qt)	1 quart = 2 pints	= 2.273 10^{-3} m^3	= 2.273 litres
	gallon (gal)	1 gal = 4 quarts	= 4.546 10^{-3} m^3	= 4.546 litres
Density kilogramme per metre^3 (kg/m^3)	pound/cu foot	1 lb/ft^3	= 16.02 kg/m^3	1 tonne /m^3
	pound/cu inch	1 lb/in^3 = 1728 lb/ft^3	= 27.68 t/m^3	= 10^3 kg/m^3
	ton/cu yard	1 ton/yd^3 = 82.97 lb/ft^3	= 1.329 t/m^3	= 1 gm/cm^3
Mass flow rate kilogramme per second (kg/s)	pound per hour	1 lb/h		
	pound per sec	1 lb/s = 3600 lb/h	= 1.26 10^{-4} kg/s	
	ton per hour	1 ton/h = 6.222 lb/s	= 0.2822 kg/s	

Table A.1 *contd*

Dimension primary SI unit	*Imperial unit*	*Conversion between IMP units*	*Nearest pref. SI unit*	*Other metric units*
Volume flow rate				
cu metre/sec	cu feet/hour	ft^3/h	= 7.868 10^{-6} m^3/s	
(m^3/s)	cu feet/sec	ft^3/s = 3600 ft^3/h	= 0.02832 m^3/s	
	gal/sec	gal/s = 577.94 ft^3/h	= 4.546 10^{-3} m^3/s	
	gal/min	gal/min = 9.632 ft^3/h	= 7.577 10^{-5} m^3/s	
Mass				
kilogramme (kg)	ounce (oz)	1 oz	= 28.35 gramme (g)	
	pound (lb)	1 lb = 16 oz	= 0.4536 kg	
	stone (st)	1 stone = 14 lb	= 6.350 kg	
	hundredweight (cwt)	1 cwt = 8 stone	= 50.80 kg	
	ton	1 ton = 20 cwt	= 1016 kg	1 tonne (t) = 10^3 kg
Force				
newton (N)	poundal (pdl)	1 pdl	= 1.383 10^{-1} N	1 dyne = 10^{-5} N
	pound-force (lbf)	1 lbf = 32.174 pdl	= 4.448 N	kilogramme-force
	ton-force	1 tonf = 2240 lbf	= 9.964 kN	1 kgf = 9.807 N
		1 kip = 1000 lbf		1 kgf = 1 kp
Torque				
newton metre	pound-force inch	1 lbf in	= 0.113 Nm	
(Nm)	pound-force foot	1 lbf ft = 12 lbf in	= 1.356 Nm	
Pressure				1 bar = 100 kPa
pascal (Pa)	lbf/in^2 (psi)	1 psi	= 6.894 kPa	atmosphere (atm)
	kilo-lbf/in^2 (ksi)	1 ksi = 1000 psi	= 6.894 MPa	atm = 1.013 bar
	$tonf/in^2$	1 tf/in^2 = 2240 psi	= 15.44 MPa	1 atm = 760 torr
		1 ft H_20 = 0.443 psi	= 2.989 kPa	1 mm Hg = 1 torr
Energy, Work				
joule (J)	foot pound-force	1 ft lbf	= 1.356 J	kWatt hour (kWh)
				1 kWh = 3.6 MJ
Power[5]				
watt (W)	foot pound-force/ sec	ft lbf/s	= 1.355 W	
	horsepower (hp)	1 hp = 550 ftlbf/s	= 0.7457 kW	

Notes:
1. US units for liquid measure only, 1 US gallon = 231 cubic inches, hence 1 Imperial gallon = 1.201 US gallons.
2. The litre was defined in 1901 as the volume occupied by 1kg of pure water, hence 1 litre = $1000.028cm^3$.
3. In 1964 the name litre was adopted for the cubic decimetre giving 1 litre = $1000cm^3$.
4. The Imperial gallon is defined as the space occupied by 10 pounds of distilled water.
5. Metric horsepower (CV or PS) = 75 kgf m/s = 735.5 W = 0.9863 hp.

A.2.3 Unity brackets

A common error when converting between units arises when the decision to divide or multiply by a conversion factor has to be made. If you find this a problem, do not despair, but instead turn to unity brackets for help. The principle is simple: the value of an expression will be unchanged if it is multiplied by one (i.e. unity).

If, for example, we wish to convert between metres and millimetres, we can create a unity bracket (i.e. an expression whose value is one).

$$\left[\frac{1000mm}{m}\right]$$

As 1000mm = 1m this expression has a value of unity; we can use it to multiply another expression without changing its value. Then, cancelling equal units in the numerator and denominator, we arrive at the correct conversion:

$$15m = 15m \times \left[\frac{1000mm}{m}\right] = 15000mm$$

This is obviously a very simple example; the true benefits arise with more complex conversions. Take as an example a pressure quoted in Imperial units as 50psi that you wish to convert to the SI unit of the pascal. First we use a unity bracket to allow the lbf to be eliminated.

$$15psi = \frac{15lbf}{in^2} \times \left[\frac{4.45N}{lbf}\right] = 15 \times 4.45\frac{N}{in^2}$$

Then we use a further bracket to convert the inches to metres, noting in this case that the unity bracket is squared.

$$15 \times 4.45\frac{N^2}{in} = \frac{15 \times 4.45N^2}{in} \times \left[\frac{39.37in}{m}\right]^2$$

$$= 15 \times 4.45 \times 39.37^2\frac{N}{m^2}$$

The conversion is now complete. As you will have no doubt recognised, a pascal is the SI name for N/m^2.

Hence 15 psi = 103.4 kPa.

In this example it was assumed that we knew the conversion factor directly between newtons and pound-force. If the direct conversion factor is not known, unity brackets can be used to generate the conversion from first principles, as follows.

Using the definition of the pound-force, we have

$$1lbf = lb \times 32.174\frac{ft}{s^2}$$

Multiplying by appropriate unity brackets, we obtain

$$1lbf = lb \times 32.174\frac{ft}{s^2} \times \left[\frac{kg}{2.204lb}\right] \times \left[\frac{0.3048m}{ft}\right]$$

Cancelling the lb and ft, we have

$$1lbf = \frac{32.174 \times 0.3048}{2.204} \times \frac{kg\ m}{s^2}$$

i.e.

$$1lbf = 4.45kg\frac{m}{s^2} = 4.45N.$$

A.2.4 Dimensional analysis

Sometimes when deriving engineering formulae it is useful to invoke a technique called dimensional analysis to check for any fundamental errors in the formula. The technique will check that the mix of units is at least feasible. It involves reducing each of the elements to its fundamental dimensions. Most expressions in mechanics can, for example, be given in units of mass (M), length (L) and time (T).

As an example, we will check that Newton's second law of motion is dimensionally correct. It can be written as

$$Ft = mv - mu$$

Now if we examine the left-hand side of the equation, we know from the definition of force that it comprises mass × acceleration. As acceleration is velocity/time, and velocity is length/time, force has the dimensions [M] [L] [T]$^{-2}$. Impulse, Ft is force × time, so the left-hand side must have dimensions [M] [L] [T]$^{-1}$.

Next we will examine the right-hand side of the equation. Velocity as above has the dimensions length/time, i.e. [L] [T]$^{-1}$. Hence momentum, *mv* and mu, must have the dimensions [M] [L] $[T]^{-1}$ These dimensions are then compared with those found for the first side. If they are not the same, we have a problem.

This simple check does not prove Newton's second law of motion, but it does show that the equation is dimensionally correct. An equation that does not balance dimensionally contains an error.

A.3 Equations of motion

Linear	*Angular*
$s = \left(\frac{u+v}{2}\right) \times t$	$\theta = \left(\frac{\omega_1 + \omega_2}{2}\right) \times t$
$v = u + at$	$\omega_2 = \omega_1 + \alpha t$
$s = ut + \frac{at^2}{2}$	$\theta = \omega_1 t + \frac{\alpha t^2}{2}$
$v^2 = u^2 + 2as$	$\omega_2^2 = \omega_1^2 + 2\alpha t$

Where
- s = distance
- u = initial velocity
- v = final velocity
- a = linear acceleration
- t = time

Where
- θ = angle rotated
- ω_1 = initial angular velocity
- ω_2 = final angular velocity
- α = angular acceleration
- t = time

If the angular motion is around a circle of radius r, then:

$$\theta = \frac{s}{r}$$

and

$$v = r\omega$$

A.4 Force, motion and work

A.4.1 Newton's laws of motion

1. An object will remain at rest or in uniform motion unless acted on by a force.
2. When acted on by a force, an object will accelerate at a rate dependent on the force and the mass of the object.
3. For every action there is an equal and opposite reaction.

The second law can be expressed as Force = mass × acceleration, i.e.

$$F = m \times a$$

or that Force is proportional to the rate of momentum change that it produces, i.e.

$$Ft = mv - mu$$

Centripetal force, the force constraining an object to move along a circular path radius r is given by $F = mr\omega^2$.

A.4.2 Work, energy and power

When the point of application of a force moves, work is done. The work is measured by the magnitude of the force and the distance its point of application moves, i.e.

$$Work = Force \times Distance$$

Energy can be considered as the capacity to do work. That is, an object containing energy can do work as the energy is converted into another form.

Potential energy of mass m stored at a height h is given by:

$$PE = mgh$$

Kinetic energy of mass m travelling at velocity v is given by

$$KE = \frac{1}{2}mv^2$$

Note that energy can only be changed in format, it cannot be created or destroyed.

Power is the rate of doing work, i.e.

$$Power = \frac{Work}{Time}$$

Torque is the moment of force (i.e. the twisting effect of a force).

$$Torque = Force \times Radius\ of\ application$$

Power is a measure of the rate at which the torque can be applied, hence

$$Power = Torque \times Speed\ of\ rotation$$

It is the *torque* generated by a car engine that, when applied to the driving wheels, causes the vehicle to accelerate. It is the *power* generated by the engine that determines the maximum speed of a given vehicle.

Appendix B Data Tables

B.1	Surface roughness	302
B.2	Relative costs of processes	303
B.3	Selected ISO fits – hole basis	304
B.4	Selected ISO fits – shaft basis	305
B.5	Properties of sections	306
B.6	Deflections and moments for simple beams	308
B.7	Friction coefficients	309
B.8	Corrosion: Electrochemical series	310
B.9	Materials data	311
	Steels	311
	Non-ferrous metals	312
	Non-metals	313

Table B.1 *Guide to surface roughness*

R_A 0.001mm	*Grade no.*	*Description, applications*	*Typical processes*
25	N11	A very basic rough finish only suitable for clearance surfaces where appearance is not important.	Flame cutting, sawn, hand grind, fettling.
12.5	N10	Moderately smooth surface without burrs. Suitable for clearance surfaces or those not requiring uniform contact. Can be used for bolted or riveted joints on structural work.	Sawing, planing, shaping, sand casting, hot rolling, forging.
6.3	N9	Reasonable appearance, mating surfaces not requiring a close fit, such as motor feet.	Sawing, planing, shaping, drilling, chemical milling, milling, turning, forging.
3.2	N8	Normal machined surface, surfaces requiring good appearance, maximum level of roughness to minimise fatigue stress concentrations. Locating surfaces using gaskets, drilled holes.	Planing, shaping, drilling, milling, chemical milling, broaching, reaming, turning, permanent mould casting, investment casting, cold rolling.
1.6	N7	Good commercial machined finish, used for locating surfaces, slow light duty plain bearings, interference and key fits.	Shaping, milling, broaching, reaming, turning, extruding, drawing, die casting.
0.8	N6	High quality finish where smoothness and freedom from scratches is important, e.g. O ring sealing faces. Suitable for bearing housings, shaft seats for bearing races, running and sliding fits at high speeds.	Turning , barrel finishing, grinding, honing, polishing.
0.4	N5	Ball bearing rings and seatings for precision fits. Suitable for cylinder bores and sealing surfaces, gear teeth.	Fine turning, barrel finishing, cylindrical and surface grinding, honing, lapping.
0.2	N4	Suitable for highly rated journal bearings, hydraulic ram surfaces, ball races.	High class centreless cylindrical and surface grinding, honing and lapping.
0.1	N3	Very high quality finish, suitable for very high speed spindles, rams, pistons, shop gauges.	Burnishing, high-quality grinding, lapping.
0.05	N2	Reference gauges.	Honing, series grinding, lapping.
0.025	N1	Master reference gauges, optical flats, slip gauges	Lapping, superfinishing.

B.2 Relative costs of processes

The cost of a machined component will depend on a variety of factors. These will include the type of feature being machined, the size and shape of the component, the tightness of the dimensional tolerance, the surface finish, and the component material. The chart below provides an approximate relationship between the cost and achievable tolerance attainable for common machining processes.

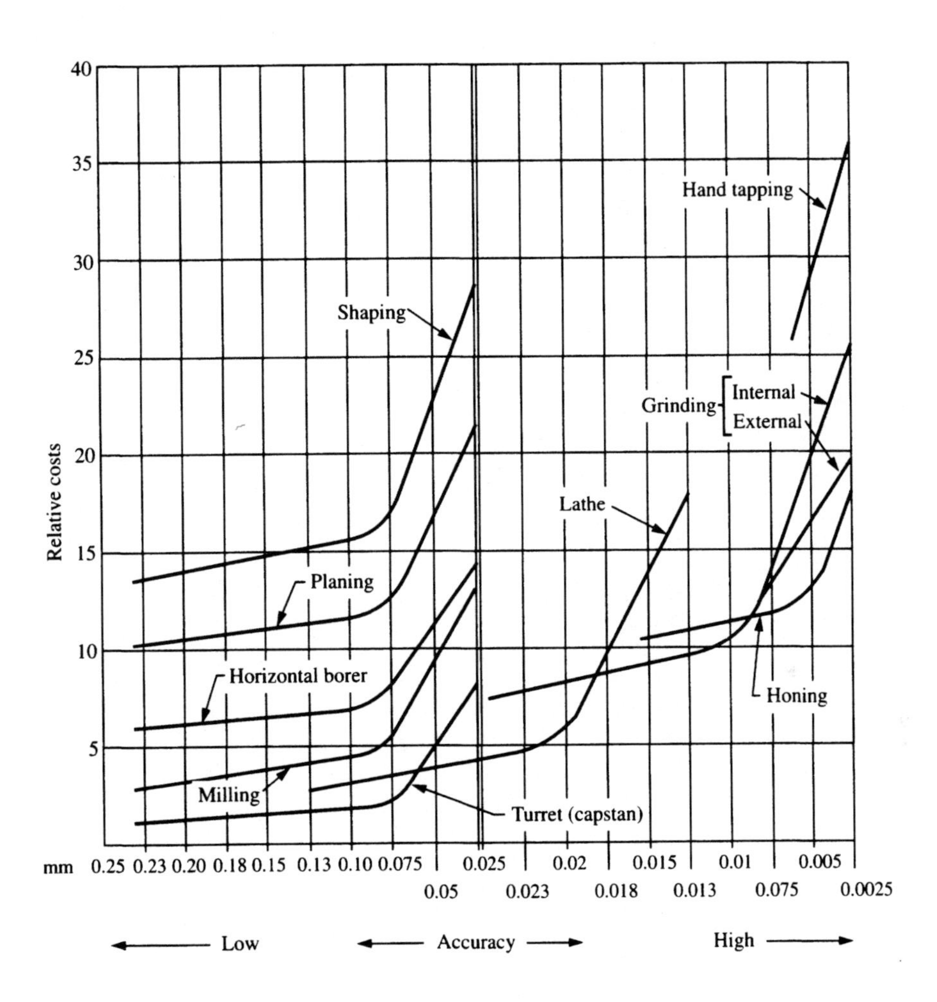

Source: BSI chart A.3.1: PD 6470; 1981.

Figure B.2 *Cost of various manufacturing processes for achieving set tolerances*

Table B.3 *Selected ISO fits – hole basis*

Diagram to scale for 25 mm. diameter: + / 0 / −

Clearance fits: H11 / c11; H9 / d10; H9 / e9; H8 / f7; H7 / g6; H7 / h6

Transition fits: H7 / k6; H7 / n6

Interference fits: H7 / p6; H7 / s6

Holes; Shafts

Nominal sizes		Clearance fits												Transition fits				Interference fits				Nominal sizes	
		Tolerance		Tolerance		Tolerance		Tolerance		Tolerance		Tolerance		Tolerance		Tolerance		Tolerance		Tolerance			
Over	To	H11	c11	H9	d10	H9	e9	H8	f7	H7	g6	H7	h6	H7	k6	H7	n6	H7	p6	H7	s6	Over	To
mm	mm	0·001 mm	0·001 mm	0·001 mm	0·001 mm	0·001 mm	0·001 mm	0·001 mm	0·001 mm	0·001 mm	0·001 mm	0·001 mm	0·001 mm	0·001 mm	0·001 mm	0·001 mm	0·001 mm	0·001 mm	0·001 mm	0·001 mm	0·001 mm	mm	mm
—	3	+ 60 0	− 60 − 120	+ 25 0	− 20 − 60	+ 25 0	− 14 − 39	+ 14 0	− 6 − 16	+ 10 0	− 2 − 8	+ 10 0	− 6 0	+ 10 0	+ 6 + 0	+ 10 0	+ 10 + 4	+ 10 0	+ 12 + 6	+ 10 0	+ 20 + 14	—	3
3	6	+ 75 0	− 70 − 145	+ 30 0	− 30 − 78	+ 30 0	− 20 − 50	+ 18 0	− 10 − 22	+ 12 0	− 4 − 12	+ 12 0	− 8 0	+ 12 0	+ 9 + 1	+ 12 0	+ 16 + 8	+ 12 0	+ 20 + 12	+ 12 0	+ 27 + 19	3	6
6	10	+ 90 0	− 80 − 170	+ 36 0	− 40 − 98	+ 36 0	− 25 − 61	+ 22 0	− 13 − 28	+ 15 0	− 5 − 14	+ 15 0	− 9 0	+ 15 0	+ 10 + 1	+ 15 0	+ 19 + 10	+ 15 0	+ 24 + 15	+ 15 0	+ 32 + 23	6	10
10	18	+ 110 0	− 95 − 205	+ 43 0	− 50 − 120	+ 43 0	− 32 − 75	+ 27 0	− 16 − 34	+ 18 0	− 6 − 17	+ 18 0	− 11 0	+ 18 0	+ 12 + 1	+ 18 0	+ 23 + 12	+ 18 0	+ 29 + 18	+ 18 0	+ 39 + 28	10	18
18	30	+ 130 0	− 110 − 240	+ 52 0	− 65 − 149	+ 52 0	− 40 − 92	+ 33 0	− 20 − 41	+ 21 0	− 7 − 20	+ 21 0	− 13 0	+ 21 0	+ 15 + 2	+ 21 0	+ 28 + 15	+ 21 0	+ 35 + 22	+ 21 0	+ 48 + 35	18	30
30	40	+ 160 0	− 120 − 280	+ 62 0	− 80 − 180	+ 62 0	− 50 − 112	+ 39 0	− 25 − 50	+ 25 0	− 9 − 25	+ 25 0	− 16 0	+ 25 0	+ 18 + 2	+ 25 0	+ 33 + 17	+ 25 0	+ 42 + 26	+ 25 0	+ 59 + 43	30	40
40	50	+ 160 0	− 130 − 290																			40	50
50	65	+ 190 0	− 140 − 330	+ 74 0	− 100 − 220	+ 74 0	− 60 − 134	+ 46 0	− 30 − 60	+ 30 0	− 10 − 29	+ 30 0	− 19 0	+ 30 0	+ 21 + 2	+ 30 0	+ 39 + 20	+ 30 0	+ 51 + 32	+ 30 0	+ 72 + 53	50	65
65	80	+ 190 0	− 150 − 340																	+ 30 0	+ 78 + 59	65	80
80	100	+ 220 0	− 170 − 390	+ 87 0	− 120 − 260	+ 87 0	− 72 − 159	+ 54 0	− 36 − 71	+ 35 0	− 12 − 34	+ 35 0	− 22 0	+ 35 0	+ 25 + 3	+ 35 0	+ 45 + 23	+ 35 0	+ 59 + 37	+ 35 0	+ 93 + 71	80	100
100	120	+ 220 0	− 180 − 400																	+ 35 0	+ 101 + 79	100	120
120	140	+ 250 0	− 200 − 450	+ 100 0	145 305	+ 100 0	− 84 − 185	+ 63 0	− 43 − 83	+ 40 0	− 14 − 39	+ 40 0	− 25 0	+ 40 0	+ 28 + 3	+ 40 0	+ 52 + 27	+ 40 0	+ 68 + 43	+ 40 0	+ 117 + 92	120	140
140	160	+ 250 0	− 210 460																	+ 40 0	+ 125 + 100	140	160
160	180	+ 250 0	− 230 480																	+ 40 0	+ 133 + 108	160	180
180	200	+ 290 0	− 240 530	+ 115 0	− 170 − 355	+ 115 0	− 100 − 215	+ 72 0	− 50 − 96	+ 46 0	− 15 − 44	+ 46 0	− 29 0	+ 46 0	+ 33 + 4	+ 46 0	+ 60 + 31	+ 46 0	+ 79 + 50	+ 46 0	+ 151 + 122	180	200
200	225	+ 290 0	260 550																	+ 46 0	+ 159 + 130	200	225
225	250	+ 290 0	280 570																	+ 46 0	+ 169 + 140	225	250
250	280	+ 320 0	300 620	+ 130 0	− 190 − 400	+ 130 0	− 110 − 240	+ 81 0	− 56 − 108	+ 52 0	− 17 − 49	+ 52 0	− 32 0	+ 52 0	+ 36 + 4	+ 52 0	+ 66 + 34	+ 52 0	+ 88 + 56	+ 52 0	+ 190 + 158	250	280
280	315	+ 320 0	330 650																	+ 52 0	+ 202 + 170	280	315
315	355	+ 360 0	360 − 720	+ 140 0	− 210 − 440	+ 140 0	− 125 − 265	+ 89 0	− 62 − 119	+ 57 0	− 18 − 54	+ 57 0	− 36 0	+ 57 0	+ 40 + 4	+ 57 0	+ 73 + 37	+ 57 0	+ 98 + 62	+ 57 0	+ 226 + 190	315	355
355	400	+ 360 0	400 760																	+ 57 0	+ 244 + 208	355	400
400	450	+ 400 0	440 840	+ 155 0	− 230 − 480	+ 155 0	− 135 − 290	+ 97 0	− 68 − 131	+ 63 0	− 20 − 60	+ 63 0	− 40 0	+ 63 0	+ 45 + 5	+ 63 0	+ 80 + 40	+ 63 0	+ 108 + 68	+ 63 0	+ 272 + 232	400	450
450	500	+ 400 0	480 880																	+ 63 0	+ 292 + 252	450	500

Source: Data sheet BS 4500A, 1970.

Table B.4 *Selected ISO fits – shaft basis*

Nominal sizes		Clearance fits												Transition fits				Interference fits				Nominal sizes	
Diagram to scale for 25 mm diameter		h11 / C11		h9 / D10		h9 / E9		h7 / F8		h6 / G7		h6 / H7		h6 / K7		h6 / N7		h6 / P7		h6 / S7		Holes / Shafts	
		Tolerance		Tolerance		Tolerance		Tolerance		Tolerance		Tolerance		Tolerance		Tolerance		Tolerance		Tolerance			
Over	To	h11	C11	h9	D10	h9	E9	h7	F8	h6	G7	h6	H7	h6	K7	h6	N7	h6	P7	h6	S7	Over	To
mm	mm	0·001 mm	0·001 mm	0·001 mm	0·001 mm	0·001 mm	0·001 mm	0·001 mm	0·001 mm	0·001 mm	0·001 mm	0·001 mm	0·001 mm	0·001 mm	0·001 mm	0·001 mm	0·001 mm	0·001 mm	0·001 mm	0·001 mm	0·001 mm	mm	mm
—	3	0 −60	+120 +60	0 −25	+60 +20	0 −25	+39 +14	0 −10	+20 +6	0 −6	+12 +2	0 −6	+10 0	0 −6	0 −10	0 −6	−4 −14	0 −6	−6 −16	0 −6	−14 −24	—	3
3	6	0 −75	+145 +70	0 −30	+78 +30	0 −30	+50 +20	0 −12	+28 +10	0 −8	+16 +4	0 −8	+12 0	0 −8	+3 −9	0 −8	−4 −16	0 −8	−8 −20	0 −8	−15 −27	3	6
6	10	0 −90	+170 +80	0 −36	+98 +40	0 −36	+61 +25	0 −15	+35 +13	0 −9	+20 +5	0 −9	+15 0	0 −9	+5 −10	0 −9	−4 −19	0 −9	−9 −24	0 −9	−17 −32	6	10
10	18	0 −110	+205 +95	0 −43	+120 +50	0 −43	+75 +32	0 −18	+43 +16	0 −11	+24 +6	0 −11	+18 0	0 −11	+6 −12	0 −11	−5 −23	0 −11	−11 −29	0 −11	−21 −39	10	18
18	30	0 −130	+240 +110	0 −52	+149 +65	0 −52	+92 +40	0 −21	+53 +20	0 −13	+28 +7	0 −13	+21 0	0 −13	+6 −15	0 −13	−7 −28	0 −13	−14 −35	0 −13	−27 −48	18	30
30	40	0 −160	+280 +120	0 −62	+180 +80	0 −62	+112 +50	0 −25	+64 +25	0 −16	+34 +9	0 −16	+25 0	0 −16	+7 −18	0 −16	−8 −33	0 −16	−17 −42	0 −16	−34 −59	30	40
40	50	0 −160	+290 +130																			40	50
50	65	0 −190	+330 +140	0 −74	+220 +100	0 −74	+134 +60	0 −30	+76 +30	0 −19	+40 +10	0 −19	+30 0	0 −19	+9 −21	0 −19	−9 −39	0 −19	−21 −51	0 −19	−42 −72	50	65
65	80	0 −190	+340 +150																	0 −19	−48 −78	65	80
80	100	0 −220	+390 +170	0 −87	+260 +120	0 −87	+159 +72	0 −35	+90 +36	0 −22	+47 +12	0 −22	+35 0	0 −22	+10 −25	0 −22	−10 −45	0 −22	−24 −59	0 −22	−58 −93	80	100
100	120	0 −220	+400 +180																	0 −22	−66 −101	100	120
120	140	0 −250	+450 +200	0 −100	+305 +145	0 −100	+185 +85	0 −40	+106 +43	0 −25	+54 +14	0 −25	+40 0	0 −25	+12 −28	0 −25	−12 −52	0 −25	−28 −68	0 −25	−77 −117	120	140
140	160	0 −250	+460 +210																	0 −25	−85 −125	140	160
160	180	0 −250	+480 +230																	0 −25	−93 −133	160	180
180	200	0 −290	+530 +240	0 −115	+355 +170	0 −115	+215 +100	0 −46	+122 +50	0 −29	+61 +15	0 −29	+46 0	0 −29	+13 −33	0 −29	−14 −60	0 −29	−33 −79	0 −29	−105 −151	180	200
200	225	0 −290	+550 +260																	0 −29	−113 −159	200	225
225	250	0 −290	+570 +280																	0 −29	−123 −169	225	250
250	280	0 −320	+620 +300	0 −130	+400 +190	0 −130	+240 +110	0 −52	+137 +56	0 −32	+62 +17	0 −32	+52 0	0 −32	+16 −36	0 −32	−14 −66	0 −32	−36 −88	0 −32	−138 −190	250	280
280	315	0 −320	+650 +330																	0 −32	−150 −202	280	315
315	355	0 −360	+720 +360	0 −140	+440 +210	0 −140	+265 +125	0 −57	+151 +62	0 −36	+75 +18	0 −36	+57 0	0 −36	+17 −40	0 −36	−16 −73	0 −36	−41 −98	0 −36	−169 −226	315	355
355	400	0 −360	+760 +400																	0 −36	−187 −244	355	400
400	450	0 −400	+840 +440	0 −155	+480 +230	0 −155	+290 +135	0 −63	+165 +68	0 −40	+83 +20	0 −40	+63 0	0 −40	+18 −45	0 −40	−17 −80	0 −40	−45 −108	0 −40	−209 −272	400	450
450	500	0 −400	+880 +480																	0 −40	−229 −292	450	500

Source: As Table B.3.

B.5 Properties of sections

Table B.5

Section	Cross section area (A)	Moment of inertia about axis xx (I_x)	Radius of Gyration (k)	Polar moment of inertia about centroidal axis (J)
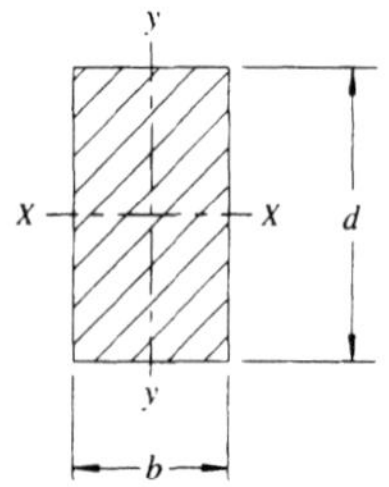 Rectangle (y, y, x, x, d, b)	bd	$\frac{bd^3}{12}$	$\frac{d}{\sqrt{12}}$	$\frac{bd}{12}(b^2+d^2)$
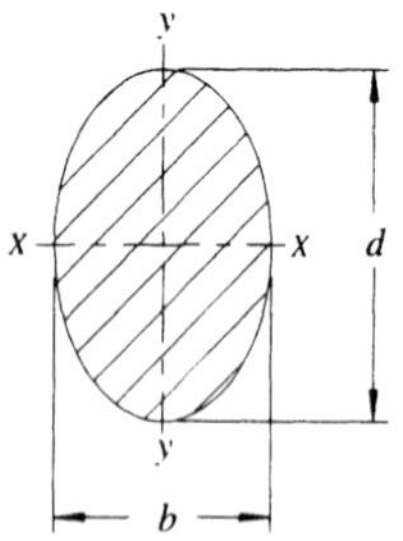 Ellipse (y, y, x, x, d, b)	$\frac{\pi bd}{4}$	$\frac{\pi bd^3}{64}$	$k_y=\frac{b}{4}$ $k_x=\frac{d}{4}$	$\frac{\pi bd}{64}(b^2+d^2)$
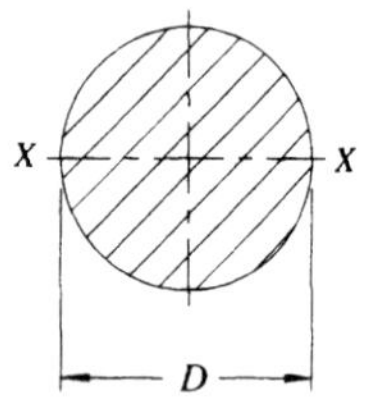 Circle (x, x, D)	$\frac{\pi D^2}{4}$	$\frac{\pi D^4}{64}$	$\frac{D}{4}$	$\frac{\pi D^4}{32}$

Table B.5 *continued*

Section	*Cross section area (A)*	*Moment of inertia about axis xx (I_x)*	*Radius of Gyration (k)*	*Polar moment of inertia about centroidal axis (J)*
X X ID OD	$\frac{\pi}{4}(OD^2 - ID^2)$	$\frac{\pi}{64}(OD^2 - ID^2)$	$\sqrt{\frac{OD^2 + ID^2}{16}}$	$\frac{\pi}{32}(OD^4 - ID^4)$
g/2 g/2 X X h H G	$GH - gh$	$\frac{(GH^3 - gh^3)}{12}$	$\sqrt{\frac{1}{12}\left(\frac{GH^3 - gh^3}{GH - gh}\right)}$	
a h/2 h H e b/2 b/2 X X c d t t/2 B Centroid	$Bt + a(H - t)$	$\frac{Bt^3}{12} + (Bt)d^2 + \frac{ah^3}{12} + (ah)e^2$	$\sqrt{\frac{I}{A}}$	Centriod location $c = \frac{aH^2 + bt^2}{2(aH + bt)}$

B.6 Deflections and moments for simple beams

The following table lists some simple beam cases. For the point load cases, the applied load is F newtons, but for the uniform load cases, the load is w newtons per metre length, where the beam length is L metres (i.e. the load F will be wL). E is the modulus of elasticity, I the moment of inertia of the beam section, y the deflection, F_s the shear force and M the moment at a distance x from the end of the beam. θ is the beam slope in radians.

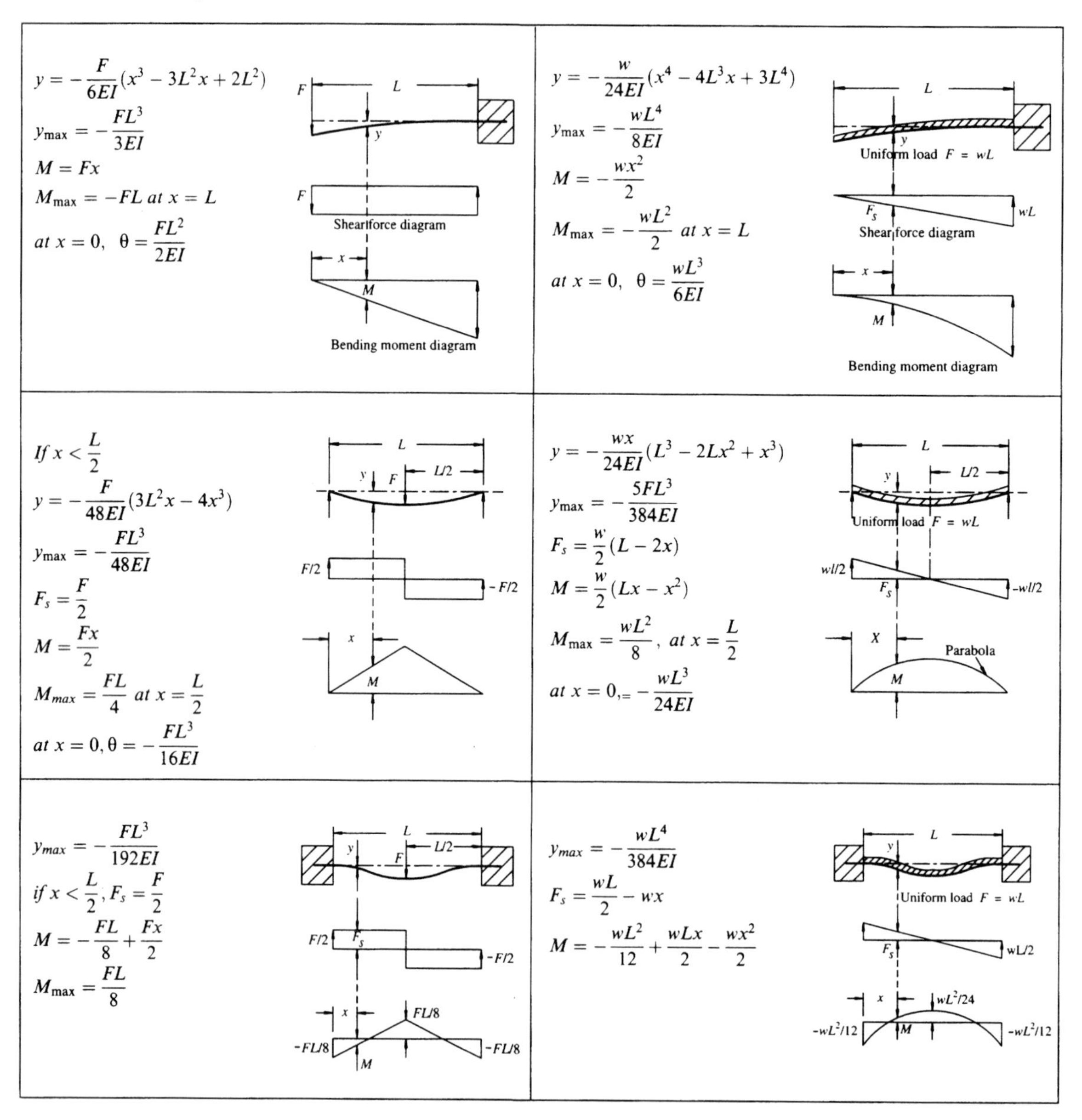

Figure B.6 *Simple beam cases*

B.7 Friction coefficients

These figures have been taken from a number of references, but as the values are highly dependent on the operating conditions they should only be used as an approximate guide. Material suppliers' data sheets are normally the best source for further information.

Table B.7 *Friction coefficients*

Materials in contact	*Condition*	*Friction coefficient (μ)*
Metal on metal	dry	0.15 to 0.25
Metal on metal	oiled	0.10 to 0.20
Sintered metal on cast iron	dry	0.20 to 0.40
Sintered metal on cast iron	submerged in oil	0.05 to 0.08
Brake lining on metal	dry	0.50 to 0.70
Asbestos in resin binder, on metal	dry	0.30 to 0.40
Asbestos in resin binder, on metal	oiled	0.10
Asbestos, flexible woven on metal	dry	0.35 to 0.45
Asbestos, flexible woven on metal	oiled	0.12
Rubber on asphalt	dry	0.50 to 0.80
Rubber on asphalt	wet	0.25 to 0.75
Rubber on concrete	dry	0.60 to 0.85
Rubber on concrete	wet	0.45 to 0.75
Timber on timber	dry, static	0.25 to 0.50
Timber on timber	wet, static	0.20
Timber on metal	dry, static	0.20 to 0.60
Timber on metal	wet, static	0.20

B.8 Corrosion: electrochemical series

Galvanic corrosion can occur when dissimilar metals make electrical contact in a design. The higher metal in the table will become the anode, and will corrode away, leaving the lower metal, the cathode, protected. In general, the greater the separation of the materials in Table B.8, the greater will be the severity of the reaction.

Table B.8 *Standard oxidation potential of the elements (volts)*

Element	Potential (volts)
Lithium	3.05
Potassium	2.93
Barium	2.90
Calcium	2.87
Sodium	2.71
Magnesium	2.37
Aluminium	1.66
Zinc	0.76
Sulphur	0.48
Iron	0.44
Cadmium	0.40
Nickel	0.25
Tin	0.14
Lead	0.13
Hydrogen	0
Copper	−0.34
Iodine	−0.54
Mercury	−0.79
Silver	−0.80
Bromine	−1.07
Platinum	−1.20
Chlorine	−1.36
Gold	−1.50
Fluorine	−2.65

B.9 Materials data

BS 970 covers wrought steels for mechanical and allied engineering purposes. Within the standard, steels must be ordered to one of the following conditions, identified by the fourth character of the identification number.

A: denotes that the steel will be supplied to close limits of chemical composition
H: to the requirements of hardenability
M: to specified mechanical properties
S: indicates a stainless steel alloy.

The first three digits indicate the family of steels to which the alloy belongs:

000 to 199 Carbon and carbon manganese steels. The figures indicate 100 × the mean manganese content.
200 to 240 Free cutting steels, where the second and third digits indicate 100 × the minimum or mean sulphur content.
250 Silico-manganese spring steels (EN45 and EN45A).
300 to 499 Stainless, heat resisting and valve steels.
500 to 999 Alloy steels, in groups or multiples of groups of ten according to alloy type.

The fifth and sixth digits represent 100 × the mean carbon content, e.g. 080A40 denotes a carbon manganese steel supplied close to limits of chemical composition, containing 0.7 to 0.9% manganese and 0.38 to 0.40% carbon.

Table B.9 *Materials data*

Common Steels									
BS Specification		*Tensile stress*	*Yield stress*	*Young's modulus*	*Density*	*Thermal expan. Coeff.*	*Common form*	*General comments*
EN No	*BS 970*	*MN/m²*	*MN/m²*	*GN/m²*	*kg/m³*	*K×10⁻⁶*		
EN1A	220M07 230M07	370 to 480	215 to 400	208	7833	12.6	various	Low carbon mild steel, soft and easily machined. Not suitable for highly stressed parts.
EN3A	070M20	450 to 560 cold drawn	325 to 440	208	7861	12.4	various	General purpose mild steel for welded or riveted structures, forgings, machined parts, hot pressings etc.
EN8	080M40	570 to 660 cold drawn	430 to 530	208	7860	11.5	various	Medium carbon steel suitable for the manufacture of forgings and general engineering parts. Extensively used on transmission components, shafts, gears, flanges etc.
EN9	070M55	670 to 770 cold drawn	530 to 610	208	7860	11.5	various	Similar to EN8 but offering higher strength and better wear properties. Harder to machine.
EN58M	303S41	540 to 695	232 to 278	193	7910	17.5	round bar	Free machining variant of stainless steel.

Table B.9 *contd*

Common Steels *Contd*									
BS Specification		*Tensile stress*	*Yield stress*	*Young's modulus*	*Density*	*Thermal expan. coeff.*	*Common form*	*General comments*
EN No.	*BS970*	MN/m^2	MN/m^2	GN/m^2	kg/m^3	$K \times 10^{-6}$		
EN58J	316S16	630	300	204	7972	17.0	plate sheet round hex bar	High corrosion resistance, but poor machining properties. Extensively used for food processing equipment.
	BS 4360 ref 43C	430 to 540	275	207	7861	12.0	angle, round or square hollow sections	Easily weldable structural steel often used for rolled hollow sections.
EN16	605M36	695 to 933	605 to 846	210	7833	12.4	hex or round bar	Light duty, hardenable steel available as hex or round bar.
EN24	817M40	823 to 1550	690 to 1127	213	7889	11.2	round bar	Heavy duty hardenable

Note that the full range of steels available is much wider than those shown above. The mechanical properties shown are typical and depend on the treatment the material has been subjected to. For further details refer to the relevant BS and material stockist's data sheets.

Non-Ferrous Metals								
Metal	*BS*	*ISO*	*Tensile stress* MN/m^2	*Young's modulus* GN/m^2	*Density* kg/m^3	*Therm expan coeff.* $K \times 10^{-6}$	*Cost factor relative to mild steel cost by mass*	*General comments*
Aluminium	BS 1200	ISO Al99.0	70 to 150	70	2710	23	2.6	Low-cost, low-strength grade. Good for welding.
Aluminium	BS 5154A	ISO Al Mg3.5	215 to 325	70		23	2.6	Good general purpose grade. Good for welding.
Brass		Cu-Zn 70/30	400	110	8400	19	2.7	Copper zinc alloy, good conductor of heat and electricity. Widely used for marine and electrical components.
Copper	BS 2870 to 2875 BS 6931	Pure	400	142	8900	17	3.5	Good conductor of both heat and electricity. Widely used as electrical wiring and pipework for heating installations.
Lead	BS 1178 BS 3332	Pure	40	14	11000	29	1	High density, used as radiation shield, additives for various materials.
Phosphor bronze		Spring temper	800	110	8800	17	6.3	Corrosion resistant material for springs, bushes and switch parts.

Table B.9 *contd*

Non-Metals							
Material	*Type or grade*	*Tensile stress* MN/m^2	*Young's modulus* GN/m^2	*Density* kg/m^3	*Therm expan coeff.* $K \times 10^{-6}$	*Cost factor relative to mild steel cost by mass*	*General comments*
Acrylic	Perspex	75	3.1	1200	90	4.9	Good weather resistance. Used for outdoor signs, cloches, glazing.
Nylon	66	76	1.5	1150	150	6.5	Widely used, but moisture absorption will cause dimension changes.
Poly-carbonate		66	2.2	1200	70	2.3	Strong, rigid thermoplastic. Used for safety shields, lenses, glazing, gears, etc.
Poly-ethylene	Polythene	38	1.0	970	130	2.4	Low friction, no moisture absorption. Used for packaging, toys, bottles, etc.
Structural foam	Poly-phenylene oxide	32	1.7	900			Easy to mould to complex shapes. Used on car fascia panels, television cabinets, etc.
Rubber	Natural	20	0.01 to 0.05	900		2.4	Used in tyre treads, rubber bands, seals, etc.
Rubber	Neoprene	24	0.01	1250			A water and weather resistant synthetic rubber. Used for belt drives.
Rubber	Nitrile	15	0.005	1000			An oil-resistant rubber suitable for seals, but of low strength, and poor weather resistance.
Wood	Soft (pine)	6	7	400			Low-cost, general-purpose wood used on doors, window frames, skirting boards, etc.
Wood	Hard (oak)	10	12	720			High-cost wood with better strength and weather resistance. Used for furniture and specialist doors, etc.
Plywood		6	12	560			Laminates of soft wood available in sheet form, for general purpose cladding etc. Normally damaged by weather, unless marine ply grade specified.
Carbon fibre + epoxy	Composite	1500	170	1700			High strength-to-weight ratio, with ability to produce complex profiles. Used on aircraft and racing cars.
Glass	Soda/lime/silica	50	70	2500			Domestic grade glass used for glazing.
Glass	Lithium al. silicate	120	85	2500			Specialist material used in high voltage insulators.
Cement	Portland	3.4 (tensile) 20 (comp-ression)	45	2200		1.5	Civil engineering and general construction.

Appendix C Costing Calculations

Comparing financial returns of engineering alternatives can prove a lengthy and very tedious process if done manually by simply using the appropriate formulae. Financial tables are often used. These list data such as present-worth factors against the number of years for different interest rates. In many cases they make the calculations much easier. However, a more modern tool, the computer, with a spreadsheet, can also provide a simple means to such problems.

The example given in Chapter 17, comparing internal rates of return for a given payback period, only involves the use of a simple formula. Solving on a trial and error basis, with the use of a spreadsheet, is very straightforward and gives an insight into the effects of different internal rates of return. Figure C.1 shows how a Lotus compatible spreadsheet could be used. Formulae need only be entered in 3 cells, B10, C9, and C10, and the rest can be copied from these. Enter the initial cost in C3, and the expected annual savings in C4. Now the trial and error part simply involves entering different values for the IRR in cell C5, until the value in C12 reaches zero. That is, the payback has been achieved by the end of year 3.

	A	B	C
1		Machine A	
2			
3		Initial cost =	
4		Annual savings =	
5		Internal rate of return =	
6			
7	Year	Present worth of savings	Net cashflow
8			
9	0		–$C$3
10	1	+$C$4/(1+$C$5/100)^A10	+C9+B10
11	2	+$C$4/(1+$C$5/100)^A11	+C10+B11
12	3	+$C$4/(1+$C$5/100)^A12	+C11+B10

Figure C.1 *Use of a Lotus compatible spreadsheet*

If an initial cost of £5000 is used, with annual savings of £2300, trial and error will lead to a value of 18.01 for the IRR, to give a calculated net cashflow of zero after 3 years, as shown in Figure C.2 below.

7	Year	Present worth of savings	Net cashflow
8			
9	0		−5000.0
10	1	1949.0	−3051.0
11	2	1651.5	−1399.5
12	3	1399.5	0.0

Figure C.2 *Spreadsheet*

Appendix D Solutions

Answers to most of the questions at the end of each chapter can be found by reference to the relevant section within the chapter. However, there are a number of instances, such as the drawing exercises in Chapter 2, where solutions are provided, as follows.

Chapter 2

Q8

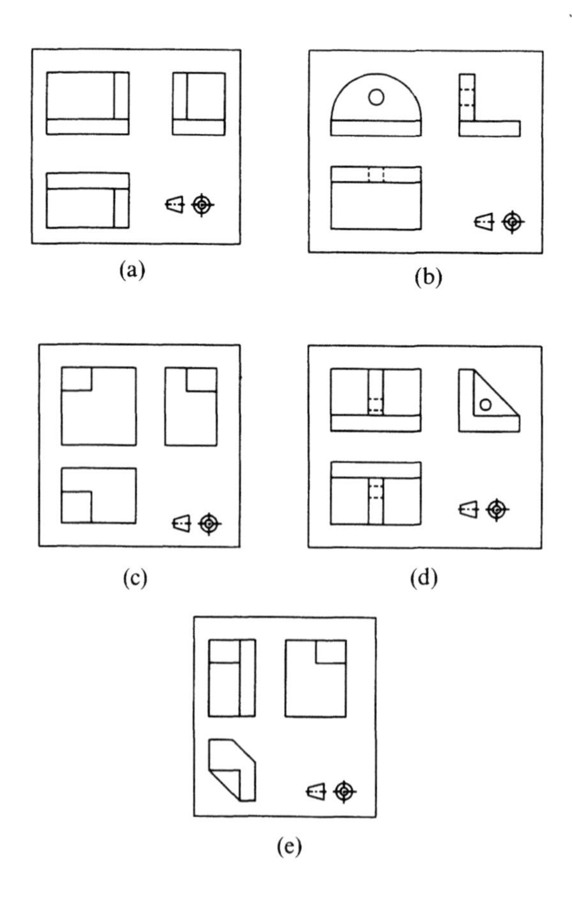

Figure *D.1*

Q 9

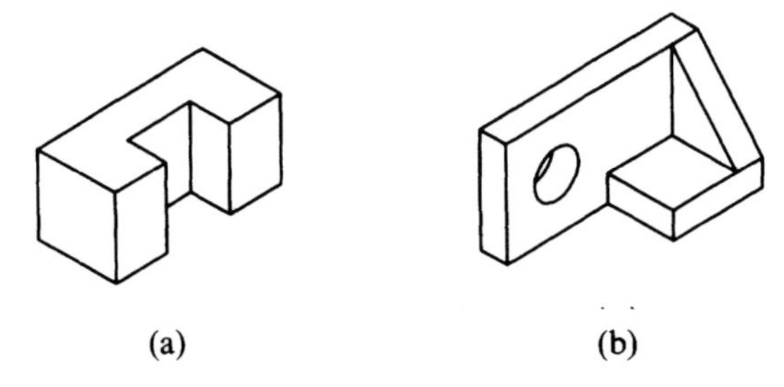

Figure *D.2*

Q 10

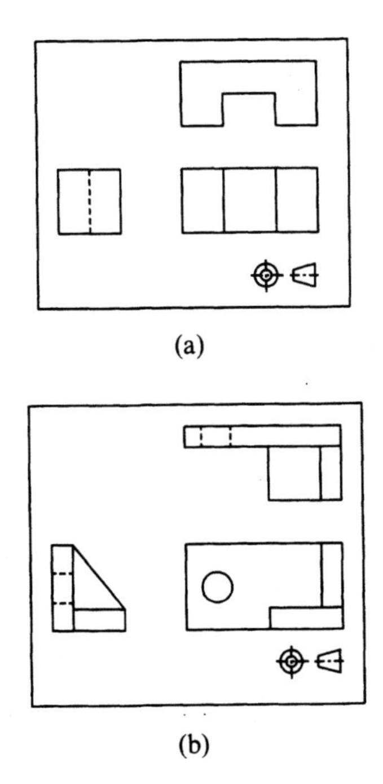

Figure *D.3*

Q 11

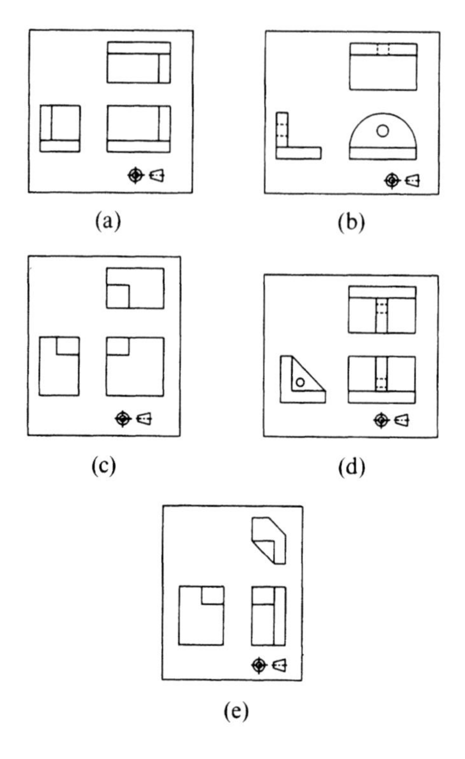

Figure *D.4*

Q 15

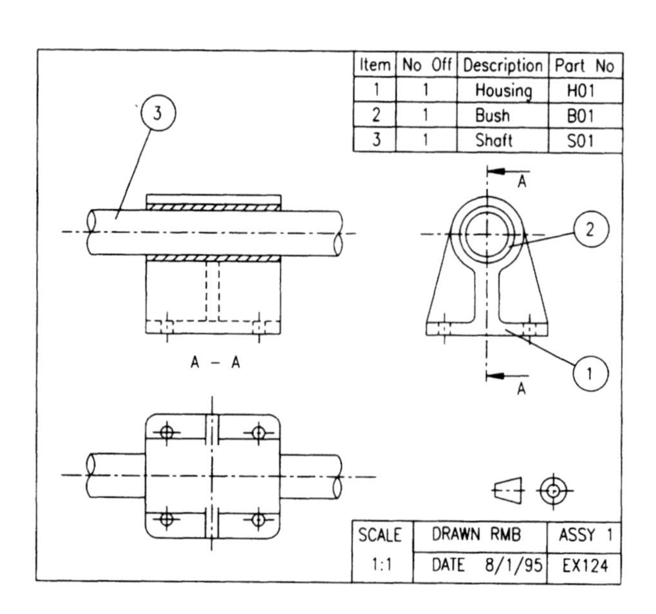

Figure *D.5*

Chapter 3

Q 3 Use hole base, close running fit is H7g6
Hole 50.025, 50.000
Shaft 49.991, 49.975.

Q 7 To within 0.075 means ± 0.075 so the tolerance is 0.15. See chart A2.1 in PD6470 (1981) (a BS publication). This gives cost factors regarding a plain face to a datum. Changing from 0.15 to 0.075 has a factor of 1.46. Machining cost is estimated to be a little over £29.

Q 10 The size and shape indicate that the component can be turned. Tolerance is on an external diameter, so chart A2.3 in PD 6470 applies. Reducing the tolerance from 0.15 to 0.04 gives a cost factor of 3.29. Machining cost will rise to £32.90, giving a total estimated cost of £37.90.

Chapter 7

Q 5 Feed rate = 0.2 mm/rev (should be in the range 0.1 to 0.4 mm/rev).

Assume cutting speed 215 m/min (should be in the range 180 to 250 m/min). Hence calculated cutting time will be 9.6 seconds.

Assuming a specific power of 55×10^{-3} W/mm^3/min gives an estimated power requirement of 24 kW.

Chapter 17

Q 7 Capital cost of £10 000, generating an annual saving of £4500, will require an internal rate of return of 16.65% to give a 3-year payback.

Capital cost of £13 000, generating an annual saving of £6000, will require an internal rate of return of 18.22% to give a 3-year payback.

Appendix E Equivalent Standards

Within the text a number of references are made to British Standards. The table below cross references these with equivalent or similar International Standards. The standards are listed in the order in which the appear in the text.

British Standard	Title	Equivalent or similar International Standard
BS308 Pt1	Engineering Drawing Practice	ISO2162/1, ISO2203,, ISO5455-, ISO6410, ISO6413, ISO6433,ISO8826, ISO9222/1
BS308 Pt 2	Dimensions and tolerancing	ISO129, ISO406, ISO1302
BS308 Pt3	Engineering Drawing Practice:- Geometric Tolerances	ISO1101-, ISO2692, ISO5458 ISO5459-, ISO7083-
PP7308	Engineering Drawing Practice for Schools and Colleges	
PD6470	The management of design for economic production	
BS4500	ISO System of Limits and Fits	ISO/R286, BS EN20286
BS1134	Assessment of Surface Texture	ASTM D466, ISO468, ISO4827/1
BS4235	Specification for Keys and Keyways	ISO773, ISO774, ISO3912
BS2059	Specification for Straight Parallel Sided Splines	ISO14-
BS3550	Specification for Involute Splines	ANSI B92, DIN5480, ISO4156
BS3790	Specification for Endless Wedge Belt Drives and Endless V Belt Drives	ISO155-, ISO1813, ISO254, ISO4183, ISO4184, ISO5292
BS436	Spur and Helical Gears	ISO1328-, IS1340-, ISO1341-, ISO53, ISO54 ISO6336/1/2/3
BS4518	O ring seals	ISO3601
BS343	Isometric Threads	ISO261-, ISO262-, ISO68-ISO724 -1978, ISO965/1-, ISO965
BS4620	Rivets for General Engineering Purposes	ISO/R1051-
BS3673	Carbon Steel Circlips	DIN471, DIN472
BS1574	Split Pins	ISO/R1234
BS1804	Steel Dowel Pins	ISO2338, DIN7
BS4311	Gauge Blocks	ISO3650
BS6000/1/2	Sampling Procedures	ISO2859, ISO8422, ISO3951, ISO8423
BS970	Steels for mechanical and allied engineering purposes	ISO683/13, ISO683/17, Euronorm90
BS4360 (BS7613) (BS668)	Specifications for weldable steel structures	EN113 - 72, EN155 - 80 Euronorm137
BS2870 (BS6931)	Specification for copper and copper alloys	ISO197
BS1178	Lead sheet	AS1804
BS3332	White metal bearing alloy ingots	ISO4381

Bibliography and Further Reading

Ashby, M. F., *Materials Selection in Mechanical Design*, 1st edn (Oxford: Pergamon Press, 1992).

Ashby, M. F. and Jones, D. R. H., *Engineering Materials, An Introduction to their Properties and Applications*, 1st edn (Oxford: Pergamon Press, 1987).

Collet, C. V. and Hope, A. D., *Engineering Measurements*, 2nd edn (London: Pitman, 1983).

Corbet, J., Dooner, M., Meleka, J. and Pym, C., *Design for Manufacture Strategies, Principles and Techniques*, 1st edn (Wokingham: Addison-Wesley, 1991).

Cross, N., *Engineering Design Methods, Strategies for Product Design*, 2nd edn (Chichester: John Wiley & Sons, 1994).

Cullum, R. D., *Handbook of Engineering Design*, 1st edn (London: Butterworth, 1988).

DeGarmo, E. P., Temple Black, J. and Kohser, R. A., *Materials and Processes in Manufacturing*, 7th edn (New York: Macmillan, 1988).

Deiter, G. E., *Engineering Design, A Materials and Processing Approach*, 2nd edn (New York: McGraw-Hill, 1991).

Derby, B., Hills, D. and Ruiz, C., *Materials for Engineering, A Fundamental Design Approach*, 1st edn (Harlow: Longman Scientific and Technical, 1992).

Easterling, K. E., *Advanced Materials for Sports Equipment*, 1st edn (London: Chapman & Hall,, 1993).

Engineering Adhesives for Structural, Multi Resistant Bonding, A Simple Guide (3M, 1990).

Ertas, A. and Jones, J. C, *The Engineering Design Process*, 1st edn (New York: John Wiley & Sons, 1993).

Faires, V. M., *Design of Machine Elements* 4th edn (New York: Macmillan, 1966).

Farag, M. M., *Selection of Materials and Manufacturing Processes for Engineering Design*, 1st edn (Hemel Hempstead: Prentice Hall International (UK), 1989).

Galyer, J. F. W. and Shotbolt, C. R., *Metrology for Engineers*, 5th edn (London: Cassell Publishers, 1990).

Hales, C., *Managing Engineering Design*, 1st edn (Harlow: Longman Scientific and Technical, 1993).

Hawkes, B. R. and Abinet, R. E., *The Engineering Design Process* (London: Pitman, 1984).

Hurst, K., *Rotary Power Transmission Design* (Maidenhead: McGraw-Hill, 1994).

Insitution of Production Engineers *Guide to Design for Production A* (London: Insitution of Production Engineers, 1984).

Lindberg, R. A., *Processes and Materials of Manufacture*, 4th edn (Boston, Mass.: Allyn & Bacon, 1990).

Lissaman, A. J. and Martin, S. J., *Principles of Engineering Production*, 2nd edn (London: Hodder & Stoughton, 1982).

Mucci, P., *Handbook for Engineering Design Using Standard Materials and Components*, 4th edn (Milton Keynes: BSI Standards, 1994).

Parker, M., *Manual of British Standards in Engineering Drawing and Design*, 1st edn (London: British Standards Institute in association with Hutchinson, 1984).

Pugh, S., *Total Design, Integrated Methods for Successful Product Engineering*, 1st edn (Wokingham: Addison-Wesley, 1990).

Radford, J. D. and Richardson, D. B., *Production Engineering Technology*, 3rd edn (London: Macmillan, 1980).

Schey, J. A., *Introduction to Manufacturing Processes*, 2nd edn (New York: McGraw-Hill, 1987).

Shotbolt, C. R., *Technician Manufacturing Technology 3*, 1st edn (London: Cassell, 1980).

Walton, J., *Engineering Design: From Art to Practice*, 1st edn (St Paul, Minno.: West Publishing, 1991).

Walton, J., *Essentials of Engineering Design*, 1st edn (St Paul, Minno.: West Publishing, 1991).

Williams, E. H., *Designing in Metals*, 1st edn (London: Iliffe Books, 1968).

Index

ABS 198
Acceleration 293
Acetylene 159
Acoustic testing 178
Acrylic 198
Added value 235
Adhesive bonding 164
  advantages and disadvantages 165
  applications 202, 207
  surface preparation 164
  types of adhesives 164
Aesthetics 220, 223
Aims 250
Analogy 106
Annealing 170
Anode 229
Anodising 172
ANSI 55
Appearance 220
Arc welding 159, 203
  flux core welding 160
  MIG 160
  plasma arc 161
  pulsed TIG 161
  stick welding 159
  stud welding 161
  submerged arc welding 160
  TIG 161
ASME 55
ASTM 55
Attribute listing 105
Austenite 169
Autoclave 154, 199
Automated assembly 216
Auxiliary projection 19

Barrel finishing 171
Batch sizes 287
Beam welding 163
  electron beam welding 163
  laser beam welding 163
Bearings 74
  angular contact 77
  bearing selection 79
  bushes 75
  cylindrical roller 77
  deep groove 77
  dynamic load 80
  hydrodynamic 76
  hydrostatic 76
  journal 75
  life 80
  liners 75
  lubrication 76, 78
  needle roller 78
  rolling element 77
  sliding contact 74
  static load 80
  taper roller 78
  thrust 74, 78
Belt drives 58
  flat belts 60
  raw edge belt 61
  V belt 60
  wedge belt 60
Bill of materials 272
Billets 148
Blind holes 215
Blooms 147
Blow moulding 152
Bodyshell 225
Bolt 86
Bolt loading 212
Brainstorming 106
Braze 202
Brazing 158, 202, 207
  applications 207
  dip brazing 158
  furnace brazing 158
  induction brazing 158
  resistance brazing 158
  torch brazing 158
Brisch number 257
Brittle 228
Broaching 119
BS 55, 320
BS4500 38
Bushes 75

CAD 266
CAD (Computer Aided Design) 267
  2D 268
  3D 268
  FEA (finite element analysis) 268
  kinematics 268
  solid modeller 268
  surface modelling 268
  wire frame 268
CADCAM 266
  integration 269
  post processor 270
  CAM (Computer Aided Manufacture) 268
  CNC (Computer Numerically Controlled) 269
  DNC (Direct Numerical Control) 269
  DNC (Distributed Numerical Control) 269
  CAPP (Computer Aided Process Planning) 271
  CIM (Computer Integrated Manufacture) 273
Capital 286
Carburising 171
Case hardening 171
  carburising 171
  gas carburising 171
  nitriding 171
Cashflow 281, 283
Casting 139, 187, 235
  ceramic moulding 142
  CO_2 moulding 142
  design points 189
  defects 235
  die 143, 188
  draft 140
  dry sand moulding 142
  dry skin moulding 142
  full mould 143
  gate 140
  gravity die casting 143
  green sand moulds 142
  high pressure die casting 145
  investment casting 143, 188
  lost foam 143
  low pressure die casting 145
  mould 140
  pattern 140
  permanent mould casting 143
  plaster moulds 142
  riser 140
  runner 140
  sand casting 141, 188, 235
  shell moulding 142, 188
  sprue 140
  when to use 187
Cathode 229
Cemented carbides 125

Cementite 169
Centrifuge 1
Ceramic 199
Ceramics 126, 153, 199
  crystalline ceramics 153
  glass 153
Chain drives 62
CIP 233
Circlips 88
CLA 42
Clearance 38
Clip 202
Clutches 73
  centrifugal 73
  dog clutch 73
  friction plate 73
  over-running 74
  sprag 74
CMM 175
CNC 135
Coated carbides 125
Coating 172
  anodising 172
  electroplating 172
  friction surfacing 172
  hot dip coatings 172, 229
Codes of practice 225
Cold working 146
Common database 273
Company aims, objectives and strategy 250
Competition 284
Component manufacture 253
Composites 154, 199, 225
  design points 200
  when to use 199
Compression moulding 151
Compromise 236
Computer Aided Process Planning (CAPP) 271
Computer Integrated Manufacture (CIM) 273
Conflict 236
Control charts 182
Control system 6
Control systems 261
Coordinate measuring machine 175
Corrosion 222, 229
COSHH 164
Cost 98, 239, 261
Costing 275
  capital 286
  cost variance 279
  direct costs 262, 277
  fixed costs 276
  income 275
  indirect costs 262, 277
  internal rate of return 283
  investment 281
  labour 278
  materials 277
  net sales value 275
  outgoings 276
  overhead 279
  perceived value 275
  present worth 282
  prime cost 280
  revenue 286
  savings 281
  standard costing 277
  variable costs 276
Cotter pins 88
Couplings 71
  chain 72
  flexible rubber disc 72
  Oldham 72
  rigid 71
  spider 72
Crevice corrosion 229
Cubic boron nitride 126
Customer orders 251
Cutting economics 128
Cutting fluids 126
Cycles 228
Cycloid 64

Datum 33
Delivery date 257
Density 293
Design detail 3
Design engineering 251
Design process 91
  definition 92
  product design specification 92
Deviation 39
Dial indicator gauge 174
Die casting 143, 188
Dimensional analysis 198
Dip brazing 158
Direct workers 262, 277
Distortion 205
Dog clutch 73
Dowel 88
Draft 140
Drawing layout 12
Drawing number 12, 257
Drilling 194
Ductility 228
Duty cycle 97
Dye penetrant 177, 178, 245

EBQ 278
EBW 163, 205
ECM 137
Economic batch quantity 278
ECR 261
EDM 137
Efficiency 235, 241, 262
Elasticity 227
Electric motor 69
  AC motor 70
  DC motor 69
  DP 71
  environment 70
  FP 71
  series 69
  servo motors 70
  shunt 69
  stepper motors 70
  TEFC 71
Electrical capacitance 295
Electrical potential 295
Electrical resistance 295
Electrochemical corrosion 229
Electrochemical series 310
Electrochemical machining (ECM) 137
Electron beam welding 163, 205, 231
Electroplating 172
Electrostatic spraying 172
Energy 295, 300
Engineering change number 262
Engineering changes 261
Engineering drawing 11
  abbreviations 25
  assembly drawings 22
  auxiliary projection 19
  BS308 11, 27
  detail drawings 21
  dimensions 23
  drawing identification 26, 257
  first angle 15
  hatching 18
  hidden detail 18
  layout 12
  orthographic projection 15
  parts list 22
  PP7308 11, 27
  scale 12
  sections 18
  standard components 25
  third angle 16
  title block 12
Engineering model 3
Environment 95, 99, 117, 229, 230
  hygienic 233

Environment (*cont.*)
  underwater 232
  vacuum 231
EPC 103
Epoxy 198
Equations of motion 299
Ergonomics 222
European patent convention 103
Extrusion 118, 148
  backward 148
  impact 148
  moulding 152
  wire drawing 148

Fabrication 235
Factory layout 264
Fail safe 226
Fatigue 227, 228
Ferrite 169
Field trials 4, 285
Final assembly 254
Finance 254
First angle 15
Fits 38
Fixed gauges, Taylor's principles 176
Flame hardening 171
Flash 151
Flashing 147
Flow turning 150
Flux 158
Fluxes 207
Foam plastics 153
Force 295, 297
Forging 146, 190, 222
  closed die 147, 191
  cold heading 147, 191
  design points 191
  drop forging 147
  machining allowance 192
  open die 147, 191
  parting line 192
  press forging 147
  upset 191
  upset forging 147
  when to use 190
Frequency 294
Friction surfacing 172
Friction welding 163
Functional dimensions 32
Functional efficiency 241
Functional requirements 238
Functionality 222
Furnace brazing 158

Galvanising 229, 230
Gas carburising 171
Gas welding 159, 203
Gate 140
Gauge blocks 174
Gear pump 68
Gears 62
  circular pitch 65
  cycloid 64
  diametral pitch 65
  double helical 66
  gear forces 67
  involute 64
  Lewis form factor 68
  method of operation 62
  modes of failure 67
  module 65
  pitch circle 64, 67
  pressure angle 64, 67
  single helical 66
  spiral bevel 66
  spur gear 65
  standard gears 68
  straight bevel 66
  tooth profile 63
  tooth size 65
  worm gears 66
Geometric tolerances 44
  datum identification 45
  examples 46
  feature identification 44
  maximum material condition 50
  symbols 45
  syntax 44
  tolerance frame 44
  when to use 49
Glass 153
Glass fibre 222, 224
GO gauge 176
Grinding 136, 194
  centreless grinding 137
  cylindrical grinding 136
  surface grinding 136
Group technology 265, 271

Hardening 171
Hatching 18
Health and Safety at Work Act 225
Heat exchanger plates 244
Heat treatment 169
  annealing 170
  case hardening 171
  normalising 170
  stress relieving 170
  surface hardening 171
  tempering 171
HIP 146, 193
Hole base 40
Honing 137
Hot dip coatings 172
  hot dip galvanising 172
  tin plating 172
Hot isostatic pressing 146, 193
Hot strip mill 148
Hot working 146
HSS 125

Idea evaluation 108
Idea generation 103
Inclusions 235
Indirect workers 262, 277
Inductance 294
Induction brazing 158
Induction hardening 171
Industrial engineers 279
Information flows 256
Information sources 101
Ingots 147
Injection moulding 151
Inspection equipment 174
  coordinate measuring machine 175
  dial indicator gauge 174
  fixed gauges 176
  gauge blocks 174
  micrometer 174
  sine bar 174
  slip gauges 175
Interest rate 283
Interference 38, 157, 202, 209
Internal rate of return 283
Inventory 287
Investment 281, 283
Investment casting 143, 188
Invoicing 254
Involute 64
ISO 55, 320
Isometric 12

JIS 55
JIT 287
Joining devices 84
  locking devices 86
  non-threaded fasteners 87
  threaded fasteners 84
Joining process 121, 202
Joints 157, 166
  adhesive bonding 164, 207
  bolted 84, 212
  brazing 158, 207
  interference 157, 202, 209
  rivet 87, 157, 216
  solder 158, 207
  welding 159, 203

Just in Time 287

Kevlar 154
Key 57
Keyboard 222
Keyway 57

Lapping 137
Laser beam welding 163
Laser profiling 138
Lathe 132
Lead time 195, 235, 287
Leak detection 178
Legal 255
Legislation 27, 99, 225
Lewis form factor 68
Library searches 101
Life 96
  duty cycle 97
  product life 96
  service life 97
  shelf life 97
Limits and fits 38
  BS4500 38
  clearance 38
  deviation 39
  hole based 39, 40
  interference 38
  maximum material condition 39
  selected fits 40
  shaft based 39, 40
  syntax 39
  tolerance grade 38
  transition 38
Lost wax 143

Machine tools 132
  drilling machine 133
  lathe 132
  machining centre 135
  milling machine 134
  multi-spindle automatic lathe 134
  turning centres 136
Machine utilisation 262
Machining 122, 193, 234
  boring 119, 133
  broaching 119
  chip formation 123
  climb milling 135
  cutting economics 128
  cutting fluids 126
  cutting forces 129
  cutting mechanism 123
  design points 195
  down milling 135
  drilling 119, 133, 194
  electrical discharge machining (EDM) 137
  electrochemical machining (ECM) 137
  electrochemical turning 137
  end milling 135
  facing 132
  finishing cut 128
  form turning 133
  grinding 136, 194
  honing 137
  lapping 137
  laser profiling 138
  machine tools 132
  milling 119, 134, 194
  MRR 129
  multi-point cutting 119, 120
  parting 133
  planing 119, 132
  ream 120, 133
  roughing cut 128
  shaping 119, 132, 194
  single-point cutting 119, 123, 129
  slab milling 135
  spark erosion 137
  specific power 129, 131
  superfinishing 137
  taper turning 133
  Taylor tool life 128
  tool geometry 124
  tool materials 125
  tool wear 127
  turning 132, 194
  up milling 135
  which machining process 194
  when to 193
Magnetic flux 295
Magnetic flux density 295
Manufacturing control 270
Manufacturing drawings 21, 258
Manufacturing engineering 251
Manufacturing process 117, 249
MAP (Manufacturing Automation Protocol) 273
Marketing 251
Marketing appraisal 284
Martensite 171
Mass 297
Master schedule 272
Material selection 115
Material data 311
Material stock movements 264
Maximum material condition 32, 39, 50
  virtual size 50
Mechanical joints 208
  bolt loads 214
  interference 157, 209
  joint loads 212
  riveted joint 87, 157, 202, 216
  threaded 212
  tightening torque 214
  see also joints
Melamine 198
Metal cutting 122
Micrometer 174
MIG 160
Milling 119, 134, 194
MMC 32, 39, 50
Module 65
Modulus of elasticity 227
Moment of inertia 293
Momentum 293
Morphological analysis 105
MRP (Manufacturing Resource Requirements 272
MRP (Materials Requirements Planning) 272
MRPII 272
MRR 128, 129

New product appraisal 286
New product introduction 1, 284
  control system 6
  review point 7, 286
  technical feasibility 7, 285
Newton's Laws of Motion 300
NIH 103
Nitriding 171
NO/GO gauge 176
Non-destructive testing 177
  acoustic 178
  dye penetrant 177
  leak detection 178
  magnetic particle 177
  proof testing 177
  radiographic 178
  ultrasonic 178
  visual inspection 177
Non-threaded fasteners 87, 216
  circlips 88
  dowels 88
  rivets 87
  rollpins 88
  split cotter pins 88
Normal distribution curve 180
Normalising 170
Nylon 198

O ring seal 88, 218, 233
Objectives 250
Operation time 278

Order number 257
Order processing 254, 257
Orthographic projection 15
Overhead costs 279, 287

Packing and despatch 254
Painting 172
  electrostatic spraying 172
  phosphate dip 172
  spraying 172
Parametric 272
Part number 257
Patent Cooperation Treaty 103
Patents 102
Pattern 140, 141
Payback 281, 283
PCT 103
PDS 2, 5, 92, 94, 112, 236
Pearlite 169
Personnel 254
Phosphate dip 172
Pig iron 169
Pitch circle 64, 67
Planing 132
Planning sheets 258, 271
Plasma arc welding 161
Plastic 227
Plastic deformation 146
Plastics 151, 197
  blow moulding 152
  compression moulding 151
  design points 197
  extrusion moulding 152
  foam plastics 153
  injection moulding 151
  properties 198
  thermoplastic 198
  thermosetting 198
  vacuum forming 153
  when to use 197
  which plastic 197
PM 145
Poisson's Ratio 210
Polishing 171, 221, 234
Polycarbonate 198
Pop rivet 87, 157, 202, 216
Post processor 270
Powder forming 145
Powder metallurgy 145, 192
  compaction 145
  hot isostatic pressing 146, 193
  powder manufacture 145
  sintering 146
  when to use 192
  design points 192
Power 294, 300
Pre-production 4, 285
Prepreg 154
Present worth 282
Pressings 150, 221
Pressure 294, 295
Pressure angle 64, 67
Primary tasks 249
Process capability 179
Product design specification 92
  customer 99
  life 96
  operation 94
  origin of need 94
  producer 98
  society 99
Product liability 225
Product planning 251
Production control 252
Projection welding 162
Proof testing 177
Protective paints 229
Prototype 4, 227, 285
Pulsed TIG 161
Purchasing 253

Quality 174, 255
  statistical process control 181
  control charts 182
  inspection equipment 174
  non-destructive testing (NDT) 177
  process capability 179
  traceability 178
Quantity 98, 223, 287
Quotations 257
QWERTY 222

R_a 42
R&D 251
Ranking order 110
Ream 119
Redundancy 226
Reliability 97, 179
Research and development (R&D) 251
Resistance brazing 158
Resistance welding 162, 206
  projection welding 162
  seam welding 162
  spot welding 162
Revenue 286
Riser 140
Risk 285
Rivet 87, 157, 202, 216
Rolling 147
Rollpins 88
Routing sheets 258, 271
Run time 278
Runner 140

SAE 55
Safety 225
Safety factors 227
Sand casting 141, 187, 235
Scheduling 271
Screw 84, 202, 202
Seals 82
  dynamic 82
  lipseal 82
  O ring 83
  static 82
Seam welding 162, 206
Service 97, 285
Set up time 278
Shaft base 39, 40
Shafts 56
Shaping 132, 194
Shear 118, 150
Sheet metal processes 148
  bending 148
  blanking 150
  deep drawing 149
  flow turning 150
  piercing 150
  shearing 150
  spinning 150
  stretch drawing 149
  stretch forming 149
Shell moulding 142, 188, 235
Shot blasting 171, 228
Shot peening 171
Simulation 271
Sine bar 174
Sintering 146, 193
Sketching 12
Slabs 147
Slip gauges 175
Solder 158, 202, 207
Spark erosion 137
Special order 256
Specials 256
Specific power 129
Specification 2, 93
Spinning 150
Splines 58
Split cotter pins 88
Spot welding 162, 206, 222, 224
Spraying 172, 246
Sprue 140
Standard costing 277
Standard deviation 180
Standard product order 259
Standards 54, 101
  ANSI 55
  ASME 55

ASTM 55
BS 55, 320
company standards 55
ISO 55, 320
JIS 55
national standards 55
SAE 55
Statistical process control 181
Steel 169, 311
austenite 169
carbon 169
cementite 169
ferrite 169
martensite 171
pearlite 169
Stick welding 159
Stock control 263
Stock levels 287
Stores 252
finished components store 252
intermediate stores 253
raw materials store 252
Strain 227
Strategy 250
Strength 227
Stress 227, 294
Stress relieving 170
Structural foam 2
Structure of steel 169
Stud 86
Stud welding 161
Stylist 220
Submerged arc welding 160
Superfinishing 137
Surface finish 41, 233, 236
assessment 42
CLA 42
cost 43
R_a 42
syntax 42
Surface finishing 120, 171
friction surfacing 172
polishing 171
shot blasting 171
shot peening 171
tumbling 171
see also coating
Surface hardening 171
flame hardening 171
induction hardening 171
Suspension 221
Synthetics 278

Taylor tool life 128
Taylor's principles 176
Technical feasibility 284, 286
Tempering 171
TEST 108
Thermoplastic 151, 197
Thermosetting 151, 197
Third angle 16
Threaded fasteners 84, 212
bolt 86
locking devices 86
screw 86, 202
self-tapping 85
strength grades 86
stud 86
thread form 84
TIG welding 161, 203, 234
Tightening torque 214
Time and attendance 271
Time study 279
Tin plating 172, 229
Tolerance 30, 180
allocation 34
bi-lateral 32
cost 31
datum 33
definition 30
dependent dimension 34
direct 32
grade 38
maximum material condition 32, 39, 150
presentation 32
uni-lateral 32
virtual size 50
see also limits and fits
Tool materials 125
carbon steel 125
cemented carbides 125
ceramics 125
coated carbides 125
cubic boron nitride 126
diamond 126
high speed steels 125
Tool wear 127
Torch brazing 158
Torque 294, 300
Traceability 178
Transition 38
Transmission and components 55
bearings 74
belts 58
chain drives 62
clutches 73
couplings 71
electric 69
flat belts 60
gears 62
hydraulic 68
key 57
keyway 57
seals 82
shafts 56
splines 58
timing belts 61, 62
V belt 60
wedge belt 60
Tumbling 171
Tup 147
Turning 132, 194

Ultimate tensile stress 228
Ultrasonic inspection 178
Units 293
conversion 296
derived unit 294
dimensional analysis 298
imperial system 295
multipliers 293
si system 293
unit names 293
unity brackets 298
Unmanned submersible 232
Unsprung weight 221
UTS 228

Vacuum forming 153
Vacuum furnace 158
Value 239
Value analysis 238
Valve 233
Variance 261
Velocity 293
Vernier callipers 174
Virtual size 50

Washers 215
Wave soldering 159
Weight 295
Weighting factor 110
Weld 203, 221
butt 203, 205
edge preparation 203
electron beam 205
fillet 203
sizing 204
throat area 204
Welding 159, 202, 203
arc welding 159, 203
beam welding 163, 205, 231
continuous 203
design points 205
distortion 205
friction welding 163
gas 203
gas welding 159
resistance welding 162, 206

Welding (*cont.*)
  seam 206
  spot 206
WIP (Work In Progress) 195, 235, 272
Wire drawing 148
Work 294, 300
Work study 279
Working file 26
Works engineers 255
Works order number 257

Yield point 228
Yield stress 228
Young's Modulus 200, 210, 204